Science of Lakes - Multidisciplinary Approach

Edited by Ali A. Assani

Published in London, United Kingdom

Science of Lakes - Multidisciplinary Approach
http://dx.doi.org/10.5772/intechopen.108096
Edited by Ali A. Assani

Contributors
Hassan Abdel Motagly Abdel Mougod Gouda, Abdel-Aal Hassan Moubasher, Mady Ahmed Ismail, Nammat Abd el Gowad Hussein, Job O. Omweno, Reuben Omondi, Fredrick M. Ondemo, Argwings Omondi, Maximilien Bernier, Andrzej Górniak, Adam Więcko, Aina O. Adeogun, Azubuike V. Chukwuka, Jinliang Liu, Huaiping Zhu, Ziwang Deng, Chris Charron, Francois Hategekimana, Digne Edmond Rwabuhungu Rwatangabo, Theophile Mugerwa, Young-Seog Kim, Lukman Lukman, David Avalos-Cueva, Anatoliy Filonov, Iryna Tereshchenko, Cesar Monzon, María del Refugio Barba-López, Mariana Belferman, Amotz Agnon, Jeanna A. Balonishnikova, Oleg D. Tregubov, Pavel Ya. Konstantinov, Vladimir V Shamov, Konstantin K. Uyagansky

First published in London, United Kingdom, 2024 by IntechOpen
IntechOpen is the global imprint of INTECHOPEN LIMITED, registered in England and Wales, registration number: 11086078, 5 Princes Gate Court, London, SW7 2QJ, United Kingdom
Printed in Croatia

British Library Cataloguing-in-Publication Data
A catalogue record for this book is available from the British Library

Additional hard and PDF copies can be obtained from orders@intechopen.com

Science of Lakes - Multidisciplinary Approach
Edited by Ali A. Assani
p. cm.

This title is part of the Environmental Sciences Book Series, Volume 14
Topic: Water Science
Series Editor: J. Kevin Summers
Topic Editor: Yizi Shang

Print ISBN 978-1-83769-041-1
Online ISBN 978-1-83769-042-8
eBook (PDF) ISBN 978-1-83769-043-5
ISSN 2754-6713

For EU product safety concerns:
IN TECH d.o.o., Prolaz Marije Krucifikse Kozulić 3, 51000 Rijeka, Croatia,
info@intechopen.com or visit our website at intechopen.com.

IntechOpen Book Series

Environmental Sciences

Volume 14

Aims and Scope of the Series

Scientists have long researched to understand the environment and man's place in it. The search for this knowledge grows in importance as rapid increases in population and economic development intensify humans' stresses on ecosystems. Fortunately, rapid increases in multiple scientific areas are advancing our understanding of environmental sciences. Breakthroughs in computing, molecular biology, ecology, and sustainability science are enhancing our ability to utilize environmental sciences to address real-world problems.

The four topics of this book series - Pollution; Environmental Resilience and Management; Ecosystems and Biodiversity; and Water Science - will address important areas of advancement in the environmental sciences. They will represent an excellent initial grouping of published works on these critical topics.

Meet the Series Editor

J. Kevin Summers is a Senior Research Ecologist at the Environmental Protection Agency's (EPA) Gulf Ecosystem Measurement and Modeling Division. He is currently working with colleagues in the Sustainable and Healthy Communities Program to develop an index of community resilience to natural hazards, an index of human well-being that can be linked to changes in the ecosystem, social and economic services, and a community sustainability tool for communities with populations under 40,000. He leads research efforts for indicator and indices development. Dr. Summers is a systems ecologist and began his career at the EPA in 1989 and has worked in various programs and capacities. This includes leading the National Coastal Assessment in collaboration with the Office of Water which culminated in the award-winning National Coastal Condition Report series (four volumes between 2001 and 2012), and which integrates water quality, sediment quality, habitat, and biological data to assess the ecosystem condition of the United States estuaries. He was acting National Program Director for Ecology for the EPA between 2004 and 2006. He has authored approximately 150 peer-reviewed journal articles, book chapters, and reports and has received many awards for technical accomplishments from the EPA and from outside of the agency. Dr. Summers holds a BA in Zoology and Psychology, an MA in Ecology, and Ph.D. in Systems Ecology/Biology.

Meet the Volume Editor

Professor Ali A. Assani holds a master's degree in Geology and a Ph.D. in Physical Geography. He also completed postdoctoral training in physical geography. He teaches climatology, hydrology, and climate change. He is the author and co-author of around 100 articles published in peer-reviewed journals and 10 chapters in collective works. His research is in the fields of hydrology, climatology, climate change, fluvial geomorphology, and ecohydrology. He is a guest editor of several special journal issues and a reviewer of journal articles. Dr. Assani is also a journal editorial board member.

Contents

Preface

Lakes (natural and artificial) are main components of the Earth's landscape. They influence biodiversity as well as biogeochemical and hydrological cycles on the continents. They constitute one of the largest reserves of surface fresh water in the world. They also influence, to varying degrees, socioeconomic and recreational tourism activities in many regions of the planet. The evolution of their physicochemical characteristics results from the interaction (climate changes, tectonic movements, evolution of slopes, etc.) of natural and anthropogenic factors (deforestation, reforestation, agriculture, urbanization, dams, reservoirs, etc.). This book presents 12 studies on these different interactions carried out in intertropical and extratropical regions. The studies are grouped into three sections. The first section, made up of four chapters, is devoted to the impacts of climate change and other natural factors of the evolution of the physicochemical characteristics of lakes. Chapter 1 analyzes the potential impacts of climate change on the water levels of the Great Lakes using high-resolution regional climate models. Chapter 2 studies the impacts of global warming on the hydrological regime of thermokarst lakes in the sub-arctic region. Chapter 3 traces the limnological changes undergone by a lake in a temperate continental region in relation to changes in land use and land cover (urbanization) in its watershed. Finally, chapter 4 analyzes the impacts of earthquakes on the evolution of the water levels of a lake through geological periods.

The second section of the book brings together five studies devoted to the interactions between anthropogenic activities and lake ecosystems. Chapter 5 summarizes these interactions. Chapter 6 analyzes the impacts of anthropogenic activities on the morphology of a lake in a tropical region. Chapter 7 focuses on estimating the quantity of water in a lake to meet the current and future needs of agricultural development in its watershed. Chapter 8 addresses the impacts of the introduction of new species of fish and aquatic plants on the ecology and fishing of a tropical lake. Chapter 9 analyzes the intersection of the hydrodynamics of lake waters and the economic development of their territories.

The third section consists of three chapters devoted to the limnological characteristics of lakes. Chapter 10 focuses on the thermal regime and the water balance of two volcanic crater lakes. Chapter 11 is dedicated to the characterization of the chemical waters of a tropical lake and its environmental impacts. Finally, Chapter 12 lists the different species of fungi that colonize several lakes in a very saline environment.

Ali A. Assani
Department of Environmental Sciences,
University of Quebec at Trois-Rivières,
Trois-Rivières, Quebec, Canada

Section 1

The Impacts of Climate Change and Other Natural Factors on the Evolution of the Physicochemical Characteristics of Lakes

Chapter 1

High-Resolution Regional Climate Projections for Ontario and the Canadian Great Lakes Basins

Jinliang Liu, Huaiping Zhu, Ziwang Deng and Chris Charron

Abstract

The IPCC-endorsed multiple model/scenario approach and a state-of-the-science combined downscaling methodology were applied to project the future climate changes over Ontario and the Canadian Great Lakes basins. Significant warming is expected across the province under all RCPs. Relative to 1986–2005 averages, the highest temperature rise is projected to occur in Ontario's Far North, 7.3°C warmer by the 2080s. The temperature over the Great Lakes Basin is projected to increase by 1.3–5.7°C. Ontario's annual total precipitation is projected to increase 86.9 mm (11%) by the 2080s under RCP 8.5, while summer precipitation is projected to decrease by 32.9 mm (12%) and winter precipitation to increase by 52.4 mm (48%). In the Great Lakes Basin, the greatest increase in annual average temperature (1.7–5.3°C) is projected to occur in the Lake Superior sub-basin by the 2080s. Winter warming is projected to exceed summer warming in all sub-basins. Annual total precipitation is projected to increase in all five sub-basins, with the largest increase in the Lake Superior sub-basin. Summer (winter) is projected to be drier (wetter) across the entire Great Lakes Basin. These projected changes could have implications on future water levels in the Great Lakes and many aspects over the study area.

Keywords: high-resolution regional climate projections, dynamical downscaling, combined downscaling, Canadian Great Lakes basin and sub-basins, Ontario, climate change impacts, water levels

1. Introduction

The impacts of climate change are broad in scope (both location-wise and sector-wise) and unprecedented in magnitude. The 5th assessment report (AR5) of the Intergovernmental Panel on Climate Change (IPCC) concludes that global warming leads to an intensification of the water cycle and attendant effects on heavy and extreme precipitation events and is likely to lead to more frequent daily precipitation extremes [1]. These changes have significant potential impacts on floods, erosion, infrastructure, agriculture, water resources, ecosystems, human health and ultimately the economy. IPCC's latest 6th assessment report (AR6) found that the warming and associated impacts are more severe than previously estimated [2]. Therefore, it is critical for governments, practitioners, and the public to be aware of future climate changes at local scales where adaptation/mitigation activities are most effective. High-resolution climate projection data at multiple time scales (e.g., annual, monthly, daily, even sub-daily) is necessary for governments to develop climate adaptation/mitigation strategies to assess and address local impacts. The geography of Ontario is unique; it is bordered by the Great Lakes in the south and Hudson Bay and James Bay in the North. Within Ontario, there are more than 250,000 inland lakes, large areas of uplands, particularly within the Canadian Shield which traverses the province from northwest to southeast and above the Niagara Escarpment in the south. Southern Ontario (south of Lake Nipissing) belongs to the Great Lakes/St. Lawrence River Basins [3, 4]. To generate projections of climate change in Ontario and the Great Lakes basin, the impacts of all these geophysical features specific to the study area should be considered [5]. IPCC has been promoting, in its recent reports, using novel methods to combine multi-model projections with observational constraints to provide robust projections of geographical patterns of climate change [2]; we are ahead of this game and developed a novel combined downscaling method which combines the Ensemble Optimal Interpolation (EnOI) and bias correction techniques [6]. The resulting high-resolution regional climate projections specific to Ontario and the Great Lakes basin are disseminated through the user-friendly Ontario Climate Data Portal [7–9], which has been supporting many practitioners in assessing climate change impacts and developing climate change pertaining policies and plans to mitigate and adapt to the changing climate (e.g., [10–13]).

To further help practitioners (including all levels of government) to mitigate and adapt to the changing climate, this chapter summarizes trends and spatial patterns of projected changes in annual and seasonal mean temperature and precipitation over Ontario, its three major watersheds (the Hudson Bay Basin, the Nelsen River Basin and the Great Lakes Basin) and the five sub-basins of the Canadian Great Lakes basin (Lake Superior, Lake Huron, Lake Erie, Lake Ontario and the Ottawa River).

The rest of this chapter is organized as the following. Section 2 describes the data and methodology used in this study; Section 3 are the results by basin and sub-basin; and conclusions and discussion are in Section 4.

2. Data and methodology

2.1 Data

Downscaling is used to increase the spatial resolution (i.e., to smaller grid sizes) from global climate model output (usually at much larger grids). Multiple sources can be found for downscaled climate projections, but most (if not all) of them are results of statistical downscaling directly from the global climate models (GCMs). In this study, a super ensemble of high-resolution regional climate projections using a state-of-the-science combined downscaling methodology [6–9] was used. Using the multiple model/scenario approach endorsed by the IPCC to address uncertainties in future projections, we developed a super ensemble (collectively 209 members) of high-resolution (~10 km × ~10 km) regional climate projections for Ontario and the Great Lakes basin based on all available creditable data sources at the time this study was carried out (including York University [6–9], the Pacific Climate Impacts Consortium (PCIC) [14], North America—Coordinated Regional Downscaling Experiment (NA-CORDEX) [15], University of Toronto [16, 17], and the University of Regina [18]).

There have been examples of papers/reports that have synthesized and reported on future climate projection information relevant to the study area [6, 19–22]. While previous studies have provided a wealth of information about climate change over the study area, this chapter is intended to provide an updated high-level comprehensive summary based on the future climate projections specific to the study area with the following most recent regional climate modeling advancements incorporated [6–9]:

- *Improved robustness*: to better consider uncertainties in future projections, a super-ensemble of Ontario-specific high-resolution climate projections has been developed based on 209 projections of 42 global climate models (GCMs) and 7 regional climate models (RCMs) which is far greater than the number of models used in other studies;

- *Improved downscaling techniques*: the super-ensemble of climate projections is based on state-of-the-science combined downscaling methodologies [6] with improved physical comprehension, in other words, improved consideration of the impacts of local geophysical features such as the Great Lakes, the Niagara Escarpment, and the Hudson Bay, which are very critical to Ontario's local weather and climate, especially extreme events (see **Figure 1**); and,

- *Improved consistency* in time frames with IPCC's AR5 (i.e., reference/base time period and future time periods).

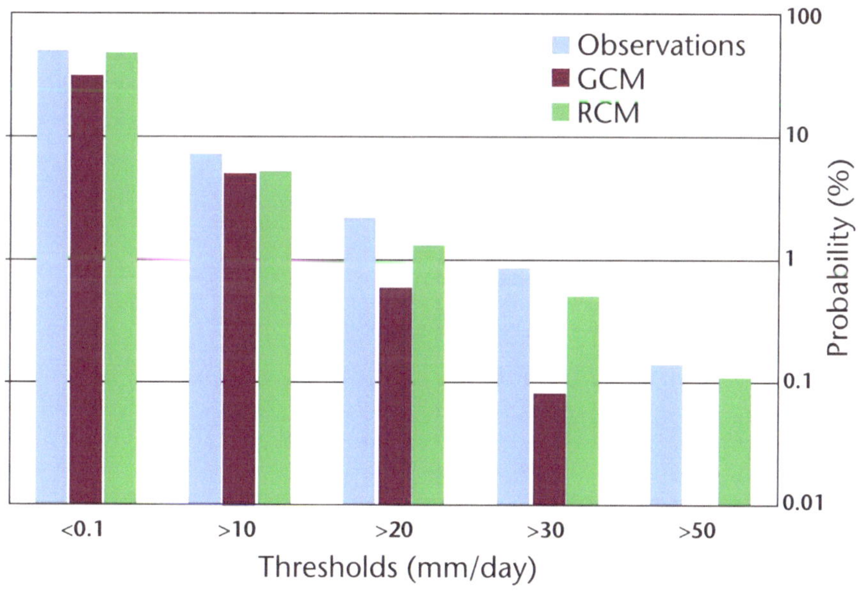

Figure 1.
Capability comparison between low-resolution global climate models (GCMs) and high-resolution regional climate models (RCMs) in simulating rainfall events in the Alps (source: [23]). This bar chart clearly demonstrates that RCMs are much more capable of predicting extreme rainfall events due to their much higher resolution which helps them to better consider the impacts from local geophysical features such as the Great Lakes and the Niagara Escarpment in Ontario.

2.2 Methodology

Projected changes in Ontario's climate are presented for three time periods (2030s: 2020–2049, 2050s: 2040–2069, and 2080s: 2070–2099) under three IPCC-defined greenhouse gas emission scenarios or representative concentration pathways (RCP 2.6, RCP4.5, and RCP8.5, see the following definitions and visual in **Figure 2**). The use of a more advanced combined downscaling methodology and more comprehensive historical meteorological data sets (based on both observations and modeled data) better accounts for the impacts of Ontario's local geophysical features such as the Great Lakes, Niagara Escarpment, Hudson Bay and James Bay on local weather and climate. Lastly, bias correction was applied to the final product. Bias correction is widely used in climate modeling. It aims to adjust selected statistics of a climate model's results so that they better match observed statistics over a present-day reference period which are then applied to correct the future projections. More details about the state-of-the-science combined downscaling methodology can be found in [6–9].

- RCP8.5: This is often referred to as the "business-as-usual" greenhouse gas emission scenario, meaning no emission reductions, emissions continue to rise throughout the 21st century;
- RCP4.5: with more vigorous greenhouse gas reductions, emissions peaks around 2040 and then stabilize;
- RCP2.6: with the most vigorous greenhouse gas reductions, emissions peak around 2020 and then decline.

This chapter provides a high-level summary of annual and seasonal averages of daily mean temperature and daily total precipitation for Ontario's three primary basins (Hudson Bay Basin, the Nelsen River Basin and the Great Lakes Basin), as well as the

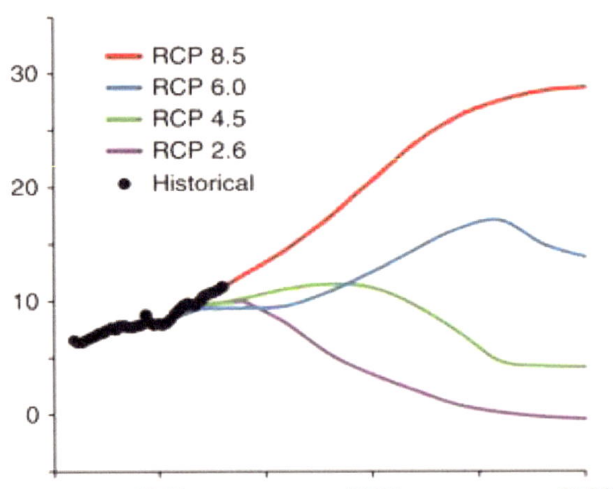

Figure 2.
Greenhouse gas emission scenarios (also known as representative concentration pathways or RCPs) are defined in the IPCC's AR5. In the vertical axis is the global carbon emissions in petagrams of carbon or 10^{15} grams of carbon per year (PgC/year) (source: [24]).

sub-basins of the Great Lakes Basin (Lake Superior, Lake Huron, Lake Erie, Lake Ontario and the Ottawa River) (**Figure 3**). Projections are, however, available for 42 climate variables, both typical variables (e.g., long-term averages of temperature and precipitation) and extreme variables (e.g., heat wave-related variables and extreme rainfall events) through York University's Ontario Climate Data Portal (ODCP) [7–9]. Readers are referred to the OCDP [7] and the references listed on the Portal for these additional variables and more details on the development of the super-ensemble of high-resolution (10 km × 10 km) climate projections specific to the study area. A high-level overview of the OCDP can also be found in [8] and more details regarding the super-ensemble projection downscaling methodology in [6, 7, 9, 24].

In this chapter, we present projected changes relative to their averages during the reference time period (1990s: 1986–2005) for the entire Province of Ontario, each of its three primary basins and the five sub-basins of the Great Lakes basin (see **Figure 3**), including the mean (average of each basin per period), standard deviation (spatial variation in the basin per period) and range (minimum and maximum across the basin per period). Projected changes in long-term averages of temperature and precipitation are presented in tabular, bar-chart and map formats at annual (January to December), summer (June to August) and winter (December to February) temporal scales. All maps of future climate are shown as the change in temperature and precipitation compared to the reference time period (1990s: 1986–2005). Generally speaking, the most dramatic changes are projected to occur in the winter months, and we have therefore chosen to focus on

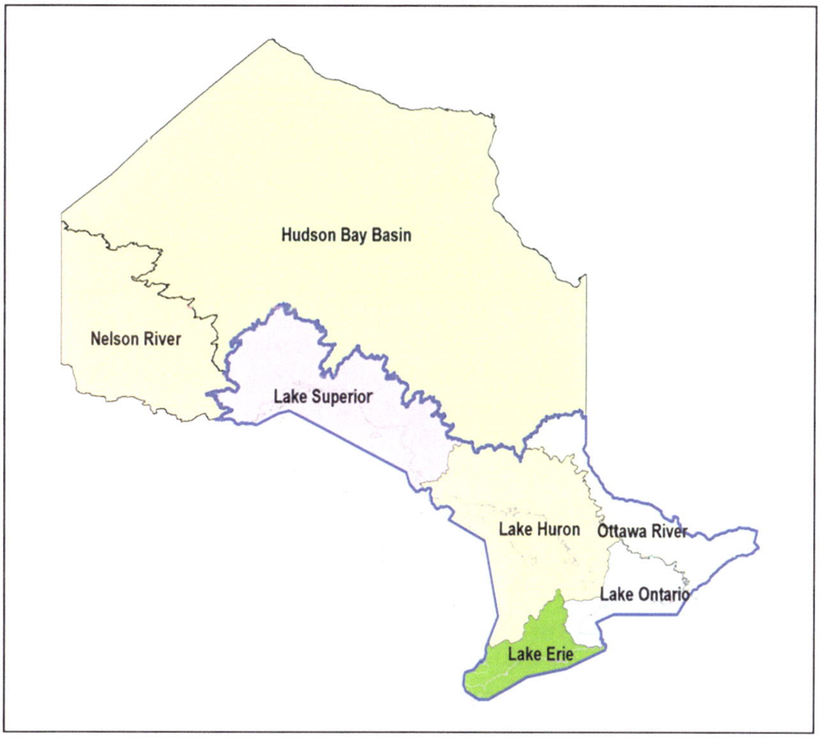

Figure 3.
Map of the Province of Ontario and its seven basins/sub-basins. The five sub-basins within the Great Lakes Basin are outlined by the blue polygon.

the summer and winter seasons to illustrate seasonal climate change extremes. It is important to note that this high-level summary focuses on long-term (annual and seasonal) averages over large areas (the entire province and its large basins) which is very different from extreme events (e. g., flooding, heat wave) that usually occur at much shorter temporal scales at specific locations. Readers are referred to the OCDP [7] for extreme variables.

3. Results

3.1 Historical climate trends in Ontario

Ontario's annual average air temperature has been warming at an average rate of 1.3°C per 100 years from 1901 to 2018 [25, 26]. This is consistent with the trend reported for 1948–2016 in Chapter 8 of the 2019 Canada's Changing Climate Report (CCCR2019) [19]. Annual mean daily temperatures in Ontario increased by 0.5°C to 1.5°C from 1950 to 2010, with the most dramatic increases occurring south of the Nelson River Basin and west of the Hudson Bay Basin [27]. Annual total precipitation has been increasing by 11.8% per 100 years from 1901 to 2016 [26, 28]. Ontario's annual total precipitation increased by 7.6% from 1979 to 2016; where summer precipitation increased by 4.3% and winter precipitation increased by 24% [28]. Weather station data in Ontario shows that rainfall has generally increased in all seasons with the most pronounced increases (by up to 50%) observed in northwestern Ontario (i.e., the vicinity of Thunder Bay) during the spring months. On average, the north shores of the Great Lakes have seen the most dramatic increases in precipitation, while the southern portion of the province near the Great Lakes snow-belt areas has seen the most significant winter snowfall increases (by 10–30%).

3.2 Future climate projections for Ontario

The annual average temperature is projected to increase across the entire province in this century when considering all RCPs (**Figures 4** and 5). The most dramatic warming is projected to occur in the Far North which could be up to 7.3°C warmer by the 2080s under RCP 8.5 (**Table 1**; **Figure** 5) while the southern most parts of Ontario are projected to see a 4.0°C rise in temperature over the same time period. When considering both the RCP 2.6 and RCP 8.5 emissions scenarios, the projected average annual temperature across the province is expected to increase by 1.1–2.3°C by the 2030s, 1.4–4.1°C by the 2050s, and 1.3 –7.3°C by the 2080s (**Table 1**; **Figures 4** and 5). Seasonally, warming is expected to be greater in winter than in summer (**Figures 4** and **5**) with the average winter temperature projected to increase by 1.3–11.7°C (**Table 1**, **Figures 4** and **6**). Lastly, the average summer temperature is projected to increase by 1.1–6.3°C (**Table 1**; **Figures 4** and 7).

Provincial average annual precipitation is projected to increase up to 18.6 mm (2.4%) by the 2030s, 47 mm (6%) by the 2050s and 86.9 mm (11%) by the 2080s when considering all RCPs. Although average precipitation over the entire province is projected to increase for the rest of this century, the projected changes in precipitation are much more complicated than temperature and vary dramatically from region to region. For example, some locations in the Hudson's Bay Basin and the Great Lakes Basin are projected to become wetter with projected increases in

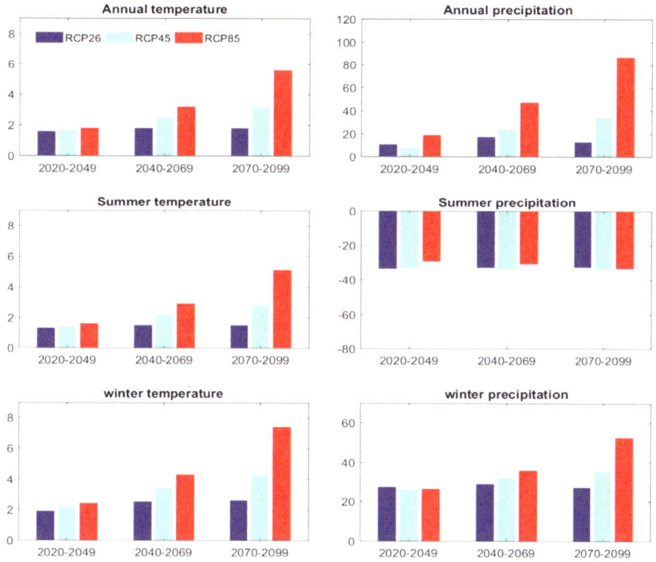

Figure 4.
Projected changes in Ontario's average annual temperature (°C) (upper-left), summer temperature (middle-left), winter temperature (lower-left), annual precipitation (mm) (upper-right), summer precipitation (middle-right), and winter precipitation (lower-right) relative to the 1986–2005 reference period, under representative concentration pathways (RCPs) 2.6, 4.5 and 8.5.

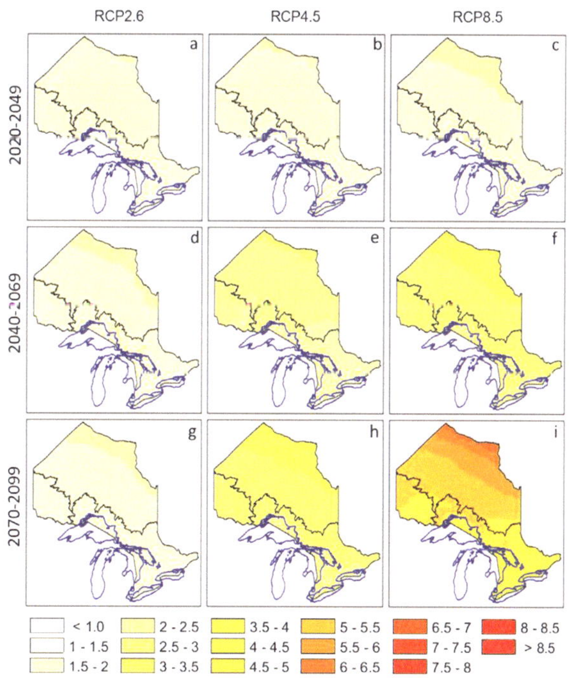

Figure 5.
Projected changes in Ontario's average annual temperature (°C) relative to the 1986–2005 reference period for representative concentration pathways (RCP) 2.6 (left column), 4.5 (middle column), and 8.5 (right column), over three 30-year time frames 2030s (2020–2049, top row), 2050s (2040–2069, middle row), and 2080s (2070–2099, bottom row). Maps reflect the 50th percentile of the super ensemble.

Change from 1986–2005 baseline		2020–2049			2040–2069			2070–2099		
		RCP2.6	RCP4.5	RCP8.5	RCP2.6	RCP4.5	RCP8.5	RCP2.6	RCP4.5	RCP8.5
Annual	Temperature (°C)	1.6(0.1)	1.7(0.1)	1.8(0.1)	1.8(0.1)	2.5(0.2)	3.2(0.3)	1.8(0.2)	3.2(0.3)	5.6(0.6)
		1.1–2.0	1.2–2.0	1.4–2.3	1.4–2.4	1.8–3.2	2.3–4.1	1.3–2.4	2.3–4.2	4.0–7.3
	Precipitation (mm)	10.5(33.6)	7.4(31.5)	18.6(25.8)	17(32.7)	23.9(32.4)	47(27.9)	12.6(37.1)	34.5(30.7)	86.9(29.8)
		−91 to 97	−73 to 92	−51 to 90	−93 to 108	−62 to 111	−27 to 121	−83 to 112	−50 to 122	7–177
Summer	Temperature (°C)	1.3(0.1)	1.4(0.1)	1.6(0.1)	1.5(0.1)	2.2(0.1)	2.9(0.1)	1.5(0.1)	2.8(0.1)	5.1(0.3)
		1.1–1.7	1.2–1.6	1.3–1.9	1.2–2.1	1.8–2.6	2.3–3.4	1.1–2.0	2.3–3.5	4.0–6.3
	Precipitation (mm)	−33.4(24.8)	−33.1(24.6)	−28.8(22.5)	−32.7(23.8)	−33.9(24.9)	−30.5(23.8)	−32.3(24.5)	−33.6(24.7)	−32.9(24.8)
		−111 to 17	−111 to 19	−98 to 19	−108 to 19	−113 to 19	−106 to 19	−109 to 26	−115 to 17	−112 to 16
Winter	Temperature (°C)	1.9(0.3)	2.2(0.4)	2.4(0.5)	2.5(0.4)	3.4(0.6)	4.3(0.8)	2.6(0.5)	4.3(0.8)	7.4(1.5)
		1.3–3.2	1.3–3.5	1.4–3.9	1.5–4.2	1.9–5.3	2.3–6.7	1.5–4.2	2.4–6.9	3.8–11.7
	Precipitation (mm)	27.3(9.0)	26.1(8.8)	26.4(7.7)	28.8(9.1)	31.7(9.0)	35.9(9.0)	27(8.6)	35.7(9.5)	52.4(10.1)
		−47 to 68	−26 to 60	−11 to 57	−51 to 65	−19 to 65	−2 to 71	−46 to 59	−21 to 74	15–89

For each entry, the first row is the provincial average, followed by the standard deviation (SD) in brackets. The second row is the range across the province (minimum to maximum).

Table 1.
Projected changes in temperature and precipitation relative to the 1986–2005 reference period for the entire province of Ontario, under three IPCC-defined representative concentration pathways (RCP2.6, RCP4.5, and RCP8.5), and for three time periods (2030s: 2020–2049, 2050s: 2040–2069, 2080s: 2070–2099).

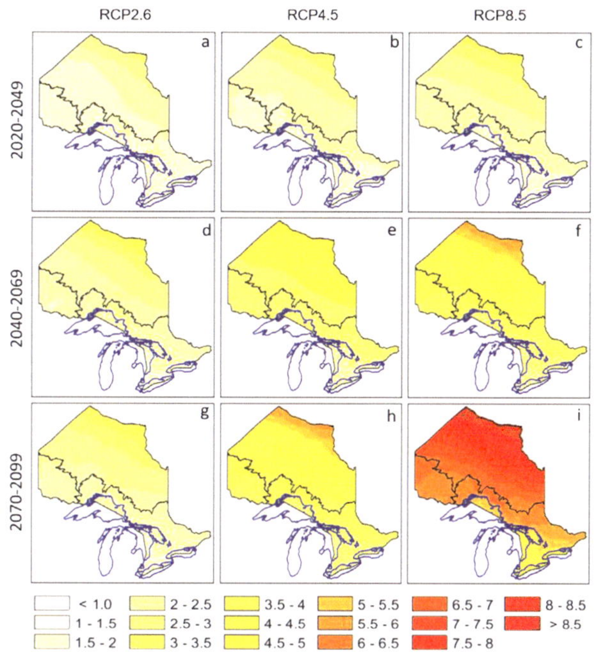

Figure 6.
*Same as in **Figure 5** but for winter temperature (°C).*

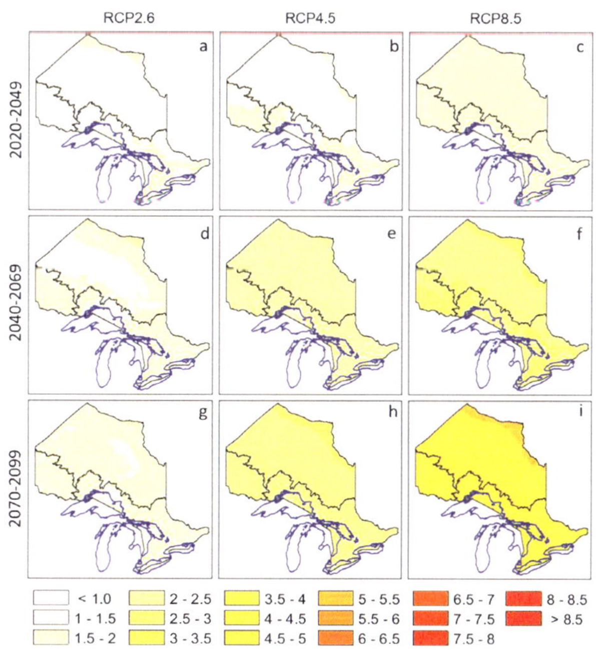

Figure 7.
*Same as in **Figure 5** but for summer temperature (°C).*

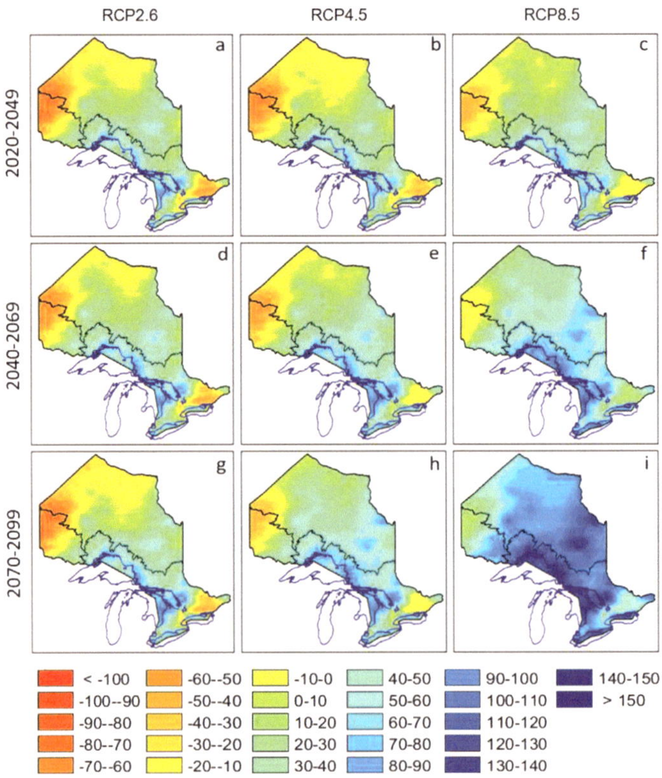

Figure 8.
*Same as in **Figure** 5 but for annual total precipitation (mm).*

precipitation of more than 150 mm (**Table 1**; **Figures 4** and **8**), whereas the Nelsen River Basin and the western Hudson Bay Basin are projected to become drier with projected precipitation decreases of up to 93 mm relative to the 1986–2005 reference period (**Table 1**; **Figures 4** and **8**).

Provincial average summer precipitation is projected to decrease up to 33.4 mm (12.1%) by the 2030s, 33.9 mm (12.3%) by the 2050s and 33.6 mm (12.2%) by the 2080s when considering all RCPs; while provincial average winter precipitation is projected to increase up to 27.3 mm (25%) by the 2030s, 35.9 mm (33%) by the 2050s and 52.4 mm (48%) by the 2080s. Summer precipitation is projected to decrease up to 115 mm over the Nelson River basin by the 2080s (**Table 1**; **Figures 4** and **9**); while winter precipitation is projected to increase up to 89 mm over some Great Lakes' eastern coastal regions relative to the reference period (**Table 1**; **Figures 4** and **10**).

3.3 Climate projections for the Hudson Bay Basin

Mean annual air temperature is projected to increase over the Hudson Bay Basin for the rest of this century when considering all RCPs. Temperature increases are projected to range from 1.4 to 7.3°C across the basin relative to the 1986–2005 reference period under all RCPs (**Table 2**; **Figures 5** and **11**). The northern portion of the basin is projected to warm faster than the southern portion (**Figure 5**). The average annual temperature across the basin is expected to increase by 1.4–2.3°C by the 2030s, 1.7–4.1°C by the 2050s, and 1.7–7.3°C by the 2080s (**Table 2**; **Figures 5** and **11**).

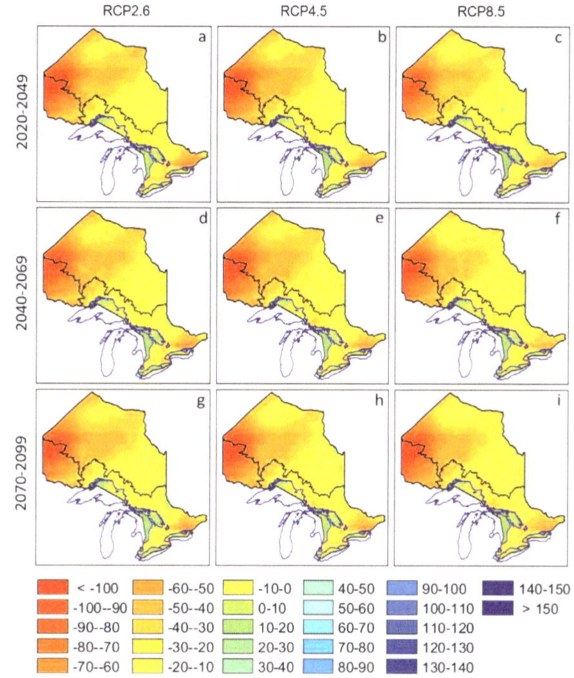

Figure 9.
*Same as in **Figure 8** but for summer total precipitation (mm).*

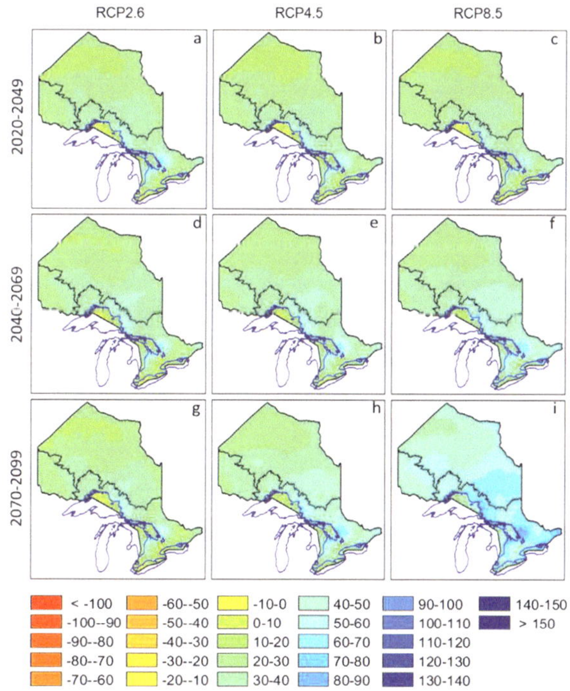

Figure 10.
*Same as in **Figure 8** but for winter total precipitation (mm).*

Change from 1986–2005 baseline		**2020–2049**			**2040–2069**			**2070–2099**		
		RCP2.6	**RCP4.5**	**RCP8.5**	**RCP2.6**	**RCP4.5**	**RCP8.5**	**RCP2.6**	**RCP4.5**	**RCP8.5**
Annual	Temperature (°C)	1.6(0.1)	1.7(0.1)	1.9(0.1)	1.9(0.1)	2.6(0.2)	3.4(02)	1.9(0.1)	3.4(0.2)	6.0(0.4)
		1.4–2.0	1.6–2.0	1.7–2.3	1.7–2.4	2.4–3.2	3.1–4.1	1.7–2.4	3.1–4.2	5.3–7.3
	Precipitation (mm)	5.8(25.7)	0.6(22.7)	14.8(18.2)	11.3(23.9)	17.6(23.8)	42.1(20.5)	3.6(26.7)	30.9(23.4)	84.6(23.6)
		−76 to 97	−72 to 85	−39 to 82	−64 to 103	−59 to 105	−22 to 121	−78 to 101	−43 to 122	15–177
Summer	Temperature (°C)	1.3(0.1)	1.4(0.1)	1.6(0.1)	1.5(0.2)	2.2(0.1)	2.9(0.1)	1.5(0.1)	2.8(0.2)	5.1(0.2)
		1.1–1.7	1.3–1.6	1.5–1.9	1.3–2.1	2.1–2.6	2.7–3.4	1.4–2.0	2.6–3.5	4.9–6.3
	Precipitation (mm)	−34.5(20.4)	−34.7(19)	−29(18.2)	−33.8(18.8)	−34.6(19.3)	−29.4(18.6)	−34.9(18.4)	−33(19.5)	−31.2(20.2)
		−111 to 8	−109 to 9	−95 to 11	−107 to 11	−110 to 10	−103 to 10	−108 to 7	−110 to 14	−110 to 13
Winter	Temperature (°C)	2.1(0.4)	2.4(0.4)	2.7(0.4)	2.7(0.5)	3.8(0.5)	4.8(0.6)	2.9(0.4)	4.8(0.7)	8.4(1.1)
		1.6–3.2	1.8–3.5	2.1–3.9	2.0–4.2	3.1–5.3	3.8–6.7	2.2–4.2	3.8–6.9	6.6–11.7
	Precipitation (mm)	25.2(6.2)	23.9(6)	24.9(5.2)	27.8(6.7)	29.8(6.7)	33.3(6.8)	26.2(6.1)	33.4(6.8)	50.1(8.4)
		1–41	5–43	10–39	4–47	11–47	19–51	3–44	16–51	35–69

Table 2.
*Same as in **Table 1** but for the Hudson Bay Basin.*

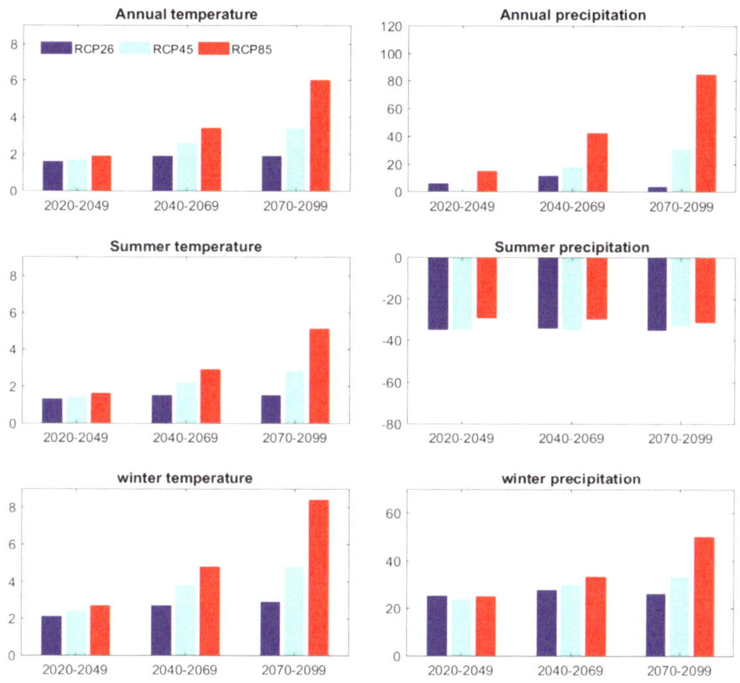

Figure 11.
*Same as in **Figure 4** but for the Hudson Bay Basin.*

Basin-averaged annual precipitation is projected to increase up to 14.8 mm (2%) by the 2030s, 42.1 mm (6%) by the 2050s and 84.6 mm (12%) by the 2080s (**Table 2**, **Figure 11**). Annual total precipitation is projected to increase in the southeastern area of the basin and decrease in the northwest (**Figure 8**).

Summer temperature across the Hudson Bay Basin is projected to increase 1.1–1.9°C by the 2030s, 1.3–3.4°C by the 2050s and 1.4–6.3°C by the 2080s when considering all RCPs (**Table 2**; **Figure 4**). Summer precipitation is projected to decline across the basin (**Table 2**, **Figures 9** and **11**), decreasing by up to 110 mm by the 2080s relative to the 1986–2005 reference period (**Table 2**; **Figure 9**). Basin-average summer precipitation is projected to decrease up to 34.7 mm (12.9%) by the 2030s, 34.6 mm (12.9%) by the 2050s and 34.9 mm (13.0%) by the 2080s when considering all RCPs (**Table 2**; **Figure 11**).

The average winter temperature across the Hudson Bay Basin is projected to increase 1.6–3.9°C by the 2030s, 2.0–6.7°C by the 2050s, and 2.2–11.7°C by the 2080s when considering all RCPs (**Table 2**; **Figure 11**). The most significant warming is expected to occur in the northern part of the province along the Hudson Bay coast under RCP 8.5 (**Figure 6**).

Basin-average winter precipitation is projected to increase up to 25.2 mm (30%) by the 2030s, 33.3 mm (41%) by the 2050s and 50.1 mm (61%) by the 2080s (**Table 2**; **Figure 11**). Average winter precipitation across the basin is projected to increase by 1–41 mm by the 2030s, 4–47 mm by the 2050s and 3–44 mm by the 2080s under RCP 2.6; and 10–39 mm by the 2030s, 19–51 mm by the 2050s, and 35–69 mm by the 2080s under RCP 8.5 (**Table 2**; **Figure 11**).

Change from 1986–2005 baseline		2020–2049			2040–2069			2070–2099		
		RCP2.6	RCP4.5	RCP8.5	RCP2.6	RCP4.5	RCP8.5	RCP2.6	RCP4.5	RCP8.5
Annual	Temperature (°C)	1.5(0.04)	1.7(0.05)	1.8(0.03)	1.8(0.05)	2.5(0.04)	3.2(0.05)	1.8(0.05)	3.2(0.06)	5.5(0.08)
		1.5–1.6	1.6–1.7	1.8–1.9	1.7–1.9	2.4–2.6	3.1–3.3	1.7–1.9	3.1–3.3	5.3–5.7
	Precipitation (mm)	−24(37.4)	−25.7(31.9)	−12.2(26.3)	−12.4(35.1)	−14.3(31.2)	15(28.8)	−24.9(37.3)	−1.6(30.5)	51.7(30.4)
		−78 to 51	−73 to 45	−51 to 48	−70 to 52	−62 to 56	−27 to 71	−83 to 51	−50 to 64	7–115
Summer	Temperature (°C)	1.3(0.05)	1.5(0.05)	1.7(0.05)	1.5(0.04)	2.2(0.04)	3.0(0.06)	1.5(0.03)	2.8(0.05)	5.2(0.07)
		1.2–1.5	1.4–1.6	1.5–1.8	1.4–1.6	2.1–2.4	2.9–3.1	1.4–1.6	2.7–2.9	5.1–5.5
	Precipitation (mm)	−69.9(24.5)	−71.8(23.3)	−65.2(20)	−69.1(21.8)	−74.7(23)	−72.3(20.7)	−69.1(23.6)	−74.8(23.9)	−75.7(22.6)
		−110 to −24	−111 to −28	−98 to −27	−108 to −29	−113 to −30	−106 to −32	−109 to −24	−115 to −31	−112 to −33
Winter	Temperature (°C)	1.6(0.1)	1.9(0.1)	2.2(0.1)	2.1(0.1)	3.2(0.1)	3.9(0.2)	2.3(0.1)	4.0(0.2)	6.8(0.3)
		1.5–1.8	1.8–2.1	2.1–2.4	1.9–2.4	3.0–3.4	3.7–4.3	2.1–2.6	3.7–4.3	6.3–7.5
	Precipitation (mm)	26.6(4.1)	26.3(3.5)	25.7(2.8)	28.4(4.1)	31.6(3.7)	33.5(2.9)	27.3(3.8)	33.5(3.9)	48.1(4)
		18–41	19–39	19–36	19–41	24–44	27–43	19–40	25–45	40–58

Table 3.
*Same as in **Table 1** but for the Nelson River Basin.*

3.4 Climate projections for the Nelson River Basin

Basin-averaged annual mean air temperature over the Nelson River Basin is projected to increase by up to 5.5°C by the 2080s (**Table 3**; **Figures 3** and **10**). Annual temperature is expected to increase, across the basin, 1.5 to 1.9°C by the 2030s, 1.7 to 3.3°C by the 2050s, and 1.7 to 5.7°C by the 2080s when considering all RCPs (**Table 3**). Although basin-wide average precipitation may increase by up to 51.7 mm (7%) by the 2080s (**Table 3**), there is significant variation across the basin. For example, total annual precipitation is likely to decrease under all RCPs for all periods in the western portion of the basin while slightly increasing in the eastern portion (**Figure 8**).

Average summer temperatures across the Nelson River Basin are projected to increase 1.2 to 1.8°C by the 2030s, 1.4 to 3.1°C by the 2050s, and 1.4 to 5.5°C by the 2080s when considering all RCPs (**Table 3**; **Figures 7** and **12**). Average summer precipitation is projected to decline over the basin throughout the century (**Table 3**; **Figures 9** and **12**) and by the 2080s, rainfall may decrease by up to 115 mm relative to the 1986–2005 reference period (**Table 3**).

Winter temperatures across the Nelson River Basin are projected to increase 1.5 to 2.4°C by the 2030s, 1.9 to 4.3°C by the 2050s, and 2.1 to 7.5°C by the 2080s when considering all RCPs (**Table 3**; **Figure 6**). Winter precipitation is projected to increase by 18–58 mm across the basin under all RCPs (**Table 3**; **Figure 10**). Basin-averaged winter precipitation is projected to increase 25.7 mm (32%) by the 2030s, 33.5 mm (42%) by the 2050s and 48.1 mm (60%) by the 2080s under the business-as-usual greenhouse gas emission scenario (RCP 8.5).

3.5 Climate projections for the Great Lakes Basin

Annual average air temperature across the Great Lakes Basin is projected to increase 1.1 to 1.9°C by the 2030s, 1.4 to 3.3°C by the 2050s and 1.3 to 5.7°C by the

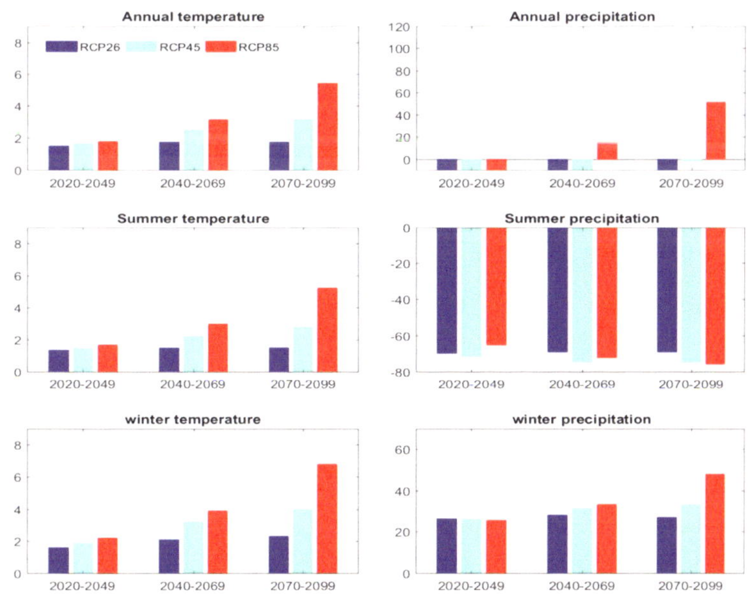

Figure 12.
*Same as in **Figure 4** but for the Nelson River Basin.*

Change from 1986–2005 baseline		2020–2049			2040–2069			2070–2099		
		RCP2.6	RCP4.5	RCP8.5	RCP2.6	RCP4.5	RCP8.5	RCP2.6	RCP4.5	RCP8.5
Annual	Temperature (°C)	1.5(0.1)	1.6(0.1)	1.7(0.1)	1.8(0.1)	2.3(0.2)	3.0(0.2)	1.7(0.1)	3.0(0.2)	5.1(0.4)
		1.1–1.7	1.2–1.7	1.4–1.9	1.4–2.0	1.8–2.6	2.3–3.3	1.3–1.9	2.3–3.3	4.0–5.7
	Precipitation (mm)	30.4(31)	30.2(27.9)	35.7(23.5)	37(32.7)	47.7(26.8)	66.5(23.8)	40.6(32.5)	53(28)	103.2(26.7)
		−91 to 97	−55 to 92	−30 to 90	−93 to 108	−49 to 111	−3 to 119	−83 to 112	−48 to 114	18–168
Summer	Temperature (°C)	1.4(0.1)	1.4(0.1)	1.6(0.1)	1.5(0.1)	2.2(0.1)	2.9(0.2)	1.5(0.1)	2.7(0.1)	4.9(0.3)
		1.1–1.7	1.2–1.6	1.3–1.8	1.2–1.8	1.8–2.4	2.3–3.2	1.1–1.7	2.3–3.0	4.0–5.4
	Precipitation (mm)	−18.8(16.4)	−16.9(15.5)	−16 (14.2)	−18.3(16.5)	−18.6(15.6)	−17.9(14.3)	−15.4(16.4)	−20.2(15.2)	−20.9(14.4)
		−72 to 17	−59 to 19	−54 to 19	−66 to 19	−63 to 19	−53 to 19	−67 to 26	−66 to 17	−57 to 16
Winter	Temperature (°C)	1.8(0.2)	1.9(0.2)	2.0(0.2)	2.2(0.2)	2.8(0.4)	3.5(0.5)	2.2(0.2)	3.5(0.4)	6.0(0.8)
		1.3–2.0	1.3–2.2	1.4–2.3	1.5–2.6	1.9–3.4	2.3–4.3	1.5–2.5	2.4–4.3	3.8–7.4
	Precipitation (mm)	31.1(12.3)	29.6(12.2)	29.2(10.9)	30.6(12.7)	35(12.1)	41(11)	28.3(12.3)	40.2(12.6)	57.6(11.8)
		−47 to 68	−26 to 60	−11 to 57	−51 to 65	−19 to 65	−2 to 71	−46 to 59	−21 to 74	15–89

Table 4.
Same as in ***Table 1*** *but for the Great Lakes Basin.*

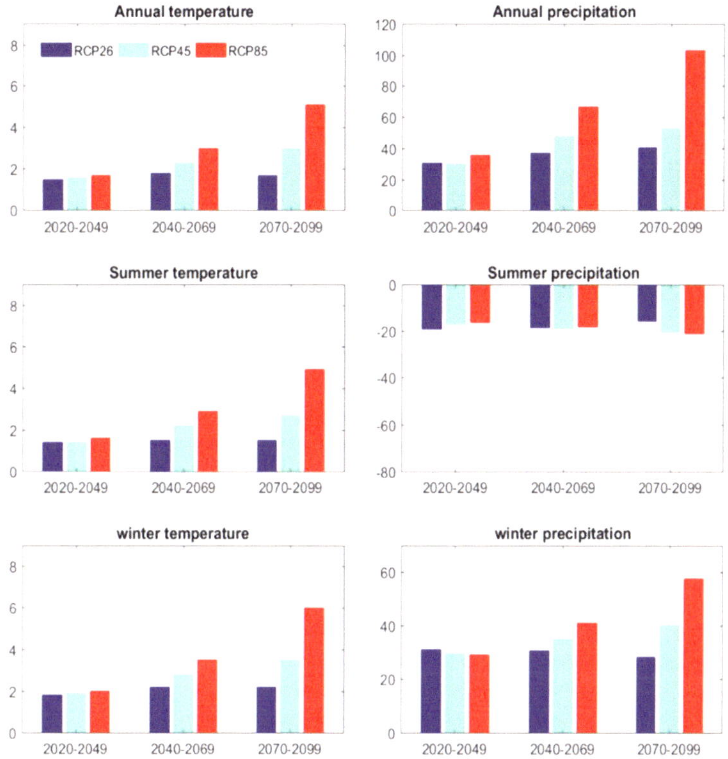

Figure 13.
Same as in ***Figure 4*** *but for the Great Lakes Basin.*

2080s when considering all RCPs (**Table 4**; **Figures 5** and **13**). Warming is projected to be more significant in the northern portion of the basin compared to its southern portion (**Figure 5**). Basin-averaged annual precipitation is projected to increase 35.7 mm (4%) by the 2030s, 66.5 mm (7%) by the 2050s and 103.2 mm (11%) by the 2080s under RCP8.5 (**Table 4**). The greatest change is expected to occur in the Lake Superior Basin (**Figure 8**). Lake effect precipitation is also evident in most future projections (**Figure 8**).

Average summer temperatures across the Great Lakes Basin are projected to increase 1.1 to 5.4°C by the 2080s (**Table 4**; **Figures 7** and **13**). Summer temperature changes are relatively uniform in space. Summer precipitation is projected to decrease over most portions of the basin except for slight increases over lakes and some over-land locations (**Figure 9**). Under RCP 8.5, basin-averaged precipitation is projected to decrease 16 mm (6%) by the 2030s, 17.9 mm (7%) by the 2050s, and 20.9 mm (8%) by the 2080s (**Table 4**).

Average winter temperature across the Great Lakes Basin is projected to increase 1.3 to 2.3°C by the 2030s, 1.5 to 4.3°C by the 2050s, and 1.5 to 7.4°C by the 2080s (**Table 4**; **Figures 6** and **13**). Winter precipitation is expected to increase across the basin under all RCPs. Basin-averaged winter precipitation is projected to increase 29.2 mm (18%) by the 2030s, 41 mm (25%) by the 2050s and 57.6 mm (35%) by the 2080s under RCP 8.5 (**Table 4**; **Figure 13**).

	Change from 1986–2005 baseline		2020–2049			2040–2069			2070–2099		
			RCP2.6	RCP4.5	RCP8.5	RCP2.6	RCP4.5	RCP8.5	RCP2.6	RCP4.5	RCP8.5
Lake Superior Basin	Annual	Temperature (°C)	1.5(0.1)	1.6(0.1)	1.7(0.1)	1.8(0.1)	2.4(0.1)	3.1(0.2)	1.7(0.1)	3.1(0.2)	5.3(0.3)
			1.3–1.7	1.4–1.7	1.4–1.9	1.5–2.0	2.0–2.6	2.6–3.3	1.4–1.9	2.6–3.3	4.5–5.7
		Precipitation (mm)	44.6(12.4)	41.1(14.8)	45.3(15)	52.3(14.9)	56(15.5)	75.3(16.3)	50.9(15.3)	65.4(14.8)	120(16.4)
			19–88	10–84	15–87	24–99	24–99	44–119	21–95	36–112	87–168
	Summer	Temperature (°C)	1.3(0.1)	1.4(0.1)	1.5(0.1)	1.4(0.1)	2.1(0.1)	2.8(0.2)	1.4(0.1)	2.7(0.1)	4.9(0.3)
			1.1–1.5	1.2–1.6	1.3–1.8	1.2–1.6	1.8–2.3	2.4–3.1	1.2–1.6	2.4–2.9	4.4–5.4
		Precipitation (mm)	−16.7(12.2)	−18.8(13)	−16.5(11.8)	−19.5(13.2)	−19.9(13.6)	−20(12.7)	−15.9(12.6)	−21.5(12.5)	−20.7(13.1)
			−43 to 14	−44 to 15	−40 to 11	−47 to 10	−47 to 11	−45 to 8	−41 to 13	−48 to 8	−47 to 9
	Winter	Temperature (°C)	1.8(0.1)	2(0.1)	2.1(0.2)	2.3(0.1)	3.1(0.2)	3.8(0.3)	2.3(0.1)	3.8(0.3)	6.6(0.6)
			1.5–2.0	1.7–2.2	1.7–2.3	2.0–2.6	2.6–3.4	3.0–4.3	2.0–2.5	3.2–4.3	5.3–7.4
		Precipitation (mm)	30.8(12.7)	30.1(14)	28.8(12.5)	33.8(12.9)	35.4(13.5)	38.1(12.2)	31.5(13.1)	38.3(13.9)	53.5(12.6)
			−1 to 53	−1 to 56	1–50	3–55	5–61	11–61	1–55	8–64	25–78
Lake Huron Basin	Annual	Temperature (°C)	1.5(0.1)	1.6(0.1)	1.7(0.1)	1.8(0.1)	2.3(0.1)	3.0(0.2)	1.7(0.1)	3.0(0.2)	5.0(0.3)
			1.2–1.6	1.4–1.7	1.5–1.8	1.5–2.0	2.0–2.5	2.5–3.3	1.5–1.9	2.5–3.2	4.3–5.5
		Precipitation (mm)	44.2(23.6)	43.9(20.8)	45.9(18)	51(24.4)	61.1(21)	75.5(19.8)	58.2(23.8)	65.7(20.3)	109(21.4)
			−33 to 97	−26 to 92	−15 to 90	−33 to 108	−2 to 111	15–119	−21 to 112	−4 to 114	36–155
	Summer	Temperature (°C)	1.4(0.1)	1.5(0.1)	1.6(0.1)	1.6(0.1)	2.2(0.1)	2.9(0.2)	1.5(0.1)	2.7(0.1)	5.0(0.3)
			1.2–1.7	1.3–1.6	1.4–1.8	1.3–1.8	1.9–2.4	2.5–3.1	1.3–1.7	2.4–3.0	4.3–5.3
		Precipitation (mm)	−12.4(13.6)	−9.7(13.6)	−9.4(12.7)	−10.2(13.9)	−11.8(14.2)	−11.2(13)	−8(14.1)	−13.4(13.5)	−15.8(14.1)
			−46 to 17	−42 to 19	−41 to 19	−45 to 19	−41 to 19	−41 to 19	−41 to 26	−44 to 17	−43 to 16
	Winter	Temperature (°C)	1.8(0.1)	1.9(0.2)	2.0(0.2)	2.2(0.2)	2.8(0.3)	3.4(0.4)	2.1(0.2)	3.5(0.3)	5.8(0.7)
			1.5–2.0	1.5–2.1	1.5–2.3	1.7–2.5	2.1–3.2	2.6–3.9	1.7–2.4	2.7–3.9	4.4–6.8
		Precipitation (mm)	37.5(10.9)	33.9(11.5)	32(10.4)	34.7(11.7)	39(11.6)	44.2(10.7)	32.6(11)	44.7(12.3)	60.6(10.9)
			12–68	7–60	7–57	8–65	13–65	19–71	7–59	17–74	35–89

	Change from 1986–2005 baseline		2020–2049			2040–2069			2070–2099		
			RCP2.6	RCP4.5	RCP8.5	RCP2.6	RCP4.5	RCP8.5	RCP2.6	RCP4.5	RCP8.5
Lake Erie Basin	Annual	Temperature (°C)	1.4(0.1)	1.5(0.1)	1.6(0.1)	1.7(0.2)	2.2(0.2)	2.8(0.2)	1.6(0.1)	2.7(0.2)	4.6(0.4)
			1.2–1.6	1.2–1.6	1.4–1.7	1.4–1.9	1.8–2.4	2.3–3.1	1.3–1.8	2.3–3.0	4.0–5.1
		Precipitation (mm)	23.5(21.4)	25(19.6)	30.2(17.8)	30.3(22.8)	45.4(21.1)	65.5(19.1)	35.8(23.9)	40.1(20.3)	89.5(20.7)
			−84 to 70	−54 to 75	−30 to 72	−78 to 82	−49 to 90	−3 to 105	−83 to 84	−48 to 88	18–132
	Summer	Temperature (°C)	1.4(0.1)	1.4(0.1)	1.6(0.1)	1.6(0.2)	2.2(0.2)	2.9(0.2)	1.5(0.1)	2.7(0.2)	4.9(0.4)
			1.1–1.6	1.2–1.6	1.4–1.8	1.2–1.8	1.8–2.4	2.5–3.2	1.2–1.7	2.3–3.0	4.1–5.3
		Precipitation (mm)	−18.3(12.1)	−15.6(11.8)	−17.2(10.5)	−15.4(12.9)	−16.7(11.6)	−15(10.2)	−10.9(10.9)	−19(10.8)	−19.5(9.8)
			−49 to 11	−44 to 14	−42 to 8	−47 to 12	−45 to 10	−40 to 7	−40 to 13	−49 to 10	−43 to 6
	Winter	Temperature (°C)	1.5(0.1)	1.6(0.1)	1.6(0.1)	1.8(0.2)	2.2(0.2)	2.7(0.3)	1.8(0.2)	2.9(0.2)	4.6(0.4)
			1.3–1.7	1.3–1.8	1.4–1.9	1.5–2.2	1.9–2.6	2.3–3.3	1.5–2.1	2.4–3.4	3.8–5.5
		Precipitation (mm)	17.5(9.6)	17.7(8.8)	18.5(6.8)	15.6(9.3)	21.8(8.7)	32.3(7.6)	14.6(9.7)	29.3(10)	48.5(8.1)
			−47 to 47	−26 to 44	−11 to 38	−51 to 45	−19 to 48	−2 to 50	−46 to 43	−21 to 55	15–67
Lake Ontario Basin	Annual	Temperature (°C)	1.4(0.1)	1.5(0.1)	1.6(0.1)	1.7(0.2)	2.2(0.2)	2.9(0.2)	1.6(0.1)	2.8(0.2)	4.8(0.4)
			1.1–1.6	1.3–1.7	1.4–1.8	1.4–1.9	1.9–2.4	2.4–3.1	1.3–1.8	2.4–3.0	4.0–5.2
		Precipitation (mm)	−10.4(30)	−6.7(24.2)	6.5(21.9)	−7.3(31.9)	14.9(25.7)	39.5(23.8)	−3.3(32.3)	16.6(27.7)	70.1(26.1)
			−91 to 55	−55 to 47	−30 to 59	−93 to 59	−49 to 71	−3 to 94	−83 to 62	−48 to 79	18–132
	Summer	Temperature (°C)	1.4(0.1)	1.4(0.1)	1.7(0.1)	1.5(0.2)	2.2(0.1)	2.9(0.2)	1.5(0.2)	2.7(0.2)	4.9(0.3)
			1.1–1.6	1.2–1.6	1.4–1.8	1.2–1.8	1.8–2.4	2.3–3.1	1.1–1.7	2.3–3.0	4.0–5.3
		Precipitation (mm)	−35.9(18.7)	−30.2(16.9)	−29.5(15.1)	−34.8(17.6)	−32(16.3)	−29.1(14.7)	−30.8(19.2)	−32.7(18.1)	−31.6(14.4)
			−72 to 6	−59 to 10	−54 to 8	−66 to 4	−63 to 6	−53 to 7	−67 to 9	−66 to 9	−57 to 2
	Winter	Temperature (°C)	1.7(0.1)	1.7(0.1)	1.8(0.2)	2.1(0.2)	2.5(0.2)	3.2(0.3)	2.0(0.2)	3.2(0.3)	5.3(0.5)
			1.4–1.8	1.5–1.9	1.6–2.1	1.7–2.4	2.1–2.9	2.7–3.6	1.7–2.3	2.7–3.6	4.4–6.1
		Precipitation (mm)	24.2(10.3)	24.4(9.3)	26.8(8.7)	21.2(10.2)	30.8(9.6)	41.4(9.2)	19.9(9.4)	38.3(9.8)	59.9(10.5)
			−40 to 55	−22 to 51	−11 to 51	−41 to 52	−15 to 58	−2 to 64	−41 to 49	−16 to 66	15–83

Change from 1986–2005 baseline			2020–2049			2040–2069			2070–2099		
			RCP2.6	**RCP4.5**	**RCP8.5**	**RCP2.6**	**RCP4.5**	**RCP8.5**	**RCP2.6**	**RCP4.5**	**RCP8.5**
Ottawa River Basin	Annual	Temperature (°C)	1.53(0.05)	1.62(0.04)	1.77(0.04)	1.90(0.02)	2.38(0.04)	3.10(0.03)	1.8(0.04)	3.0(0.05)	5.2(0.11)
			1.5–1.6	1.5–1.7	1.6–1.8	1.8–1.9	2.3–2.4	2.8–3.2	1.7–1.8	2.8–3.1	4.8–5.4
		Precipitation (mm)	−1.7(32)	2.5(29.5)	13.4(23)	3.7(32.8)	21.4(26.6)	43.4(20.3)	9.7(35.3)	28.1(30.3)	80.9(22.9)
			−66 to 62	−50 to 59	−26 to 62	−64 to 74	−31 to 78	6–84	−65 to 79	−32 to 88	35–122
	Summer	Temperature (°C)	1.43(0.06)	1.45(0.05)	1.67(0.05)	1.58(0.05)	2.2(0.03)	2.9(0.06)	1.5(0.05)	2.76(0.05)	5.1(0.06)
			1.3–1.5	1.4–1.5	1.6–1.8	1.5–1.7	2.1–2.3	2.8–3.0	1.4–1.6	2.7–2.8	4.8–5.2
		Precipitation (mm)	−28.8(18.9)	−22.6(17)	−22 (15)	−26.7(17.5)	−25.4(16)	−24.4(13.8)	−24.8(18)	−26.8(16.6)	−28.8(12.9)
			−72 to 4	−59 to 6	−54 to 4	−65 to 4	−63 to 2	−53 to −1	−65 to 7	−66 to 0	−57 to −7
	Winter	Temperature (°C)	1.9(0.1)	2.0(0.1)	2.2(0.1)	2.4(0.1)	3.0(0.1)	3.8(0.2)	2.3(0.1)	3.8(0.2)	6.4(0.3)
			1.7–2.0	1.8–2.2	1.9–2.3	2.2–2.5	2.6–3.3	3.3–4.1	2.1–2.4	3.3–4.1	5.4–7.1
		Precipitation (mm)	32.2(5.4)	31.4(4.9)	33.5(4.8)	31(5.8)	37.7(4.8)	46.2(5.7)	27.4(5.9)	44.2(5.4)	65.1(6.6)
			23–54	22–48	25–47	19–50	28–55	38–63	16–48	33–62	54–84

Table 5.
*Same as in **Table 1** but for the five Great Lakes sub-basins.*

3.6 Climate projections for the Great Lakes sub-basins

By the 2080s, the greatest increase in basin-averaged annual mean temperature is projected to occur over the Lake Superior sub-basin (1.7–5.3°C warmer relative to the 1986–2005 reference period), followed by the Ottawa River sub-basin (1.8–5.2°C warmer), the Lake Huron sub-basin (1.7–5.0°C warmer), the Lake Ontario sub-basin (1.6–4.8°C warmer), and the Lake Erie sub-basin (1.6–4.6°C warmer) when considering all RCPs (**Table 5**).

Basin-averaged annual precipitation is projected to increase over all five sub-basins under almost all RCPs (**Table 5**). By the 2080s, annual precipitation is likely to increase the most over the Lake Superior sub-basin (50.9–120 mm above 1986–2005 reference period when considering all RCPs), followed by the Lake Huron sub-basin (58.2–109 mm), the Lake Erie sub-basin (35.8–89.5 mm), the Ottawa River sub-basin (9.7–80.9 mm) and the Lake Ontario sub-basin (−3.3 to 70.1 mm) (**Table 5**).

Projected changes in basin-averaged summer mean air temperature are relatively uniform across the five sub-basins (**Figure 7**; **Table 5**). By the 2080s, the basin-averaged summer air temperature increases are projected to range from 1.5 to 4.9°C over the Lake Ontario sub-basin, 1.4 to 4.9°C over the Lake Superior sub-basin, 1.5 to 5.1°C over the Ottawa River basin, 1.5 to 5.0°C over the Lake Huron sub-basin and 1.5 to 4.9°Cover the Lake Erie sub-basin, when considering all RCPs (**Table 5**).

Summer precipitation is projected to decrease over all sub-basins (**Table 5**; **Figure 9**). By the 2080s, the greatest summer precipitation decrease is projected to occur over the Lake Ontario sub-basin (a decrease of 30.8–32.7 mm), while the least summer precipitation decrease is projected to occur over the Lake Huron sub-basin (a decrease of 8–15.8 mm) (**Table 5**; **Figure 9**) when considering all RCPs.

Significant spatial variation is found in the projected changes in winter air temperatures among the Great Lakes sub-basins. By the 2080s, the basin-averaged winter temperature is projected to be 2.0–5.3°C above the 1986–2005 reference values for the Lake Ontario sub-basin, 2.3–6.6°C for the Lake Superior sub-basin, 2.3–6.4°C for the Ottawa River sub-basin, 2.1–5.8°C for the Lake Huron sub-basin, and 1.8–4.6°C for the Lake Erie sub-basin when considering all RCPs (**Table 5**). The Lake Superior sub-basin may experience the greatest winter warming by the 2080s (up to 6.6°C warmer than the reference period under RCP8.5), followed by the Ottawa River sub-basin (up to 6.4°C warmer), the Lake Huron sub basin (up to 5.8°C warmer), the Lake Ontario sub-basin (up to 5.3°C warmer), and the Lake Erie sub-basin (up to 4.6°C warmer) (**Table 5**).

Among the five sub-basins, the Ottawa River sub-basin is projected to experience the greatest increase in winter precipitation by the 2080s (27.4–65.1 mm above the reference period), followed by the Lake Huron sub-basin (32.6–60.6 mm), the Lake Ontario sub-basin (19.9–59.9 mm), the Lake Superior sub-basin (31.5–53.5 mm), and the Lake Erie sub-basin (14.6–48.5 mm), when considering all RCPs (**Table 5**). Dramatic spatial variations are found in the projected winter precipitation changes, even within a sub-basin. For example, the lake-effect snow belt areas in the Lake Huron sub-basin could expect up to 89 mm more winter precipitation than the 1986–2005 baseline values by the 2080s under RCP8.5, much higher than the rest of this sub-basin (**Table 5**; **Figure 10**).

4. Conclusions and discussion

The purpose of this chapter is to summarize the trends and spatial patterns in projected changes in annual and seasonal mean temperature and precipitation over the entire Province of Ontario, its three primary watersheds and associated sub-basins, under three IPCC's AR5 scenarios. These projections were developed using the IPCC-endorsed CMIP5 climate projections and state-of-the-science combine downscaling methodologies [6–9, 19].

These Ontario-specific high-resolution climate projections presented in this chapter indicate that increases in average annual temperature are expected to continue through the 21st century across the Province of Ontario; the Hudson Bay Basin could experience the largest increase in temperature (6.0°C) by the 2080s as compared to the Nelson River Basin (5.5°C) and the Great Lakes Basin (5.1°C) under the business-as-usual greenhouse gas emission scenario (RCP 8.5).

Projected changes in precipitation vary more significantly from region to region than the projected changes in temperature. While annual precipitation in the Nelson River Basin may decrease, precipitation may increase in other basins. The Great Lakes Basin is expected to experience precipitation increases, particularly over and near the lakes and in the Lake Superior sub-basin, and most dramatically in winter.

More detailed analyses on projected climate changes for 42 climate variables (including extremes) over 50 regions and 150 municipalities in Ontario are available through York University's Ontario Climate Data Portal (ODCP) [7–9]. Readers are referred to the OCDP and its references for these additional variables and more details on the development of the super-ensemble of Ontario-specific high-resolution climate projections.

While this study is based on the latest available science and data to us when this study was carried out, we understand that climate science and climate modeling techniques evolve rapidly [2]; therefore, as new data becomes available, specifically new CMIP6-based high-resolution regional projections using more advanced downscaling methodologies specific to the study area, this chapter should be updated to reflect the most up-to-date science and data.

Acknowledgements

The authors would like to thank Professor William Richard Peltier at the University of Toronto, Professor Gordon Huang at the University of Regina, the Pacific Climate Impacts Consortium (PCIC), the North America—Coordinated Regional Downscaling Experiment (NA-CORDEX) for providing/allowing us to use their data for our combined downscaling in developing the Ontario-specific high-resolution regional climate projections on the Ontario Climate Data Portal, which are used in this study. The authors would also like to thank Xiaolan Zhou from York University for her significant effort in making the maps used in this chapter, Shannon Fera, Jenny Gleeson, Adam Hogg from the Ministry of Natural Resources and Forestry, and Jen Turnbull from the Ministry of Agriculture, Food and Rural Affairs for their constructive suggestions/edits in improving this chapter.

Conflict of interest

The authors declare no conflict of interest.

DOI: http://dx.doi.org/10.5772/intechopen.112416

Author details

Jinliang Liu[1*], Huaiping Zhu[2], Ziwang Deng[2] and Chris Charron[1]

1 Ontario Ministry of Environment, Conservation and Parks, Toronto, Canada

2 Laboratory of Mathematical Parallel Systems, Department of Mathematics and Statistics, York University, Toronto, Canada

*Address all correspondence to: jinliang.liu@ontario.ca

References

[1] Field CB, Barros VR, Dokken DJ, Mach KJ, Mastrandrea MD, Bilir TE, et al., editors. Summary for policymakers. In: Climate Change 2014: Impacts, Adaptation, and Vulnerability. Part A: Global and Sectoral Aspects. Contribution of Working Group II to the Fifth Assessment Report of the Intergovernmental Panel on Climate Change. Cambridge, United Kingdom and New York, NY, USA: Cambridge University Press; 2014. pp. 1-32

[2] Zhou T. New physical science behind climate change: What does IPCC AR6 tell us? Innovation (Camb). 2021;**2**(4):100173. DOI: 10.1016/j.xinn.2021.100173

[3] Environment and Climate Change Canada. Climate Trends and Variations Bulletin [Internet]. 2023. Available from: https://publications.gc.ca/collections/collection_2023/eccc/En81-23-2022-eng.pdf [Accessed: May 08, 2023]

[4] Chadwick P, Hume B. Weather of Ontario. UK: Lone Pine Publishing. p. 240

[5] Szeto KK. On the extreme variability and change of cold-season temperatures in Northwest Canada. Journal of Climate. 2008;**21**:94-113. DOI: 10.1175/2007JCLI1583.1

[6] Deng Z, Liu J, Qiu X, Zhou X, Zhu H. Downscaling RCP8.5 daily temperatures and precipitation in Ontario using localized ensemble optimal interpolation (EnOI) and bias correction. Climate Dynamics. 2018;**51**:411-431. DOI: 10.1007/s00382-017-3931-3

[7] Zhu H, Deng Z. The Ontario Climate Data Portal (OCDP) [Internet]. 2018. Available from: https://yorku.ca/ocdp [Accessed: May 08, 2023]

[8] Zhu H, Deng Z, Liu J, Qiu X, Chen X, Zhou X. A Look at Ontario's Climate of the Future with the Ontario Climate Data Portal (OCDP). Bulletin of the Canadian Meteorological and Oceanographic Society. [Internet]. 2018. Available from: https://bulletin.cmos.ca/ocdp/ [Accessed: May 08, 2023]

[9] Zhu H, Liu J, Zhou X, Chen X, Qiu X, Bello R, et al. The Ontario climate data portal, a user-friendly portal of Ontario-specific climate projections. Nature – Scientific Data. 2020;7:147. DOI: 10.1038/s41597-020-0489-4

[10] Simcoe Muskoka District Health Unit. A Changing Climate: Assessing Health Impacts and Vulnerability due to Climate Change Within Simcoe Muskoka [Internet]. 2017. Available from: https://www.simcoemuskokahealth.org/docs/default-source/topic-environment/smdhu-vulnerability-assessment-2017-finale1e3e25f97be6bc38c2dff0000a8dfd8.pdf?sfvrsn=0 [Accessed: May 08, 2023]

[11] Haliburton, Kawartha, Pine Ridge (HKPR) District Health Unit. Climate Change Health Vulnerability and Adaptation Assessment [Internet]. 2018. Available from: https://pub-kawarthalakes.escribemeetings.com/filestream.ashx?DocumentId=34379 [Accessed: May 08, 2023]

[12] York Region: Historical and Future Climate Trends in York Region [Internet]. 2016. Available from: https://climateconnections.ca/app/uploads/2015/02/Historical-and-Future-Climate-Trends-in-York-Region_Report-1.pdf [Accessed: May 08, 2023]

[13] Durham Region: Integrating Climate Change Considerations into Plans and Policies in Durham Region

[Internet]. 2018. Available from: https://climateconnections.ca/app/uploads/2018/10/Integrating-CC-in-Durham.pdf [Accessed: May 08, 2023]

[14] Pacific Climate Impacts Consortium, University of Victoria. Statistically Downscaled Climate Scenarios [Internet]. 2017. Available from: https://data.pacificclimate.org/portal/downscaled_gcms/map/ [Accessed: May 08, 2023]

[15] Mearns LO, et al. The North American CORDEX (NA-CORDEX) Program [Internet]. 2017. Available from: https://na-cordex.org/ [Accessed: May 08, 2023]

[16] Erler AR, Peltier WR, d'Orgeville M. Dynamically downscaled high resolution hydro-climate projections for Western Canada. Journal of Climate. 2015;**28**:423-450. DOI: 10.1175/JCLI-D-14-00174.1

[17] Peltier WR, d'Orgeville M, Erler AR, Xie F. Uncertainty in future summer precipitation in the Laurentian Great Lakes Basin: Dynamical downscaling and the influence of continental-scale processes on regional climate change. Journal of Climate. 2018;**31**:2651-2673. DOI: 10.1175/JCLI-D-17-0416.1

[18] Wang X, Huang G, Liu J, Li Z, Zhao S. GIS-based data portal for climate change impact assessment. Bulletin of the American Meteorological Society. 2016;**97**:17-18

[19] Cohen S, Bush E, Zhang X, Gillett N, Bonsal B, Derksen C, et al. Synthesis of findings for Canada's regions. In: Bush E, Lemmen DS, editors. Canada's Changing Climate Report. Canada: Government of Canada; 2019. pp. 424-443. Available from: https://changingclimate.ca/site/assets/uploads/sites/2/2019/01/CCCR-Chapter8-ChangesInCanadasRegionInANationalGlobalContext.pdf [Accessed: May 08, 2023]

[20] McDermid J, Fera S, Hogg A. Climate change projections for Ontario: An updated synthesis for policymakers and planners. Queen's Printer for Ontario. 2015;**2015**:40

[21] Colombo SJ, McKenney DW, Lawrence KM, Gray PA. Climate change projections for Ontario: Practical information for policymakers and planners. Queen's Printer for Ontario. 2007;**2007**:49

[22] Deng Z, Qiu X, Liu J, Madras N, Wang X, Zhu H. Trend in frequency of extreme precipitation events over Ontario from ensembles of multiple GCMs. Climate Dynamics. 2016;**46**:2909-2921. DOI: 10.1007/s00382-015-2740-9

[23] Jones RG, Noguer M, Hassell DC, Hudson D, Wilson SS, Jenkins GJ, et al. Generating High Resolution Climate Change Scenarios Using PRECIS. Exeter, UK: met Office, Hadley Centre; 2004. p. 40. Available from: https://www.metoffice.gov.uk/binaries/content/assets/metofficegovuk/pdf/research/applied-science/precis/precis_handbook.pdf

[24] Smith MR, Myers SS. Impact of anthropogenic CO_2 emissions on global human nutrition. Nature Climate Change. 2018;**8**:834-839. DOI: 10.1038/s41558-018-0253-3

[25] Deng Z, Qiu X, Zhou X, Zhu H. Ontario Climate Trends and Variations Bulletin [Internet]. 2018. Available from: https://lamps.math.yorku.ca/OntarioClimate/ClimateTrendsandVariationsBulletin/index.htm [Accessed: May 08, 2023]

[26] Ontario Climate Data Portal (OCDP), York University. Historical and Projected Climate Trends over Ontario [Internet]. 2017. Available from: https://lamps.math.yorku.ca/

OntarioClimate/assets/img/plot/
OntarioClimateChangeTrend.png
[Accessed: May 08, 2023]

[27] Vincent LA, Wang XL, Milewska EJ, Wan H, Yang F, Swail V. A second generation of homogenized Canadian monthly surface air temperature for climate trend analysis. Journal of Geophysical Research-Atmosphere. 2012;**117**:1-13. DOI: 10.1029/2012JD017859

[28] Deng Z, Qiu X, Zhou X, Zhu H. Trends of Precipitation in Ontario [Internet]. 2018. Available from: https://lamps.math.yorku.ca/OntarioClimate/PrecipitationTrends/index.htm [Accessed: May 08, 2023]

Chapter 2

The Influence of Climate Warming on the Hydrological Regime of Thermokarst Lakes in the Subarctic (Chukotka, Russia)

Oleg D. Tregubov, Pavel Ya. Konstantinov, Vladimir V. Shamov and Konstantin K. Uyagansky

Abstract

Using remote methods and materials for meteorological observations, climate changes and the area of 36 thermokarst lakes located in the Anadyr lowland in Chukotka over a 65-year period were analyzed. More than 20 lakes were studied by field methods. With an increase in the average annual air temperature by 1.8°C and an increase in the amount of annual precipitation by 135 mm, the total area of the lakes mirror decreased by 24%. Cryogenic processes have had a significant impact on the decrease in the water quantity of lakes. Thermal erosion in drainage channels has led to multiple discharges of water in abnormally warm years. The heaving of permafrost in the coastal zone affected the reduction of the lake catchment area. If the trends of climate change continue, further drainage of large lakes and an increase in the number of small sag pond is expected in the next 25 years.

Keywords: thermokarst lakes, thermal erosion, active layer, global warming, Chukotka

1. Introduction

Periglacial plains of the Late Pleistocene in East Asia currently represent the distribution areas of thermokarst-drained lake basins (DLB). The beginning of the formation of lake basins is associated with climate warming and areal thermokarst at the turn of the Late Pleistocene and Holocene [1, 2].

The ratio of the total surface of lakes to the catchment in Arctic lowlands of the cryolithozone locally reaches 60%. Secondary thermokarst lakes are mostly located within the boundaries of the DLB on the terraces of river valleys and the sea coast. Such reservoirs are formed as a result of the melting of Holocene underground ice, which is accompanied by subsidence of the surface and filling of depressions with water. The area of thermokarst lakes varies from 0.001 to 20 km^2. The depth of most lakes is 1–1.2 m and rarely exceeds 3 m [2–5]. In winter, shallow reservoirs freeze to the bottom. The age of most modern thermokarst lakes in the DLB is several 100 years, exclusively up to a 1000 years [1, 3]. The water of the lakes has a brown color, contains a significant amount of dissolved organic substances, and is not suitable for water supply [4]. At the same time, the DLB, together with lakes, swamps, and floodplains of rivers, are ecologically significant wetlands: nesting sites for migratory birds, summer feeding of freshwater fish fauna, and spring pastures of wild reindeer [6]. Having a significant total area, lakes significantly impact the microclimate: they increase air humidity and reduce seasonal and daily temperature differences. In the summer, the runoff of the water of small rivers is provided, and excessive precipitation is deposited in the flood. Despite the ubiquity and duration of existence in the cryolithozone, thermokarst lakes are quite sensitive to interannual and long-term changes in water supply and runoff conditions. This prompted researchers from various fields of science to study the influence of modern climate warming on the water content of thermokarst lakes.

Global warming is not the result of an increase in solar activity, but is associated with changes in atmospheric air circulation on a planetary scale, which lead to an increase in air temperature [7–9]. Therefore, the ongoing climate changes should be considered as a shift of the boundaries of climatic areas in the latitudinal and meridional direction. In the high latitudes of the northern hemisphere, this is a redistribution of the zones of distribution of marine, temperate, and sharply continental subarctic and Arctic climates. The change in the location of the borders of climatic areas is accompanied by both an increase and a decrease in solar insolation, average seasonal air temperatures, as well as an increase or decrease in precipitation. The most contrasting and dynamic climatic changes in the arctic and subarctic occur in the transition zone from the ocean to the continent.

Regional transformations of climatic conditions have a corresponding impact on the water balance of territories. Depending on the type of changing climate, the atmospheric nutrition of water bodies, the intensity of evaporation, and river runoff vary in different ways. In general, this allows us to assume that knowledge about the trends of regional climatic changes and the rate of waterlogging of the lake is sufficient to predict the parameters of the lake's water content. However, the validity of this approach is not confirmed by the results of numerous remote studies of the hydrological regime of lakes in the subarctic lowlands of Eastern Siberia and North America [10–12]. In the cryolithozone, in areas with different climates and trends in its changes, the same type of changes is often recorded: the drainage of large lakes and an increase in the number of small reservoirs [1, 3]. This suggests that, along with the climate, the dynamics of the lake area is influenced by transboundary factors of changes in the hydrological regime, leveling the diversity of regional climatic changes. It is obvious that such a factor, given the thermokarst origin of lakes and the spread of permafrost in the subarctic, may be cryogenic processes, which are most affected by global warming and climatic fluctuations.

In this chapter, using the example of thermokarst lakes of the Anadyr lowland of the Bering Sea coast in Chukotka, changes in the area of lakes under the influence of climatic dynamics of seasonal melting, thermokarst, permafrost heaving, and thermoabrasion are considered. The scientific work is based on the results of research of predecessors, remote comparative analysis of cartographic materials from different years and, importantly, on field measurements of lake sizes and observations of cryogenic processes.

2. Predecessor studies

The list of works devoted to the remote study of the dynamics of the lake area in the Arctic lowlands is very long. This article does not overview these publications and is not intended to provide their exhaustive analysis. We considered the works listed below as the most interesting and significant ones. It was found that during 1965–2016, the area of lakes in the Kolyma R. lowland decreased by 7%, averagely [13]. At the same time, it was noted that the interannual dynamics of climatic indicators does not affect the water capacity of the objects. In a later work, when studying the areas of distribution of rocks of the ice complex, an increase in the area of thermokarst lakes was noted in 1999–2013 by 0.89% and in 1999–2018 by 4.1% [14]. The work shows that changes in the total area of lakes within one area are subject to statistical laws [15]. Methodological aspects of remote retrospective analysis are considered in the case of the Eurasian lowlands [16]. Author notes a slight increase in the water capacity of lakes and expresses the opinion that global climate warming slightly effects the water capacity of the lowlands in the northern hemisphere. Other studies consider dynamics of areal thermokarst and number of thermokarst lakes in Western and Eastern Siberia and track the changes in the water capacity in Yamal areas caused by anthropogenic impact [10, 17–19]. Intensive remote studies of the lakes in the Arctic plains were conducted in North America. In 1948–2013, the authors noted a decrease in the area and number of lakes in northern Alaska by 30.3% and 17.1%, respectively [20]. An earlier study in western Alaska conducted in the period from 1949 to 2002 showed draining of 50 out of 7400 remotely analyzed lakes [11]. The reduction in the area of lakes in northern Canada is described by researchers [12, 21, 22]. The authors pay attention to the drainage of large lakes due to the formation of new ways of surface runoff. The problem of water discharge under abnormal weather conditions is considered in the example of lakes in northeastern Alaska [23]. Abnormal precipitation in the winter period of 2017–2018 led to erosion of the shores and a one-time discharge of 192 lakes. Changes in the water capacity of lakes were recorded in mountain permafrost conditions of China, on the QinghaiTibet Plateau [24]. Researchers noted an increase in the number of small and large thermokarst lakes in 1969–2010.

Lakes and thermokarst in Chukotka and, in particular, in the Anadyr lowland began to be actively studied by geocryologists and hydrogeologists in the last century. Among the works in which the problems of genesis and transformation of water bodies are considered in detail, the works [3–5, 25] should be noted. Lyubomirov investigates the conditions of formation and evolution of the lakes of the Anadyr lowland (1990). Krivoshchekov analyzes the experience of recultivation of lakes for the cultivation of meadows (2000). Tregubov [4] and Ruzanov [5] focus on the applied significance of lakes as sources of water supply, and analyze the genesis of water bodies and their interaction with hydrogenic talics. The results of remote sensing studies of 8305 thermokarst lakes in the Anadyr lowland are presented (2013) in the work of Rodionova [16]. According to the interpretation of Landsat satellite images taken between 2009 and 2013, the surface area of 338 water bodies (4% of the sample) decreased by 86 km^2 (3.3%). And only one lake from the surveyed reservoirs has slightly increased its water capacity. Field observations were not carried out; Hydrological processes and overgrowth of reservoirs were named as the reasons for the change in the area of lakes. Case studies of recent years of Tregubov with co-authors characterize changes in permafrost and climatic conditions of the Anadyr lowland (2020) over the past 25 years and analyze climatic changes in the water balance [26].

Thus, in most studies, a decrease in the area of water bodies is recorded, and there is no evidence of activation of thermokarst during a more than 50-year observation period [27–29]. In a short series of observations lasting about 10–20 years, the authors noted the activation of thermokarst and an increase in the proportion of lake surface in basins due to the formation of new small deepening lakes [1, 19, 28–33]. At the same time, it is confirmed that the area of large lakes is decreasing. Surface water runoff, increased evaporation, accumulation of bottom sediments and overgrowth of reservoirs, complete thermokarst, thermal abrasion, and treatment of icy soils along the shores of lakes are indicated as the reasons for the drainage of basins.

Summing up the results of the review of studies of changes in area lakes of lowlands in the high latitudes of the northern hemisphere, we note the following:

1. Long periods of observations of changes in the number and area of lakes show a tendency to drain large reservoirs. The short-term analysis gives contradictory results about changes in the water content of lakes, which is probably due to interannual fluctuations in climatic conditions.

2. In subarctic regions, where intermittent and insular permafrost is becoming common, there is an increase in the water content of lakes and swamps with an increase in climate humidity. In the areas of occurrence of rocks of the ice complex, the authors record the preservation or a slight increase in the water content of lakes over the past 15–20 years.

3. Most remote studies are not confirmed by the author's field observations of reservoirs, the establishment of the causes of degradation of lakes, or the preservation of their morphology. Conclusions about the causes of dehumidification or increase are hypothetical.

4. When interpreting the results of the assessment of the lake character of lowlands and the water content of specific lakes, materials for monitoring seasonal thawing of the active layer and field observations of the activity of cryogenic processes due to climate fluctuations were practically not used.

3. The study area

The Anadyr lowland is located on the southeastern outskirts of Chukotka and covers an area of 35,000 km^2. The climate of the territory is subarctic, moderately continental, and marine. According to the Anadyr Meteorological Station, the average annual temperature for the period 1981–2010 is −5°C. The annual precipitation is 382 mm; most of it falls in winter. The thickness of the continuous permafrost decreases from 300 to 50 m from north to south, where it becomes intermittent. The temperature of frozen soils at the bottom of the layer of annual heat turnover varies from north to south from −7.1°C to −1°C. The depth of seasonal thawing in undisturbed flat landscapes is 45–55 cm. The area occupied by lakes ranges from 20 to 60%. Lake water is characterized by a bicarbonate-sodium composition, a neutral or slightly acidic reaction, and a low salinity of 15–30 mg/l. As for chemical composition of waters of the lakes located on low sea-shore terraces near the coastline of Anadyr Estuary, the proportion of chlorides in them increases; salinity can increase from 20–50 mg/L to 1.5–2.0 g/L. The waters of the drained lake basins are brown color and contain increased concentrations of total iron.

It is important to note that old or modern thermokarst as a cryogenic phenomenon is inherent more or less in all categories of lakes. Therefore, S.V. Tomirdiaro classified most of the lowland lakes as thermokarst lakes [1]. Thermokarst lakes of the sinkhole type are located on elevated areas of the distribution of relict glacial landforms and permafrost of the late Pleistocene. These are relatively deep (3–5 m) lakes with an uneven funnel-shaped bottom, formed during the melting of ice deposits and subsidence of the surface. The most common secondary thermokarst lakes with a flat bottom and a depth of 1.2–3.0 m, occupying the bottom of DLB or flattened saddles of watersheds. The objects of our direct study included 36 lakes with an area of 0.01–0.50 km^2 located at a distance of 7–36 km south and southwest of the city of Anadyr (**Figure 1**). The reservoirs are located within the DLB, confined to the interridge watershed relief depressions (5)[1], gentle slopes (10), river valleys (15), and sea-shore terraces (6). The absolute marks of the water's edge vary from 8 to 80 m, the depth of the lakes is 1–4 m, the salinity of the waters is 15–60 mg/L. According to the pattern of water exchange, we classified all surveyed lakes into landlocked lakes and lakes with seasonal overflow (20), lakes with permanent overflow (11), and drainage lakes (5).

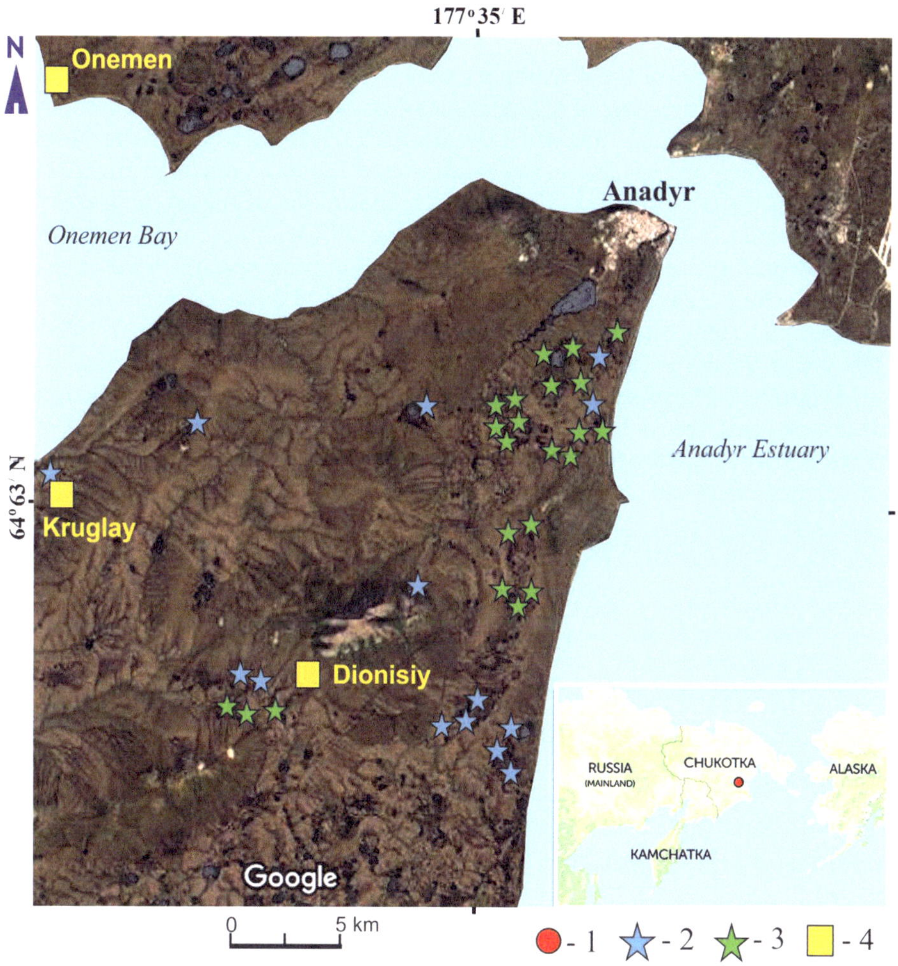

Figure 1.
Location of the study area: 1 – The study area on the inset map; 2 – Lakes studied on the basis of cartographic data; 3 – Lakes studied on the basis of cartographic data and field observations; and 4 – Monitoring areas of the active layer.

[1] Here and below, the number of lakes is indicated in brackets.

The lakes are fed by summer precipitation; snowmelt water and summer melt water of seasonal and perennial underground ice. The melt water of the underground ice enters the lakes as part of the underground runoff of the suprapermafrost waters.

The dynamics of the depth of seasonal thawing was analyzed based on the results of long-term monitoring of the active layer (AL) at the facilities of the international CALM[2] program (see **Figure 1**). This is the "Onemen" site, which occupies the flat top of a tundra ouval with a height of 26 m, covered with hummocky moss and cotton. The site of "Dionisiy", covered with a bumpy yernikov moss-grass tundra, is located on a slope of 2–3° at the foot of the mountain at an altitude of 120 m. The "Round" site was laid in 2010 on the bottom of the DLB with polygonal relief and sphagnum-sedge vegetation. The absolute height of the bottom is approximately 6 m. In general, the plots represent typical landscapes of catchment basins and the bottom of the DLB.

4. Methods

The main objectives of the research part of the work were: statistical verification of the representativeness of lake area measurement data; analysis of trends in lake water content; identification of the causes and leading factors determining the hydrological regime of lakes in the conditions of modern climate change. At the preliminary, preparatory stage of the work, the boundaries of the research area were determined, and a list of lakes that are suitable for reliable remote and direct assessment of morphology and size was compiled. The laboratory research methodology is based on a comparative analysis on the same scale (1:25,000) of the contours of 36 lakes on a topographic map compiled on the basis of aerial photography in 1953 and satellite images from the Google Maps application based on the results of the 2018 survey (**Figure 2**). Morphometric characteristics of lakes, that is, perimeter, area, and linear dimensions, were determined using the Universal Desktop Ruler V. 3.8.6498 software. Statistical parameters of the lake (arithmetic mean, asymmetry, kurtosis, and frequency) were calculated using Microsoft Excel tools.

Figure 2.
An example of a comparative analysis of lakes in the valley of the creek Promyslovyy on a topographic map (left) and on a satellite image (right).

[2] Circumpolar Active Layer Monitoring.

The field stage of the work took place in August 2020, and 22 lakes were surveyed. The technical capabilities of optical devices made it possible to study reservoirs with an area of 0.01–0.4 km^2. The transverse dimensions were measured using an RGK D1000 laser rangefinder. The measurement range was 3–1000 m; the technical accuracy at a distance of 500 m was 1.0–1.8 m (depending on weather conditions). The absolute height of the water edge of the lakes was determined by the elevation marks on topographic maps at a scale of 1:25,000. During field observations, the height of the water's edge was determined in accordance with the measurements of the GPS navigator (Garmin Legend HCx). Each reservoir was surveyed along the perimeter. We have received information about the state of coastal ledges, feeding streams, and surface runoff channels; we identified the landslip slope and determined the depth of lakes, the area of shoals, the composition of the bottom soil, groundwater outlets, as well as the salinity (electrical conductivity) and pH of lake waters. In the coastal zone, we recorded floodplain terraces, polygonal relief, ice mounds, and tand thermokarst depression lakes with a diameter of 3–15 m. The depth of seasonal thawing along the shores of lakes in swampy areas and on dry terraces was measured with a metal probe 1.2 meters long. Soil moisture was measured at a depth of 25 cm using a TK-100-01 moisture meter.

At monitoring sites measuring 100 × 100 m, the depth of thawing in the active layer was measured annually from August 25 to September 5 according to the 10 × 10 m scheme (Onemen since 1994, Dionisiy since 1996, Round since 2010).

5. Results

The results of statistical analysis of remote analysis data are shown in the diagrams below. The histogram shows the frequency distribution of lakes with different water capacity variability (**Figure 3**). Frequency distribution analysis is necessary in order to understand how homogeneous the data sample is. It was important to find out how many isolated frequency distributions were in the data array and, accordingly, the processes that determine the water content of lakes. One normal frequency distribution has been revealed, which is disrupted by a large number of reservoirs that have dried up by 90% or more. As it turned out, five of the eight such lakes were drained during land reclamation in the 1970s for the cultivation of meadows.

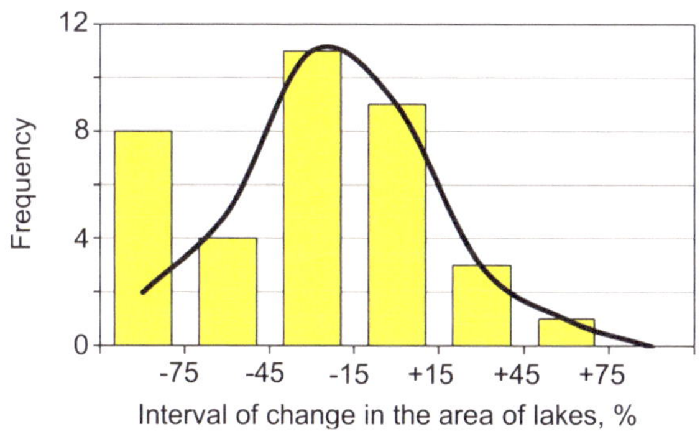

Figure 3.
Histogram of the frequency of occurrence of lakes with varying degrees of watering and drainage. The distribution graph is shown without taking into account lakes drained as a result of land reclamation.

After excluding them from the analysis, the empirical distribution acquired a normal form (see the distribution curve) with an average percentage decrease in the water surface area of −0.24 (−24%), a standard deviation of 0.36, an asymmetry of −0.31 and a peak kurtosis of 0.52.

Pie charts show the ratio of changes in the area of lakes in groups with different runoff conditions (**Figure 4**). Closed lakes that do not have surface runoff paths have preserved their water surface area to the greatest extent. Feeding and drainage such lakes occurs as part of the suprapermafrost waters of the active layer. Among the reservoirs with seasonal runoff during high water, the drained lakes are twice as large as those that have preserved or increased their area. Lakes with constant flow and flowing lakes represent a variety of changes in water content. At the same time, among the flowing lakes, there are the largest number of reservoirs that have increased their area by 45–75%, but there are no lakes with an increase in water content within 45–15%. There is also a surprisingly small amount in the drainage range of 55–25% among closed lakes and lakes with seasonal wastewater (see **Figure 4**).

A pointed diagram of the distribution of lakes by the absolute height of the water's edge shows two clouds of scattered objects separated by an interval of heights of 40–60 m (**Figure 5a**).

In the relief of the territory, this interval of heights corresponds to the foothills of hills covered with a plume of deluvial sediments and tundra ridges, remnants of the 3rd marine terrace, composed of ice-bearing glacial–marine sediments belonging to the early interglacial transgression. This explains the absence of lakes at these heights. In another dot graph, lakes are grouped according to their original size (**Figure 5b**). The dissipating cloud bounds an almost isosceles triangle with an area of 0.2 km^2 horizontally and −24% vertically. As the initial area of water bodies increases, the spread in the amount of their drainage-watering degrees decreases. This is probably

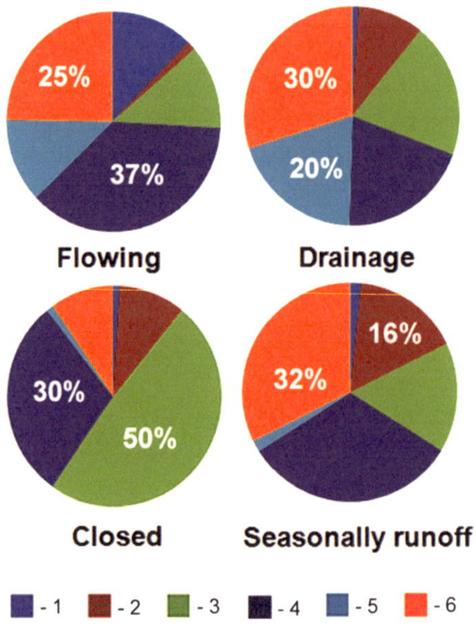

Figure 4.
The ratio of the number of watered and drained lakes in groups with different flow conditions: 1-2 — lakes whose area has increased significantly: 1 – 1,75:1,45; 2 – 1,45:1,15; 3 – lakes without significant changes 1.15:0.85; 4-6 — lakes in which there was a decrease in area: 4 – 0.85:0.55, 5 – 0.55:0.25, and 6 – 0.25:0.

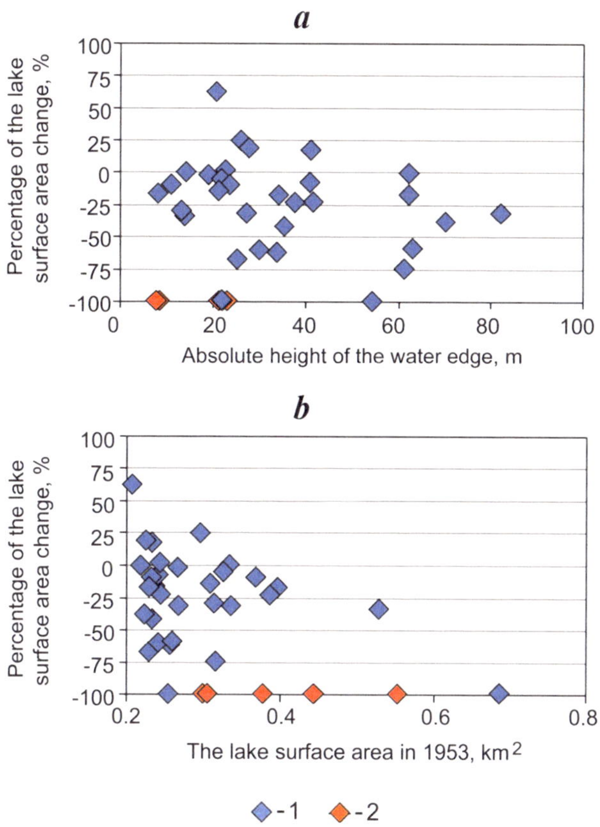

Figure 5.
Distribution of lakes with different degrees of drainage – Watering, depending on hypsometric position (a) and area of water bodies according to aero photos taken in 1953 (b): 1 – Lakes without traces of technogenic impact, and 2 – Lakes with proven facts of melioration.

because the parameters of small water bodies change rapidly and reflect current, possibly cyclical, changes. The larger the reservoir, the more resistant it is to local impact and the more slowly its parameters change.

The results of field measurements of reservoirs showed that the size of seven lakes remained virtually unchanged; five lakes were completely dry; in three lakes, the water surface area increased; and in seven lakes, it decreased. Compared with the result of the analysis of satellite images in 2018 (June), the deviation of the observed parameters, i.e., the increase in the area and drainage of the shores of the lake, in 2020 from the calculated ones was 5–10%. The morphology of the shores (open shoals or flooded shores) indicates that this is a consequence of the interannual dynamics of nutrition and runoff of water bodies. The size of open reservoirs with low, swampy shores, and small swampy catchments has decreased. The field transverse dimensions of thermokarst, seasonally open, and drainage lakes, which make up the majority, were increased compared to the 2018 image.

The results of the field survey proved the complex structure of the shore zone of the lakes. It is expressed in the presence or absence of lake terraces, the degree of development of thermal erosion, thermokarst, and thermal abrasion along the shores, the morphology of the drainways, in the material of bottom sediments of lakes and groundwater outlets. Fragments of two terraces with ledges 0.3–0.5 m high were found in the DLB of various drainage degrees (**Figure 6a**). The upper terrace is distinguished by a polygonal relief, composed of peat deposits with a thawing depth of 50–55 cm and a moisture content of 65–75% (**Figure 6b**). Vegetation cover is represented by shrub moss-and-lichen. The lower terrace is mostly boggy; areas with a polygonal, sometimes mound relief are subject to thermokarst – the intersections of polygonal wedges are filled with water. The vegetation cover varies from moss-cotton grass to forb-sedge and sedge-sphagnum. The depth of seasonal thawing is 45–50 cm; humidity is more than 80%. The shores of the lakes adjacent to the ridges' convex slopes are distinguished by solifluction sloughing and thermal erosion ditches. Drainage of the coast at the footslope causes springs with woody shrubs along the shores (**Figure 5c**). Thermoabrasive shores, due to the relatively small size of the reservoirs, are developed to a limited extent, mainly on elongated reservoirs oriented to the southeast. Among the general regularities, more or less inherent in all lakes, there is a combination of a coastline of ledges and a height of 0.3–0.5 m and boggy coastal shoals. Another regularity concerns new or renewed surface flow channels in the majority of drained lakes. These can be both rectilinear melioration canals and natural zigzag paths of the surface flow along the thawed polygonal ice wedges of

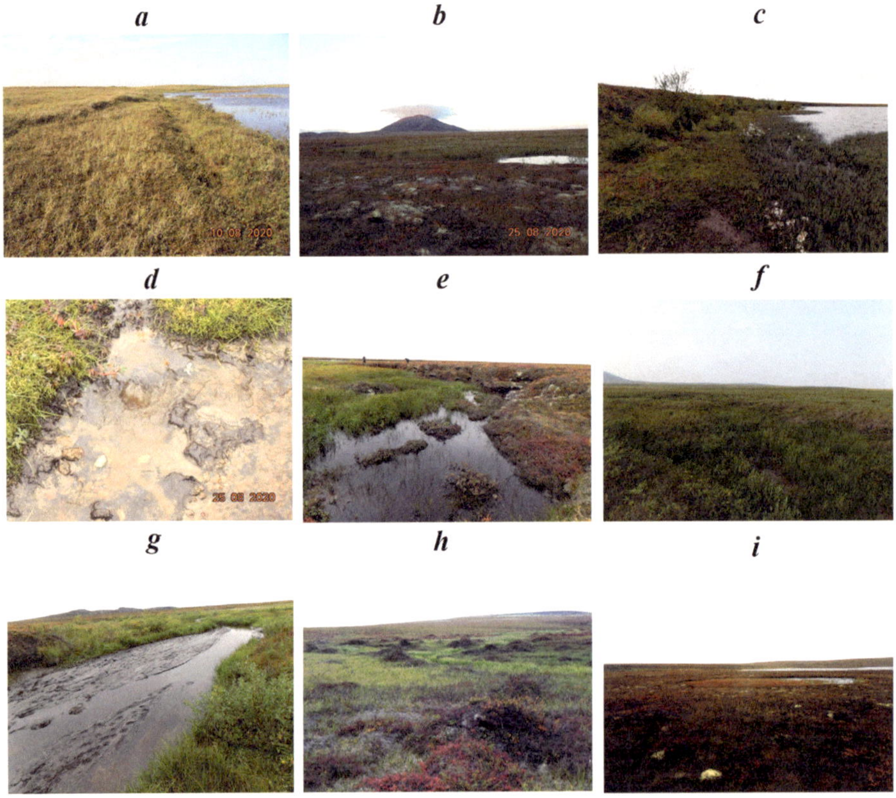

Figure 6.
Field observations of the coastal zone of the lakes of the Anadyr lowland: A – Ledges of lake terraces, b – Polygonal dwarf shrub tundra on the upper terrace of the lake, c – large shrubs along the shore at the foot of the tundra ouval, d – Sources of suprapermafrost waters on the lake shore, e – Thermokarst and thermal erosion along the drainage channel, f – Newly formed drainage channels along the thawed ice of polygons, g – Traces of water discharge from the drained lake into the channel, h – Permafrost mounds along the shores of the lake, i – and Thermokarst lake-saucer 30 m from the drying lake.

the first DLB terrace or preexisting inter-lake channels widened and deepened by thermokarst and thermal erosion (**Figure 5d** and **e**). In the coastal zone of lakes at the foot of tundra ouval, there are often mounds of permafrost heaving 1–1.5 m high (**Figure 6h**). The low shores are characterized by small thermokarst lakes-saucers 5–15 m across (**Figure 6i**).

The bottom sediments of the lakes were studied from the ice in the winter of 2021 and earlier, in 2010, when determining the depth of thermokarst lakes and the thickness of the ice [18]. The bottom sediments are represented by a layer of black organic silt 0.4–0.5 m thick. Depending on the position in the relief and the geological structure of the territory, the silt layer is underlain by peat, sandy loam, or loam.

Monitoring of the depth of seasonal thawing of the active layer of the Anadyr lowland is limited to a 29-year period (**Figure 7**). During the observation period, the depth of thawing at the Onemen control site, which occupies an autonomous position in the relief, increased by 13 cm, or 31% of the initial value. A slightly smaller increase in the depth of thawing (by 22%) was observed in the transit (trans-water) conditions of the Dionisiy site. The intermediate position is occupied by the superaqual landscape of the "Round" site. Taking into account the retrospective interpolation of the data, the increase in the thickness of AL within its limits was 9 cm, or 26%. All landscapes are characterized by fluctuations in the depth of seasonal thawing, lasting from 2.7 to 9–11 years. The average annual temperature of the active layer in the depth range of 20-50 cm at the Onemen site increased by 2.5°C over 20 years of observations. The average annual temperatures of the active layer of the Dionisiy site remained unchanged.

The result of observations of seasonal thawing at the northwest and southeast ends of the "Kruglaya" site, located at the DLB bottom, is shown in **Figure 8**. These are the shores of two secondary thermokarst lakes: gentle boggy (point 1) shore of the closed lake "Severnoye" and steep, 1.2 m high (point 2), shore of the open lake "Yuzhnoye". As can be seen, the dynamics of the active layer seasonal thawing of the two shores over the 10-year observation period is different (see **Figure 8b** and **c**).

Against the background of the general increase in the magnitude of seasonal thawing, the thickness of the thawed layer on the boggy shore decreased sharply

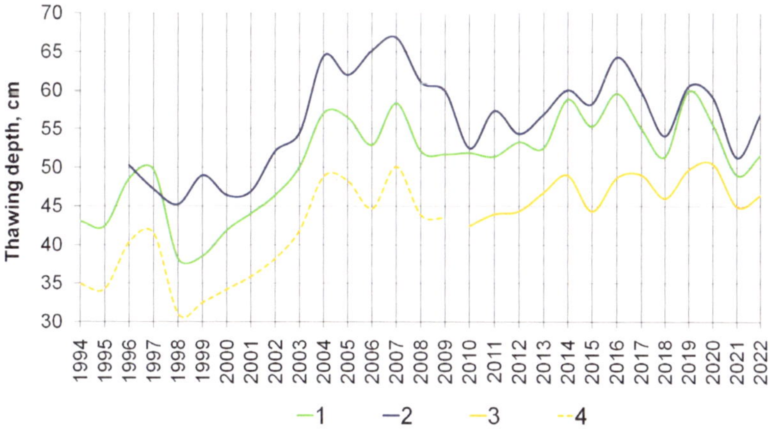

Figure 7.
Dynamics of seasonal thawing at the active layer monitoring sites CALM: 1 – Onemen (Rogozhnyy), 2 – Dionisiy, 3 – Kruglaya, and 4 – Kruglaya (retrospective).

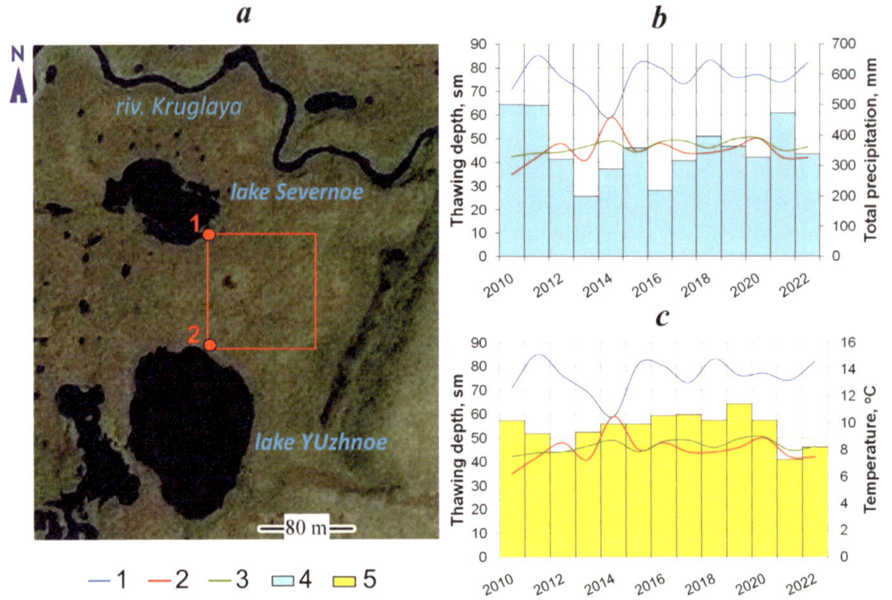

Figure 8.
The dynamics of seasonal thawing at the "Kruglaya" site (a) against the background of changes in annual precipitation (b) and summer air temperature (c): The depth of thawing at point 1 (1), at point 2 (2) on average over the area of the site (3); 4 – Annual precipitation; and 5 – Average air temperature in the frost-free period of the year.

in 2011–2015, and thawing depth decreased on the elevated dry shore of the lake in 2013–2014. Analysis of the dynamics of climatic indicators suggests that this is due to an abnormally sharp drop in precipitation in 2010–2013 (by 302 mm) in relation to an increase in the mean annual air temperature in 2012–2014 from −7.5°C to −4.5°C. Amplitude of fluctuations in precipitation in 2016–2017 is twice less (140 mm). Consequently, the thawing depth on the boggy coast slightly decreased in 2017 (see **Figure 7b**). Such phenomena, a decrease in the precipitation volume leads to drainage of the low shores of thermokarst lakes and a decrease in the depth of thawing along the shores of coastal bogs. At the same time, a decrease in the moisture content of high shores, on the contrary, contributes to an increase in the depth of seasonal thawing due to the higher intensity of heat turnover in polygonal tundra's as compared to tundra bogs.

The dynamics of air temperature and annual precipitation are shown in **Figure 8**. The graphs are compiled from observations at the Anadyr weather station. From the middle of the last century to the present, the average annual air temperature has increased by 2.1°C with fluctuations from year to year by 1.5–3°C. At the same time, the air temperature in summer increased by 1.6°C, and in the cold period - by 3.0°C. Annual precipitation increased by 61.6 mm, with the amplitude of interannual fluctuations up to 300 mm. The increase in precipitation was due to an increase in the number of snowfalls. Over the past 25 years, the air temperature has increased by 1.7°C. During this period of time, the annual amount of precipitation has practically not changed and has reduced the amplitude of fluctuations. The duration of the frost-free period in the region increased by 12 days with an average value of interannual fluctuations of 5 days.

6. Discussion

The tangible impact of global warming on the climate of the Anadyr lowland has been recorded since the late 90s of the last century. This does not contradict the known data on climate change in the Arctic and Subarctic. Climate changes in the region are expressed in the reduction of continental influence and the expansion of the boundaries of the subarctic marine climate area. This happens due to an increase in temperatures and an increase in precipitation during the cold season. Winters become warmer and snowier; the warm period increases its duration but remains cool enough with the same amount of precipitation. Such climate changes should contribute to increasing the inbound part of the water balance and preserving the water content of lakes. However, the results of remote studies and field observations reveal a different picture.

Elementary statistical analysis of lake area measurement data has shown that the size of water bodies tends to decrease. The average value of the sample is −24%; the confidence interval for a normal distribution with a 95% probability is in the range from +46.6% to −94.6% of the change in the area of lakes. That is, under the conditions of climate change over the past 65–70 years, it is a statistical norm that only 1/5 of the lakes have retained and increased their water content to 12%, and the remaining lakes have reduced the water surface area to 60%. This conclusion is not contradicted by the data shown in the pie charts (see **Figure 4**).

To analyze the possible hydrological and geomorphological causes of drainage, let us turn to the facts. The smallest losses of the water mirror area are inherent in closed lakes. The change in the area of lakes is associated with their position within the DLB and the relative excess of the water level above the erosion base. Lakes located in the upper reaches of streams or occupying the upper position in the cascade of lakes turned out to be significantly drained. These lakes are characterized by the formation of new or deepening existing surface runoff paths. In most cases, the lakes located on the edge of the DLB at the foot of the slope have retained their water content. This is due to the larger catchment area and stable supply of lakes with suprapermafrost waters. At the same time, there are exceptions to the patterns described above. For example, one of the lakes, oriented by the rose of the winds of the warm period, increased the area by 18% due to thermal abrasion of the windward shore. Another example is a small lake on a river terrace, the area of which has increased by 63% due to the activation of thermokarst along the ledge of the above-floodplain terrace with underground ice. A closed thermokarst lake with an area of 0.35 km^2, located at the bottom of the DLB, for unknown reasons reduced the water surface by 33%. The decrease in the area of a group of lakes by 17–60%, located at the foot of the slopes of tundra hummocks, cannot be interpreted. Thus, information about the conditions of the location of lakes that contribute to their drainage or preservation of water content is not enough to predict hydrological processes.

The information on evaporation capacity dynamics is not available. According to the known calculation schemes of zoning and literary sources for the warm season in the studied area, it is approximately 200–250 mm. Evaporation capacity probably increased with an increase in temperature and duration of the warm frost-free period. Open and seasonally open secondary thermokarst lakes with an increase in air temperature, and hence evaporation, will have a negative long-term water balance even with a constant or partially increasing amount of precipitation. This is due to the limited capabilities of the lake basin in the accumulation and retention of moisture and the absence of other feed sources except for atmospheric precipitation. An increase in the water content and area of such lakes is possible only due to the deepening and expansion of the bed during the development of thermokarst and thermal abrasion. A local source of replenishment of lake water and preservation of the area of water bodies can be ground ice meltwater in the composition of the increasing suprapermafrost flow. In the present case, this applies to the lakes located at the foot of extended slopes. The intensity and availability of this feed source are limited by the ice content of the permafrost roof and wedges and icy horizons' melting time. But the hydrological regime of the lakes is influenced not only by long-term changes in climate indicators but also by short-term fluctuations in their values. Often, fluctuations in the mean annual temperature and the amount of precipitation are in antiphase (see **Figure 9**). Hot dry summers and warm winters precede years with high water and summer-autumn floods, i.e. 1962–1966, 1978–1979, 1991–1994, 1996–1997, 2004–2006, 2011–2013, 2017–2018. Positive temperature extremes correspond to the maxima of interannual fluctuations in the depth of seasonal thawing. These facts, as well as the field observations of the authors, allow us to present the stages of formation of lake water discharges as follows:

1. Maximum seasonal melting in conditions of dry hot summer, long autumn, and warm snowy winter. In such years, thermokarst is activated in the channels of the channels, and in winter nonfreezing talik zones are formed.

2. With the beginning of the flood, the channels of lake channels are subjected to thermal erosion; they deepen and ensure the flow of meltwater. The massive discharge of lake water falls on summer-autumn floods caused by prolonged rains.

3. In the years following the draining of the lake, the riverbed is closed due to siltation and landslide of the eroded and thawed banks. Drainage channels become useless. With sufficient precipitation, the bottom of the lake is filled with water. In case of insufficient precipitation, the bottom of the lake becomes overgrown and turns into a tundra swamp or a grassy meadow.

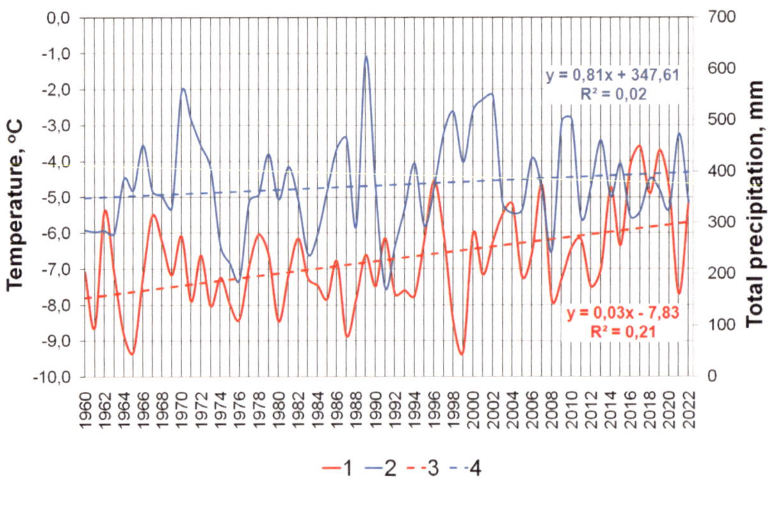

Figure 9.
Dynamics of mean annual temperature and precipitation according to the Anadyr weather station: 1 – Mean annual temperature, 2 – Total precipitation, 3 – Linear trend of temperature, and 4 – Linear trend of total precipitation.

Discharges as a reason for the drainage of the DLB were noted in [22, 23]. But the authors describe this process more as an extraordinary event and do not consider it as a natural phenomenon for the permafrost zone. According to residents of Anadyr, one of the studied lakes with a constant flow, which is often visited by hunters and gatherers of wild plants, has been drained three times since 1994 and refilled with water.

Despite the attractiveness of the salvo discharge model as a reason for the drainage of lakes, it does not explain the slow and partial decrease in the area of reservoirs. In most of the surveyed lakes, the water surface area has decreased by 15–45%, and they do not contain any newly formed elements of drainage paths. Among these reservoirs, there are many closed or seasonally open lakes, differing in catchment area. The last remark does not allow us to unambiguously associate the drying of these lakes only with an increase in evaporation, since lakes with a large catchment area should suffer less from a decrease in precipitation and an increase in average annual temperature. The results of the analysis of the dynamics of seasonal thawing of the shores of Lake Severnoye (see **Figure 8**), coupled with observations of frost heaving in the coastal strip of drying reservoirs, allowed us to put forward a cryogenic hypothesis about their partial natural drainage. The assumption is based on the interaction of lakes with the suprapermafrost DLB aquifer. **Figure 10** shows a schematic model of this interaction. With a constant amount of annual precipitation and the depth of seasonal melting, the water surface area and reservoir depth are in equilibrium (**Figure 10a**). During floods, the lake feeds the aquifer of the active layer. This happens in summer due to the reverse filtration of excess lake water through the deeply thawed active layer of the coastal strip. Reverse underground runoff into the lake occurs during the summer dry season and at the beginning of winter through when the water level in the lake decreases. In the conditions of sharp interannual fluctuations of climatic parameters, the dynamic equilibrium is disturbed. A decrease in precipitation and an increase in the average annual temperature lead to increased evaporation, a reduction in the area of the water surface, drainage of low shores and shoals. This, in turn, leads to deeper freezing and a decrease in the depth of seasonal thawing of the coastal strip (see **Figure 10b**). Therefore, for example, during the 10–year observation period, the interannual decrease in the area of Lake Severnoye reached 27% – the water retreated from the control point (No. 1) by 6 m. This happened in 2014, while the annual precipitation in 2013 decreased to 200 mm (see **Figure 8b**). When the low shores are drained, the suprapermafrost aquifer is separated from the lake by a frozen partition and forms two unequal areas – coastal and drainage. The dependence of the lake's nutrition on the catchment area decreases. As a result, drying increases and reaches a maximum in 1–2 years after a decrease in precipitation. Within the catchment area in the upper permafrost layer, with an increase in the depth of thawing, thermokarst begins along ice horizons and veins of underground ice, which is accompanied by the formation of subaerial talik zones. The subsequent freezing of thawed watered soils leads to the formation of heaving mounds (**Figure 6h**).

As a result, after the establishment of a new dynamic equilibrium, several catchment basins coexist in the DLB: primary catchments of large drained lakes, reduced in area; many small catchments of newly formed thermokarst saucer lakes; catchments of marsh streams, bounded by embankments of permafrost. That is, while maintaining the overall water balance of the territory, the appearance of lake-swamp landscapes changes to one degree or another.

The ongoing changes in the DLB have a negative impact on wetlands: newly formed shallow reservoirs are subject to interannual climate fluctuations and are not favorable for the nesting of waterfowl and feeding of ichthyofauna. Drying up reservoirs are no longer able to regulate the microclimate of the basins to the same extent.

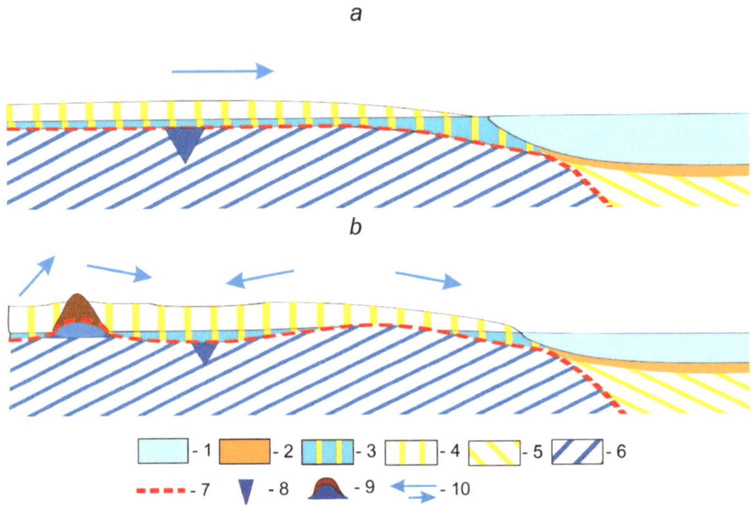

Figure 10.
Model of the interaction of lake waters and the suprapermafrost aquifer in static climatic conditions (a) and with climate warming (b): 1 – Lake; 2 – Bottom sediments; 3 – Suprapermafrost aquifer; 4 – Active layer; 5 – Talik zone under the lake; 6 – Permafrost; 7 – Roof of permafrost; 8 – A stylized image of wedge-shaped ice; 9 – Mounds of heaving permafrost; and 10 – The direction and intensity of water exchange (description in the text).

The proposed dynamic model of cryogenic processes affecting the water content of lakes and the density of their distribution is not universal. But it partially explains the paradox of the simultaneous development of thermokarst in the lower part of the DLB and the formation of new wedge-shaped ice on the upper terraces. Simultaneous weakening of thermokarst along the shores, drainage of large lakes with the formation of new local point thermokarst lakes. These, in fact, opposite natural processes have confused researchers many times and have not received a proper explanation [1, 3, 24, 30, 31, 33].

7. Conclusions

1. In the period from 1953 to 2018, the area of the water surface in the drained lake basins of the Anadyr lowland, ranging in size from 0.008 to 0.5 km^2, reduced by an average of 24%. The largest percentage (40–100%) of drainage was registered in open and flowing water bodies located at the sources of streams and cascades of lakes. The smallest decrease in the water surface (0–40%) is typical for closed water bodies located at the foot of long slopes. The area of three out of 36 lakes was increased. Field observations conducted in these lakes recorded manifestations of thermokarst and thermal abrasion, as well as the inflow of drainage water from drained lakes.

2. The reasons for the drainage of reservoirs included anthropogenic and natural processes: melioration of lakes for meadow growing (1965–1985); natural discharges of lake waters; changing conditions for feeding reservoirs with suprapermafrost waters. Discharges occur in conditions of abnormally high precipitation preceded by an increase in the depth of seasonal thawing, activation of thermokarst, and thermal erosion. Changes in the surface flow conditions of

suprapermafrost waters are caused by a differentiated change in the depth of seasonal thawing in the coastal zone of closed and seasonally drained lakes.

3. Favorable conditions for the discharge of lake waters are repeated at intervals of 3–12 years; this is typical of open lakes with an excess of the water edge over the base of erosion by 1 m or more. The area of the water surface of closed lakes located in the central part of the depressions decreases due to the weakening of the supply of groundwater from the suprapermafrost horizon. The coastal zone of drying lakes is characterized by bogging areas, frost heaving, and thermokarst, isolated from the reservoir by frozen barriers. It is assumed that this is the main mechanism of drainage of secondary thermokarst lakes, which have exhausted their expansion potential due to thermokarst and thermal abrasion. In the 20-year perspective, permafrost drainage of lakes in the bottom of the DLB is expected to be followed by expansion of the area of mound tundra bogs with numerous thermokarst lakes.

4. The undoubted advantage of this work is the emphasis on modern cryogenic processes in the coastal zone of lakes and in their catchments as a whole. This allows us to consider the impact of global warming on the lakes of the Arctic lowlands through the prism of climatic activation of cryogenic processes. Unfortunately, the authors do not yet have the opportunity to conduct coupled field and remote studies of the dynamics of water content in DLB lakes on the Bering Sea coast in North America, which would allow us to assess the degree of universality of the proposed models of exogenous cryogenic lake drainage. A logical continuation of the conducted research can be the organization of systematic instrumental observations of the water level and runoff of typical thermokarst lakes of the circumpolar lowland of the cryolithozone by analogy with monitoring of the active layer.

Acknowledgements

The work was carried out with the financial support of the CALM international program "Circumpolar Active Layer Monitoring" (NSF OPP-9732051 and OPP–0225603).

Author details

Oleg D. Tregubov[1*], Pavel Ya. Konstantinov[2], Vladimir V. Shamov[3] and Konstantin K. Uyagansky[1*]

1 North-Eastern Interdisciplinary Scientific Research Institute N.A. Shilo FEB RAS, Magadan, Russia

2 Melnikov Permafrost Institute SB RAS, Yakutsk, Russia

3 Pacific Geographical Institute FEB RAS, Vladivostok, Russia

*Address all correspondence to: tregubov2@yandex.ru and uyagansky@yandex.ru

References

[1] Tomirdiaro SV. Permafrost and the Development of Mountainous Countries and Lowlands. Magadan: Book Publishing House; 1972. p. 172

[2] Edwards M, Grosse G, Jones BM, McDowell P. The evolution of a thermokarst-lake landscape: Late quaternary permafrost degradation and stabilization in interior Alaska. Sedimentary Geology. 2016;**340**:3-14. DOI: 10.1016/j.sedgeo.2016.01.018

[3] Lyubomirov AS. Chukotka permafrost lakes. Yakutsk: Permafrost Institute of the Siberian Branch of the USSR Academy of Sciences. 1990:176

[4] Tregubov OD. Reconstruction of glacial processes and the search for water supply sources for the Zapadno-Ozernoye gas field. Bulletin of the North-Eastern Scientific Center of the Far Eastern Branch of the Russian Academy of Sciences. 2010;**4**:20-27

[5] Ruzanov VT. The nature of the lakes of the Anadyr lowland and their development. Engineering Research. 2014;7:68-72

[6] Wetland Conservation. Wikipedia. [Internet]. 2022. Available from: https://en.m.wikipedia.org/wiki/Wetland_conservation [Accessed: May 23, 2023]

[7] Foukal P, Fröhlich C, Spruit H, et al. Variations in solar luminosity and their effect on the Earth's climate. Nature. 2006;**443**:161-166. DOI: doi.org/10.1038/nature05072

[8] Davis ER. Extreme Weather, Atmospheric circulation and global warming. In: Jones L, editor. Global Warming. The Science and the Politics. Vancouver: The Fraser Institute; 1997. pp. 113-123. Available from: https://www.fraserinstitute.org/sites/default/files/GlobalWarmingExtremeWeather.pdf

[9] Reichler T. Changes in the atmospheric circulation as indicator of climate change. In: Letcher TM, editor. Climate Change. Observed Impacts on Planet Earth. Salt Lake City: Department of Meteorology, University of Utah; 2009. pp. 145-164. DOI: 10.1016/B978-0-444-53301-2.00007-5

[10] Bryksina NA, Polishchuk YM. Analysis of changes in the number of thermokarst lakes in the permafrost zone of Western Siberia based on satellite images. Cryosphere of the Earth. 2015;**XIX**(2):114-120

[11] Hinkel KM, Frohn RC, Nelson FE, Eisner WR, Beck RA. Morphometric and spatial analysis of thaw lakes and drained thaw lake basins in the western Arctic coastal plain, Alaska. Permafrost and Periglacial Processes. 2005;**16**(4):327-341

[12] Labrecque S, Lacelle D, Duguay CR, Lauriol B, Hawkings J. Contemporary (1951-2001) evolution of lakes in the old Crow Basin, northern Yukon, Canada: Remote sensing, numerical modeling, and stable isotope analysis. Arctic. 2009;**62**(2):225-238

[13] Veremeeva AA. Formation and Modern Dynamics of the Lacustrine-Thermokarst Relief of the Tundra Zone of the Kolyma Lowland According to Space Survey Data. Moscow: Institute of Geography of the Russian Academy of Sciences; 2017

[14] Veremeeva A, Nitze I, Gunther F, Grosse G, Rivkina E. Geomorphological and climatic drivers of Thermokarst Lake area increase trend (1999-2018)

in the Kolyma lowland Yedoma region, north-eastern Siberia. Remote Sensing. 2021;**13**(2):178. DOI: 10.3390/rs13020178

[15] Kapralova VN. Regularities of the Development of Thermokarst Processes within the Lacustrine-Thermokarst Plains (Based on the Approaches of the Mathematical Morphology of the Landscape). Moscow: Sergeev Institute of Geoecology of the Russian Academy of Sciences; 2014

[16] Rodionova TV. Investigation of the Dynamics of Thermokarst Lakes in Different Regions of the Permafrost Zone of Russia Using Satellite Images. Moscow: Lomonosov Moscow State University; 2013

[17] Dneprovskaya VP, Bryksina NA, Polishchuk YM. Study of changes in thermokarst in the zone of discontinuous distribution of permafrost in Western Siberia on the basis of satellite images. Earth Research from Space. 2009;**4**:1-9

[18] Salva AM. Tracking areas of thermokarst manifestations using satellite images (the case of the main water conduit route in the central Yakutia). Arctic and Antarctica. 2020;**2**:126-137

[19] Sannikov GS. Cartometric studies of thermokarst lakes on the territory of the Bovanenkovskoye field, Yamal peninsula. Earth's Cryosphere. 2012;**XVI**(2):30-37

[20] Andresen CG, Lougheed VL. Disappearing of Arctic tundra ponds: Fine-scale analysis of surface hydrology in drained thaw lake basins over a 65-year period (1948-2013). Journal of Geophysical Research – Biogeosciences. 2015;**120**:466-479. DOI: 10.1002/2014JG002778

[21] Marsh P, Russell M, Pohl S, Haywood H, Onclin C. Changes in thaw lake drainage in the Western Canadian Arctic from 1950 to 2000. Hydrological Processes. 2009;**23**:145-158

[22] Lantz T, Turner K. Changes in lake area in response to thermokarst processes and climate in old crow flats, Yukon. Journal of Geophysical Research: Biogeosciences. 2015;**120**:513-524. DOI: 10.10 02/2014JG00274 4

[23] Nitze I, Cooley SW, Duguay CR, Jones BM, Grosse G. The catastrophic thermokarst lake drainage events of 2018 in northwestern Alaska: Fast-forward into the future. The Cryosphere. 2020;**14**:4279-4297. DOI: 10.5194/tc-14-4279-2020

[24] Luo J, Niu F, Lin Z, Liu M, Yin G. Thermokarst lake changes between 1969 and 2010 in the Beilu River basin, Qinghai–Tibet plateau, China. Science Bulletin. 2015;**60**(5):556-564. DOI: 10.1007/s11434-015-0730-2

[25] Krivoshchekov VS. Land Reclamation and Land Development in Chukotka. Magadan: NEISRI FEB RAS; 2000. p. 274

[26] Tregubov OD, Uyagansky KK, Nuteveket MA. Monitoring of permafrost-climatic conditions of the Anadyr lowland. Geography and Natural Resources. 2020;**2**:143-152

[27] Romanenko FA. Dynamics of Lake Basins in the Central Yamal. Erosion Processes of Central Yamal. St. Petersburg: Publishing house of the Gomel TsNTDI; 1999. p. 350

[28] Jones BM, Grosse G, Arp CD, Jones MC, Anthony KW, Romanovsky VE. Modern thermokarst lake dynamics in the continuous permafrost zone, northern Seward peninsula, Alaska. Journal of Geophysical Research. 2011;**116**:G00M03. DOI: 10.1029/2011JG001666

[29] Jones BM, Arp CD. Observing a catastrophic Thermokarst Lake drainage in northern Alaska. Permafrost and Periglac Process. 2015;**26**:119-128. DOI: 10.1002/ppp.1842

[30] Chen M, Rowland JC, Wilson CJ, et al. The importance of natural variability in lake areas on the detection of permafrost degradation: A case study in the Yukon flats, Alaska. Permafrost and Periglacial Processes. 2013;**24**:224-240. DOI: 10.1002/ppp.1783

[31] Arp CD, Jones BM, Liljedahl AK, Hinkel KM, Welker JA. Depth, ice thickness, and ice-out timing cause divergent hydrologic responses among Arctic lakes. Water Resources. 2015;**51**:9379-9401. DOI: 10.1002/2015WR017362

[32] Boike J, Grau T, Heim B, Gunther F, et al. Satellite-derived changes in the permafrost landscapes of central Yakutia, 2000-2011: Wetting, drying, and fires. Global and Planetary Change. 2016;**2016**(139):116-127. DOI: 10.1016/j.gloplacha.2016.01.001

[33] Nesterova NV, Makarieva OM, Fedorov AN, Shikhov AN. Geocryological factors of activation of thermokarst processes in central Yakutia. In: Collection of reports of Fourth Vinogradovskie Readings. Hydrology from Knowledge to World View. Vol. 2020. St. Petersburg: LLC Publishing House VVM; 2020, 2020. pp. 739-744

Chapter 3

Wigry Lake: The Cradle of Polish Hydrobiology – A Century of Limnological Exploration

Andrzej Górniak and Adam Więcko

Abstract

Wigry Lake, located in North-Eastern (NE) Poland, has a century-old history of limnological exploration and is an excellent place to assess the impact of catchment changes caused by urbanization on the functioning of a large, polymictic, and flow-through lake. The history of prewar limnological research, the course of hydrochemical effects of urbanization in the river flowing into the lake since the 1970s, and long-term changes in the functioning of Wigry Lake are presented. The collected archival and current results indicate that the hydrochemical type of the lake's waters remained the same, and the inflow of river waters from the urban catchment strongly transformed the lake bay receiving the load. In the remaining part of the lake, the eutrophication load caused smaller changes because of a gradual reduction in the inflow of nutrients. Consequently, there was an increase in vertical differentiation of oxygen, algal biomass, and their structure. A significant share of supplying the lake with groundwater and the natural in-lake system of biotic and chemical regulations significantly reduced the effects of the eutrophication process. The existing lake biodiversity has been maintained, constituting a valuable element of the European NATURA 2000 system.

Keywords: Wigry Lake, long-term study, lake trophy, eutrophication, urbanization, history of limnology

1. Introduction

Wigry Lake in Poland is considered a historical place where an in-depth study of natural sciences was carried out, as it was here that a new, ecological approach was developed on the study of lake ecosystems. It was here that the theoretical and practical foundations of lake science were created. Intensive research of Wigry Lake, started in 1920 by Alfred Lityński, thanks to the establishment of the Hydrobiological Station on this lake, resulted in spread of the knowledge of the processes and phenomena occurring in the lake ecosystem, fundamental for limnology. They confirmed the high biodiversity and uniqueness of water ecosystems noted in earlier, random studies [1, 2]. A continuation of prewar research by teams of limnologists from several Polish academic centers since the 1970s made the collection of data and knowledge about the functioning of Wigry Lake exceptionally rich. In addition, state agency units dealing

IntechOpen

with the state of the atmosphere, water resources, and their quality systematically monitored the environment of this area. Since the 1990s of the twentieth century and the first monograph on the lakes of the Wigry National Park (WNP) [3], despite further monitoring of the water habitats of this region, no further general studies about the Wigry Lake have been made. Therefore, the aim of this study was to assess the state and functional changes of the Wigry Lake ecosystem for over 100 years from the first quantitative analyses. Particular attention was paid to extending the knowledge about the quantity and quality of lake's resources in the context of global climate change and the progressing urbanization of the catchment.

2. Wigry Lake environment

Wigry Lake, located in the north-eastern corner of Poland (54°01′09″N, 23°04′49″E), has an area of 21.7 km^2, a maximum depth of 74.2 m, water volume $352.1 \cdot 10^6 m^3$, and a catchment area of 466 km^2. It lies within the Lithuanian Lake District, which, belongs to Eastern Europe, on the course of the Czarna Hańcza River [4]. It has a characteristic location in two geographical subregions reflecting the geological sequence of the Pleistocene glaciation of Central Europe.

The northern part of the lake is located within the borders of the hilly and loamy Suwałki Lake District, while the southern part occupies a part of the outwash gravel and sandy Augustów Plain.

Intensive development of lake coastline with numerous bays indicates the polygenetic origin of this multichannel lake, with two directions of the N-S and E-W glacial channels. Detailed studies from the turn of the nineteenth and twentieth centuries indicate that the main assumptions of the lake's origin lie in the uniqueness in Poland, shallow, metamorphic Precambrian ground, and Pleistocene neotectonic activity causing cracks in the ice sheet, referring to the network of deep faults. These, in turn, shaped the main flow directions of in-glacial waters forming postglacial channels [5]. The second important moment in the genesis of the lake is the existence of two megafloods in the marginal and proximal part of the Scandinavian Ice Sheet (SIS), shortly after the ice sheet reaches its maximum extent. Two such megafloods occurring on the 19 and 16 ka (kiloannum, thousand years) ago were identified [6]. The first of them took place from the east (from the Nemunas River basin), and the second from the N-NE direction through the valley of the Czarna Hańcza River. Two prehistorical events with catastrophic loss of glacier mass had a water flow velocity of around 15–17 $m\ s^{-1}$. The large discharge of meltwater was estimated to be about $2 \times 10^6\ m^3\ s^{-1}$ and must have decisively affected the shaping of the river valley system in the European Lowland and possibly influenced global climate by flowing in significant volumes of freshwater into the North Atlantic Sea. The contemporary catchment of the Czarna Hańcza River belongs to the left-bank basin of the Niemen River, and thus to the catchment area of the Baltic Sea.

The economic history of the Polish-Lithuanian borderland, and the previous Yotvingian rule in this area, gradually changed the water ecosystems and their catchments. In particular, the activity of the Camaldolese Order from the sixteenth to nineteenth centuries led to gradual deforestation, utilization of bog ore in the surrounding peat bogs and changed the nature of the water circulation in the catchment area. Since the twentieth century, there has been a gradual increase in anthropogenic pressure on Wigry Lake as a result of urbanization in the city of Suwałki, located on the Czarna Hańcza River, the main lake tributary. The population of Suwałki in 1920

(17,000 inhabitants) slowly increased until 1939 (25,000 inhabitants). It was not until 1970 that the city's population reached its pre-World War II state. From 1974 to 1999, as the capital of the newly created voivodeship, intensive urbanization of the catchment took place. The changing economic situation and changes in the administrative structure stopped the population growth trend and since 2000, the number of inhabitants in Suwałki is approximately 70,000. It is recently estimated that in the entire catchment area of the Czarna Hańcza River, there are 90,000 people. It was not until 1986 that a new, biological wastewater treatment plant was built, and in 1994 the dephosphatation process of wastewater was started in this plant. An important action preventing the degradation of Wigry Lake was the creation of the Landscape Park in 1976, and the Wigry National Park in 1989, in accordance with the earlier proposals for its protection submitted in 1921–1924, among others by Dr. A. Lityński.

3. Limnological data collection and methods

The results presented in this paper are based on historical data, published limnological data, and own field research conducted in the years 1997–2022 by the Department of Hydrobiology of the University of Bialystok. Lake water level data and discharge in Sobolewo (1951–1992), Stary Bród (1992–2020), and Czerwony Folwark (1951–2020) stations of the Czarna Hańcza River come from the IMGW-PIB Warszawa database. Czarna Hańcza River discharge data for Sobolewo station (2000–2020) were collected from Wigry National Park database. In addition, the results of parallel, river flow measurements in 2011 and 2012 were used, the purpose of which was to determine the total amount of water flowing into Wigry Lake and the mutual relations under various hydrometeorological conditions. The missing data were supplemented by the authors using methods common in hydrology. Precipitation data for the Stary Folwark station from years 1928 to 2015 come from IMGW-PIB data, supplemented by earlier data from the Płociczno station (years 1922–1927) and data for years 2015–2021, converted from data from the Integrated Environmental Monitoring System (IEMS) in Sobolewo station. When preparing the annual lake water balances in the period 1951–2020 [7], the monthly evaporation rate was calculated using the Ivanov formula [see 8]. Results of the long-term National River Monitoring System (NRMS) provided on the Czarna Hańcza River in Sobolewo station and hydrochemical data collection from IEMS were used. Due to the state accreditation of laboratories, the collected hydrochemical data form a homogeneous set, and the methods of chemical analysis from the 1920s and 1930s are similar to those methods currently used. Own studies on Wigry Lake were carried out in the years 1997–2022 in seven companies at 6 to 13 stations (**Figure 1**), each time at the turn of July and August, when the thermocline is at its deepest [2]. Thermal and oxygen profiling were made at each station, as well as the Secchi Disk visibility noted. Water samples for a detailed chemical analysis were taken by Patalas sampler from epilimnion, metalimnion, and hypolimnion layers. Water analyses were provided in the Laboratory of the Department of Hydrobiology, University of Białystok. Standard analytical methods for water samples were used [8], but DIONEX ICS-1100 was applied for ion analysis from 2015 as well as the determinations of total chlorophyll concentrations from spectrophotometric change to fluorometric determinations. Applied FluoroProbe (bbe-Moldaenke, Kiel, Germany) identifies the four phytoplankton classes based on precalibrated excitation and emission spectra "fingerprints" programmed into the instrument, which include: (1) Green

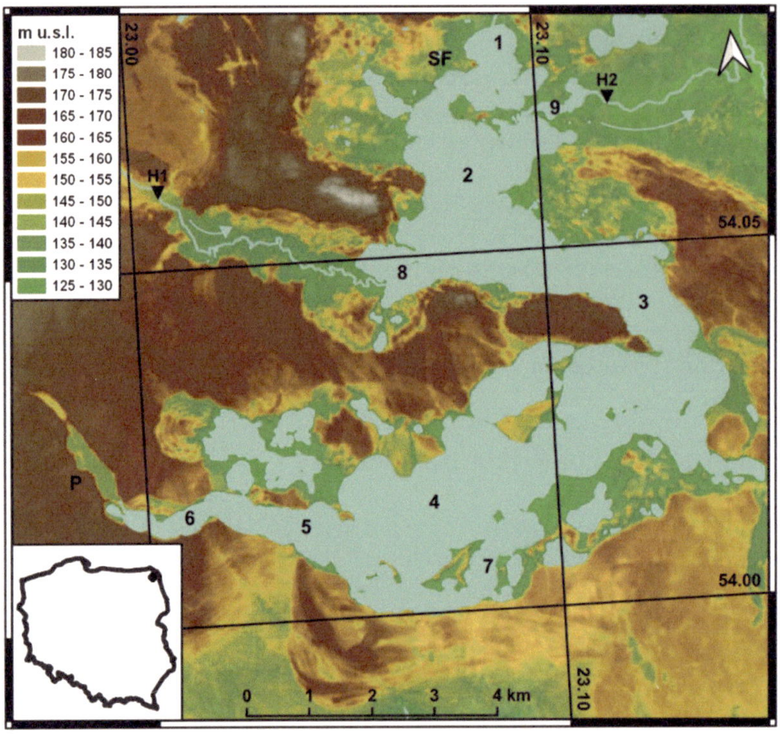

Figure 1.
Wigry Lake map with Hydrobiological Station locations (P- Płociczno, SF- Stary Folwark in 1928–1943), hydrological stations on the Czarna Hańcza River (H1- Sobolewo, H2- Czerwony Folwark), and sampling stations (1–8).

algae: Chlorophyta and Euglenophyta; (2) Cyanobacteria: phycocyanin (PC)-rich cyanobacteria; (3) Diatoms: Heterokontophyta, Haptophyta, and Dinophyta; and (4) Cryptophytes: Cryptophyta and the phycoerythrin (PE)-rich cyanobacteria. The lake's trophic state was evaluated using the Carlson formula [9]. The Mann-Kendall trend test was used to analyze the data, and the Kruskal-Wallis test for equal medians was used to demonstrate the statistical significance of differences between the study subperiods.

4. Hydrobiological Station's activity and the first period of lake exploration

In 1920, Dr. Alfred Lityński, as a young researcher of the lakes of the Tatra Mountains (Carpathians) and the lakes of Polesie Lubelskie, began limnological research at the Hydrobiological Station on the Wigry Lake (**Figure 2**).

It was initially located in the village Płociczno on the south-western shore of the lake. The Nencki Institute in Warsaw was the founder institution of the station, which was constantly supported by government subsidies and funds, obtained thanks to the activity of its founder [10]. Initially, under "spartan" conditions (**Figure 3**) and with the increasing staff of the station, the bathymetry of the lake, the structures of phytoplankton, zooplankton, and ichthyofauna were analyzed in detail. In 1921, a meteorological station began its work, which in the following year became part

Figure 2.
Alfred Lityński (1880–1945)—Source: Wigry Lake Museum archive (WNP permission).

Figure 3.
Hydrobiological Station in the first location in Płociczno (1920–1927); source: Wigry Lake Museum archive (WNP permission).

of the national network of meteorological stations in Poland. From 1922, constant measurements of the lake water level began, and from 1926, the temperature of the lake water was measured daily at 7 am. In 1928, the station and the staff of the station were moved to a newly built facility in Stary Folwark, in the north-western part of the lake (**Figure 4**). The new building, specially designed for the needs of the station, was furnished with an electrical network and direct supply of the station with lake

Figure 4.
Hydrobiological Station on Wigry Lake building in Stary Folwark. Source: National Archive of photography.

water for biological experiments. The station housed numerous laboratories, vivaria, offices, and apartments for the staff. Diverse and modern equipment of the station with the most modern limnological equipment, microscopes, measuring devices, and a motorboat were great assets of the Station. The high level of research conducted at the station, the great enthusiasm and commitment of Dr. A. Lityński were confirmed by Prof. Naumann and scientists from Lund (Sweden) during their visits to the Wigry Lake station. In the interwar period, it was one of the first four hydrobiological stations in Europe.

The first limnological studies on Wigry Lake appeared as early as 1922. Thanks to the efforts of Dr. A. Lityński, the Hydrobiological Station on Wigry River in 1926 started publishing the first Polish limnological journal—Archives d'hydrobiologie et Ichthyologie (Archives of Hydrobiology and Fisheries). The results of subsequent floristic and zoological research on Wigry Lake were noticed among limnologists around the world, and Dr. A. Lityński, as a representative of Poland, directly participated in the founding meeting of the SIL Society of Limnology and in its subsequent Congresses in Innsbruck and Kiel. The ability to cooperate, the good working atmosphere created by the founder of the station, resulted in a large group of limnologists working at the station and who became well known among water ecologists. These include Dr. J. Wiszniewski—the discoverer of the psammolitoral, Dr. Z. Koźmiński—as the first scientists who had measured chlorophyll concentrations in North American lakes, Prof. M. Stangenberg—who had documented the hydrochemical state of waters in the 1930s, and Prof. J. Wołoszyńska—phycologist— who had first described Wigry Lake phytoplankton community.

The station on Wigry lake organized summer hydrobiological schools for students and young adepts at limnology. Two doctoral dissertations on zooplankton and bacterioplankton of Wigry Lake were written by young hydrobiologists from the University of Warsaw under the supervision of A. Lityński. The Jagiellonian University in

Krakow, on the other hand, awarded him the degree of Habilitated Doctor, as the first in Poland in this field. Thanks to the Scientific Station on the Wigry Lake, a modern fish hatchery was created, supporting the fishing economy of the lakes in the entire region. Field experience, numerous scientific trips, and a rich limnological library gathered at the station enabled Dr. Lityński to create the first Polish textbook on "General Hydrobiology," published a few years after the end of World War II, thanks to the involvement of Prof. L. K. Pawłowski [11].

Unfortunately, World War II and the events related to it contributed to the end of the Station's activity, the death of Dr. A. Lityński in 1945 (near Smoleńsk, Russia) and many of his associates. Station equipment was plundered and part of it was taken away by the German occupiers to their fishing stations. The war activities of Prof. A. Neuman, a "friend" of Dr. Lityński from the prewar period, as a Nazi functionary, and his personal visits to Wigry Lake did not save the Station from its complete devastation and liquidation.

Because of its natural values, the Wigry Lake was "adopted" in 1998 by the International Association of Theoretical and Applied Limnology (SIL "Lake Adoption" Project) [12]. In 2009, thanks to the initiative of the Wigry National Park, the Museum of Wigry Lake was established in the restored building of the former Hydrobiological Station, and a monument to Dr. Alfred Lityński stands in front of its building.

5. Results and discussion

5.1 Hydrological history of the Wigry Lake in the last 100 years

Significant long-term and seasonal atmospheric dynamics is characteristic of the temperate climate zone and translates into the dynamics of the land water cycle [13]. The natural and man-induced hydroclimatic cycle of Central Europe caused changes in water resources in lake basins, observed especially in the dynamics of the water level [14, 15]. The analyzed lake is characterized by a moderate long-term fluctuation in the average annual water level of the order of 50 cm (**Figure 5**), with seasonal increase of water level during April–May and minimum in July. It is the effect of high lake capacity, moderately long (2.96 years) average time of lake water exchange. Most of the flow-through lake types have the same annual amplitude, as that observed in

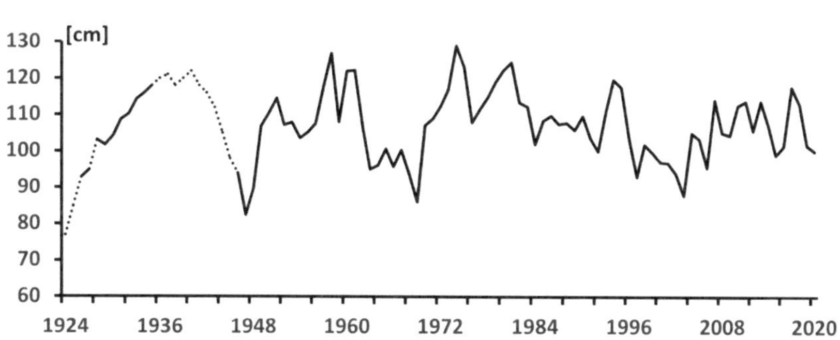

Figure 5.
Wigry Lake water level changes in years 1924–2021, gathered on the base of Hydrobiological Station observations and published IMGW-PIB data; the dotted line shows approximate data.

Wigry Lake [16]. In more than 100 years of observations, there is no significant trend of lowering or increasing the water level of the lake (Mann-Kendall test confirmed a lack of its trend), but only its periodic fluctuations. As in many other lakes in Poland, hydrological and climatic factors have a smaller impact on the long-term dynamics of lake water levels than anthropogenic changes in water circulation in the catchment area [15]. The polygenetic nature of the lake, the hydrographic network and the morphology of the lake basin cause the dominant inflow and outflow of river waters to reach the northern lake part.

The main part of surface water reaches the lake along with the Czarna Hańcza River. According to own hydrological measurements in 2011 and 2016, it currently accounts for 60% of the entire river inflow to the lake. This value is lower than the previously estimated value [7].

Riverine waters constitute 59% of the annual volume of water circulated in the lake, the underground supply vector has a share of 38%, and only 3% is the result of the vertical exchange of the lake with atmosphere (P-E). More than fourfold differences in the volume of groundwaters involved in the lake's annual water circulation volume (from $53.5{\cdot}10^6{\cdot}10^6$ m^3 to $204.6{\cdot}10^6$ m^3) were observed (**Figure 6**). The volume of water from tributaries flowing into the lake per year had a twofold smaller variability than that noted for groundwater inflows (variation coefficient, respectively, 43.8% and 21.8%). Similar estimates, but for a shorter period of analysis, were presented earlier [7].

There is a statistically significant long-term trend (confirmed by the Mann-Kendall trend test with $p < 0{,}001$) of decreasing the share of underground supply (**Figure 6**), which in recent years has been disturbed by a significant increase in the intensity of vertical lake water exchange due to progressing global warming. During the 70-year period of measuring the flow of the Czarna Hańcza River below the lake, no statistically significant trends were noted, similarly to the inflow to the lake (**Figure 7**). Mann-Kendall trend test confirmed its absence in both cases. Despite changes in the hydroclimatic balance of the catchment, gradually urbanized catchment supplies the lake with treated sewage, partly produced from groundwater.

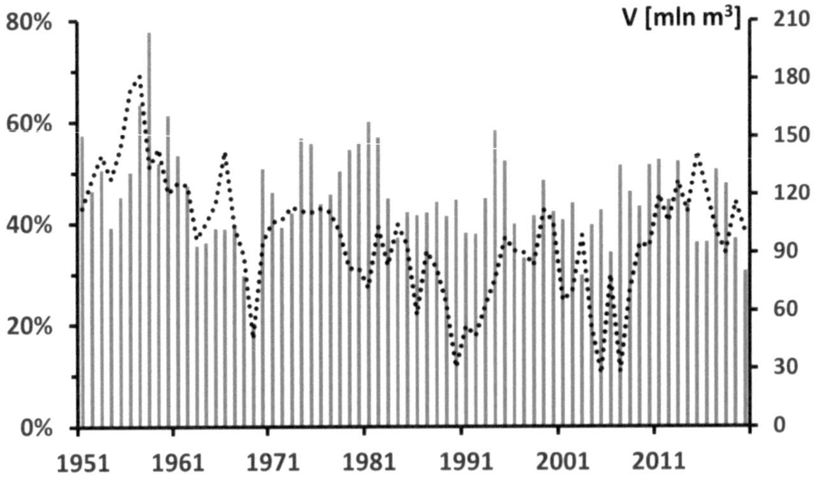

Figure 6.
Wigry Lake water volume under annual cycle in years 1951–2020 (right axis and bars) and groundwater share (in % dots, left axis) in yearly water cycle.

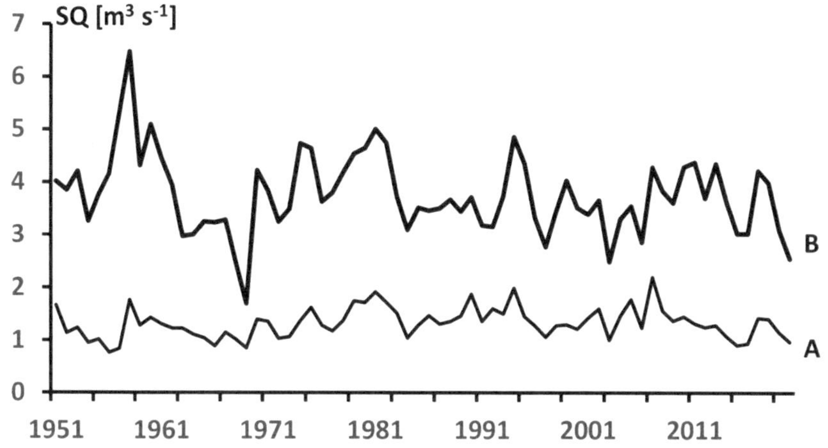

Figure 7.
Yearly (calendar) discharge of Czarna Hańcza River upstream (a) and downstream of Wigry Lake (B) in years 1951–2020; source data are IMGW-PIB and integrated environmental monitoring system (IEMS) archive with own calculations.

Before the intensive process of urbanization of the catchment, the share of inflowing sewage was small, now especially in years with low flow, e.g., in 2019, it reached 14%.

5.2 Water quality of Wigry Lake tributaries

The use structure of the Czarna Hańcza River basin upstream of the lake, its urbanization, and also tourism on the lake itself, have been changing for over 100 years, gradually affecting the quality of the hydrosphere. In addition, the effects of global warming progress, which appeared as lake epilimnion water temperature increase not only in summer period, but for mean annual values also [14]. For the first time during limnological monitoring, the ice phenomena were not observed in winter 2019/2020 at all.

The main source of long-term changes in the water quality of the analyzed lake should be seen as significant changes in the chemical composition of the Czarna Hańcza River during urbanization of the city of Suwałki and the surrounding area. The hydrochemical condition of the Czarna Hańcza River at the end of the 1980s indicated strong water pollution and the need to take protective actions for the Wigry Lake ecosystem. Collected hydrochemical data for the Czarna Hańcza River from the last 40 years document effects of wastewater treatment plant's work, including phosphorus removal technology from 1996. This led to a decrease in the average annual concentrations of total phosphorus in the river waters from 0.8–1 g $P{\cdot}m^{-3}$ to 0.05–0.1 $gP{\cdot}m^{-3}$, and total nitrogen concentrations from about 4–5 $gN{\cdot}m^{-3}$ to about 2 $gN{\cdot}m^{-3}$ (**Figure 8**).

Therefore, the total load of eutrophication (including atmospheric load) on a unit of lake surface has radically decreased. The catchment load reduction below the limit of hazardous unit values for nitrogen and phosphorus occurred at the same time, but for nitrogen periodical fluctuations were noted (**Figure 9**). At the same time, it was accompanied by a constant increase in mineral substances' concentrations that were present in river waters, which confirm the increase in water conductivity (**Table 1**). In the quality of river water, there were changes in the ionic structure, the share of

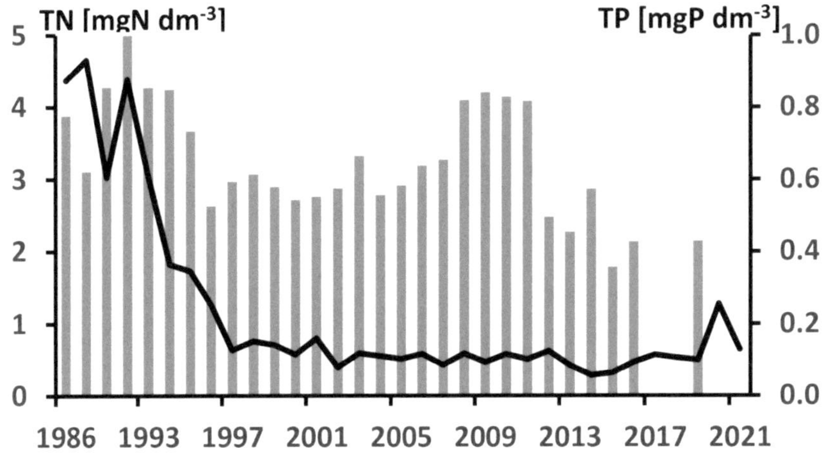

Figure 8.
Average of yearly total phosphorus (TP) and total nitrogen (TN) concentrations in the Czarna Hańcza River, upstream of Wigry Lake in years 1986– 2020; data from archives of the National River Monitoring System (NRMS) and integrated environmental monitoring system (IEMS).

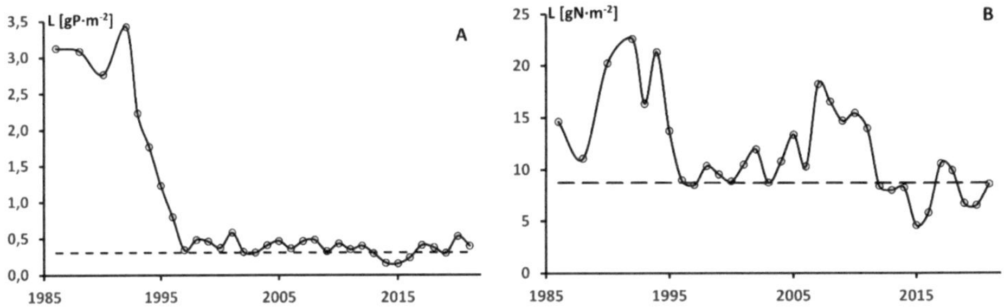

Figure 9.
Average of yearly total phosphorus (TP) and total nitrogen (TN) unit loads to Wigry Lake in years 1986–2020; the broken line indicates the level hazard unit load calculated using Vollenweider criteria [see 7].

chlorides among anions and the share of sodium and potassium ions among cations increased. These changes are statistically significant, as confirmed by the Kruskal-Wallis test for equal medians ($P < 0.0001$). Changes in the sewage management of urbanized areas and the improvement in the quality of river waters in Poland at the end of the 1990s took place a decade later than in Western Europe or the USA [3, 7].

An increase in the concentrations of basic cations and anions as well as biogenic elements in the river-lake system quickly activates eutrophication process in the receiver of river waters, which is Wigry Lake [3, 7].

5.3 Long-term hydrochemical changes of Wigry Lake

Already during the first stage of the study of Wigry Lake, it was indicated that the northern part of the lake, the shallower one with the highest flow rate, had transitional features between the state typical of oligotrophic and eutrophic waters [2]. Most of the lake's (1920s and 1930s) water mass was characterized by good oxygenation of the waters to the bottom. Qualitatively and spatially diversified bottom sediments, ranging from gravel, sandy to carbonate gytties and lacustrine chalk [17, 18], provided

EC and ions	1994–1997			2018–2021			Test for equal medians
	Average	SD	% of anions or cations	Average	SD	% of anions or cations	
Conductivity	462.6	53.8		575.7	75.2		*
HCO_3^-	305.6	81.3	82.9%	344.6	27.9	80.3%	
SO_4^{-2}	26.9	6.0	9.3%	27.8	4.2	8.2%	
Cl^-	16.8	6.7	7.8%	28.4	9.5	11.4%	*
Ca^{+2}	79.3	8.1	69.5%	77.7	5.7	63.3%	
Mg^{+2}	14.3	1.7	20.6%	14.4	1.3	19.3%	
Na^+	11.8	3.9	9.0%	23.9	8.5	16.9%	*
K^+	4.2	1.7	0.3%	7.1	2.5	0.5%	*

*Data from the National River Monitoring System (NRMS) and Integrated Environmental Monitoring System (IEMS) archives for Sobolewo station (station 8—see **Figure 1**). Statistically significant differences ($p < 0.001$) are indicated in the last column.*

Table 1.
Change in water conductivity [$\mu S\ cm^{-1}$] and ions' composition in Czarna Hańcza River upstream of Wigry Lake in the last 30 years.

effective, intralake protection against periodic eutrophic inflows from the catchment. The late summer Secchi's disk visibility was over 6 m in the first half of the twentieth century [3, 19].

The increasing pressure on the lake, mainly from the central-western direction (station 8 see **Figure 1**), since the mid-1970s has triggered the process of eutrophication of waters, resulting in a dramatic decrease in the photic zone thickness, especially in the eutrophication front zone of the lake (**Figure 10**). In parallel, oxygen orthograde, dominating in the lake, turned into a negative heterograde in the central and southern parts of Wigry Lake. Clinograde type has developed in the north lake part and Wigierki Bay (station 6 in **Figure 1**), when hypolimnion becomes anaerobic during the summer thermic stratification (**Figure 11**).

For the last 100 years, the waters of Wigry Lake have constantly maintained their bicarbonate-calcium hydrochemical type, just like most of the harmonious lakes of northern Poland [20]. Due to its extent, complex bottom shape, and water supply structure changing over time, it is characterized by periodic fluctuations of chemical composition in time and space, especially during the summer thermal stratification of waters (**Table 2**). Generally, the amount of minerals dissolved in lake water increases from the south lake part to the north of lake in the summer epilimnion; therefore, the average electric conductivity (EC) difference between them is over 20 $\mu S\ cm^{-1}$.

Along with thermal stratification, there is a chemical diversification of the chemical composition of the surface water layer (**Table 2**), as a result of biogenic utilization and geochemical processes, such as sedimentation of carbonates and tryptone to the bottom of the lake [21].

Therefore, the water is decalcified, which has been going on in this lake for several thousand years with varying intensity. This is evidenced by several-meter thick carbonate gyttja deposits with numerous underwater landslides and tectonic disturbances [18]. Significant plankton activity in the trophogenic layer is evidenced by a clear difference in the concentrations of easily utilized nutrients between the epi- and

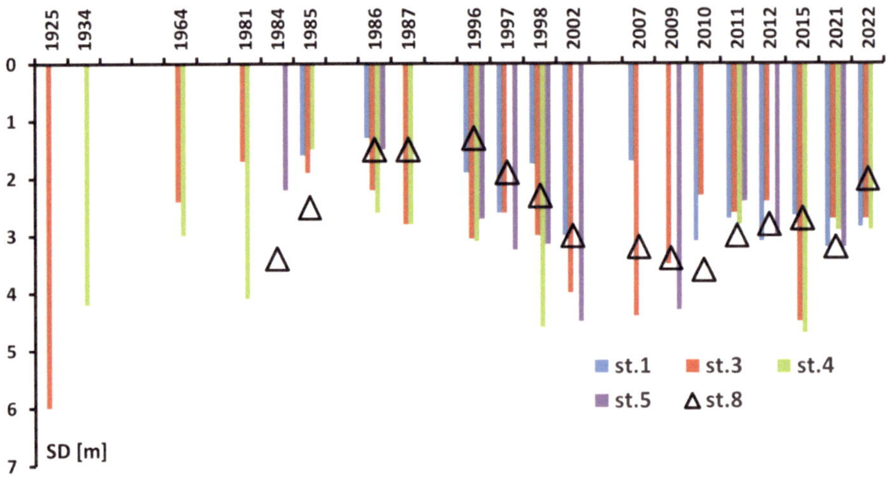

Figure 10.
Secchi disk visibility changes in Wigry Lake in years 1925–2022; data from [3] and own measurements.

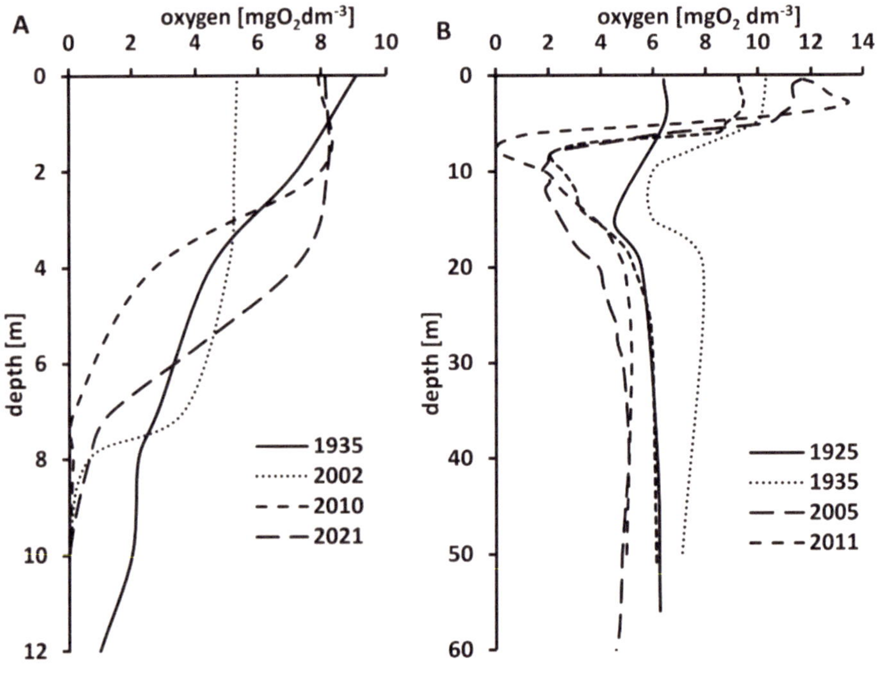

Figure 11.
Summer oxygen profiles in the shallow station 1 and deep station 3 (see ***Figure 1****) at Wigry Lake in years 1925–2021; data for 1925 from [2] and data for 1935 from [19].*

hypolimnion, as well as the effects of increased accumulation of organic matter in the epilimnion than in the hypolimnion.

With the seasonal phytoplankton development in dimictic lakes, the resources of biogenic elements in some hydrological situations may become scarce. In the analyzed lake, a gradual decrease in silicon concentrations was noted (**Table 2**), which is an impulse for spontaneous changes in the structure of algae.

Parameters	Layers	2002	2009	2012	2021	2022
Sampling stations		15	6	13	9	8
EC [μS cm^{-1}]	Epil	309.1	349.8	373.9	383.6	404.9
	Met	342.4	373.6	420.2	405.5	423.0
	Hypo	353.2	372.6	414.1	421.5	431.5
Ca [mg dm^{-3}]	Epil	51.5	59.7	61.7	42.9	50.7
	Met	60.1	64.3	60.8	49.3	54.6
	Hypo	72.9	60.1	62.8	48.1	61.7
N-NO_3 [μgN dm^{-3}]	Epil	24.0	75.5	149.5	62.4	<10.0
	Met	27.0	369.6	158.6	79.6	
	Hypo	296.9	484.7	506.3	74.3	55.6
DOC [mgC dm^{-3}]	Epil	7.32	n.d.	6.70	5.59	6.03
	Met	7.23		6.56	5.18	6.02
	Hypo	7.73		5.23	4.95	5.23
SRP [μgP dm^{-3}]	Epil	18.3	7.6	20.5	8.6	<5.0
	Met	15.2	7.2	18.7	<5.0	
	Hypo	75.0	8.5	19.2	5.5	
TP [μgP dm^{-3}]	Epil	49.1	119.4	48.6	102.1	37.1
	Met	53.8	74.5	47.3	66.4	42.7
	Hypo	117.1	164.9	47.0	92.0	42.9
SiO_2 [mgSi dm^{-3}]	Epil	1.54	0.49	1.48	0.10	0.44
	Met	1.87	1.14	1.88	0.07	0.36
	Hypo	2.54	1.39	3.38	0.09	0.24

Table 2.
Water conductivity and concentration of selected water parameters in three thermic layers of Wigry Lake in years 2002–2022.

In the analyzed lake, after a strong eutrophication impulse in the 1980s and 1990s, reduced concentrations of phosphorus and nitrogen were noted especially in the bay of the lake receiving the waters of the Czarna Hańcza River (**Table 3**).

Although the reduction in the phosphorus load on the lake was many times greater than in the case of nitrogen, the average concentrations of total phosphorus (TP) in the analyzed bay of the lake, and at the same time, the main receiver of fertilizing load, remain within the limits accepted for eutrophic waters. Results from Wigry Lake confirm observations from other European lakes that their regeneration (re-oligotrophication) is much slower in deep lakes than in shallow ones [23].

Mainly phosphorus and also nitrogen are factors that increase trophy, and thus the phytoplankton production [21, 24]. As can be seen in the example of Wigry Lake, along with the increase in catchment load, since the 1990s, the range of concentrations of chlorophyll has been increasing in the entire lake, especially in station 8 (**Figure 12**), characteristic of strongly eutrophic waters. The long-term fluctuation in algae biomass was dependent on the amount of river inflow, and the periodic decrease in the amount of algae should be partly associated with an increase in groundwater

Years	SRP	TP	TN
	[μgP dm^{-3}]		[mgN dm^{-3}]
1991–1994*	296.0 ± 87.0	424.0 ± 88.0	2.66 ± 1.27
1997	91.1 ± 11.9	124.3 ± 2.0	n.d.
2002	55.5 ± 39.4	120.3 ± 37.3	n.d.
2012	26.5 ± 6.5	56.3 ± 9.8	n.d.
2021	1.1 ± 0.7	66.8 ± 6.4	1.08 ± 0.04
2022	0.3 ± 0.1	52.1 ± 10.4	0.81 ± 0.09

*n.d. – not determined. *data for years 1991–1994 after [22].*

Table 3.
Annual concentrations of phosphorus and total nitrogen in water of the Hańczańska Bay (station 8 in ***Figure 1****) of Wigry Lake in years 1991–2022.*

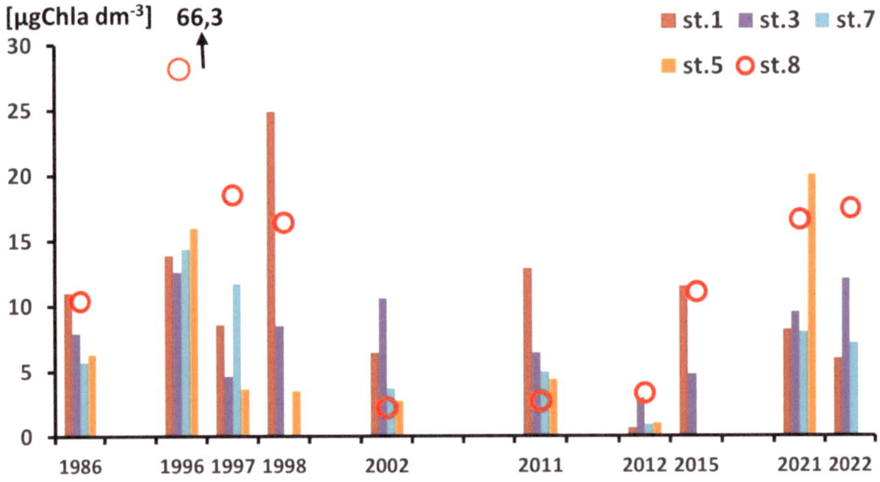

Figure 12.
Multiannual changes of summer epilimnion chloropyll "a" concentration in Wigry Lake; data for 1986 from [25].

supply. In the years 1980–2022, the previously observed [2, 3] dichotomy of the lake's pelagial in terms of phytoplankton biomass persisted, with the northern part of the lake being richer than the southern.

The increase in the lake's trophy was also accompanied by a change in the vertical distribution of algae, consisting in a stronger development of the lower epilimnion maximum of chlorophyll (**Figure 13**). More accurate *in situ* fluorometric measurements also reveal the changing taxonomic structure of phytoplankton. It turns out that in the summer epilimnion, the algal community is green-diatom-cryptophyte. In the metalimnion, the share of diatoms and green algae decreases in favor of cyanobacteria, which become absolute dominants in the hypolimnion (**Figure 13**). A characteristic feature of Wigry Lake are periodic, summer mass blooms of the genus *Ceratium* found in the lake as early as 1916 [26], with high morphological diversity, and their density is inversely proportional to TP concentrations [27].

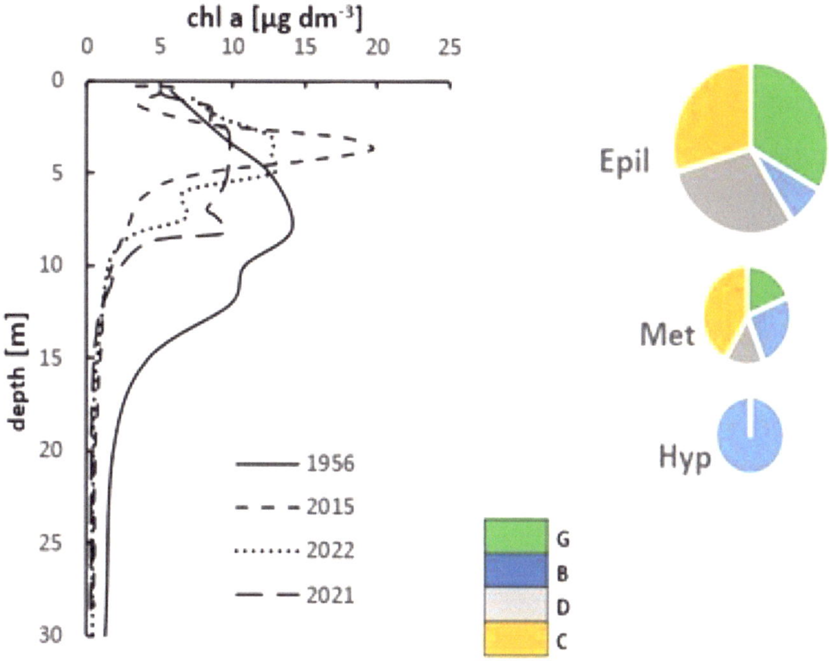

Figure 13.
Vertical differentiation of chlorophyll "a" concentration in the deepest part of Wigry Lake (station 3 in ***Figure 1****) in years 1956–2022 and phytoplankton structure (% chloropyll) in years 2015–2022; G - green algae, B - cyanobacteria, D – Diatoms, C- Cryptophytes (in situ measurement results).*

The history of hydrochemical transformations of the river flowing into Wigry Lake presented above changed the fertility of the lake to the greatest extent during the period of maximum inflow of nutrients (in 1990s).

Since the beginning of the twenty-first century, the level of trophy expressed by the Carlson index oscillates around the border between meso- and eutrophy in most parts of the lake, except for small bays and its northern part (**Figure 14**). Thanks to the natural hydrological features of the lake and the existing in-lake biogeochemical and biotic systems, the anthropogenic eutrophic impulse did not lead to a meaningful change in the trophic state. This proves that a properly functioning river-lake biogeochemical continuum is maintained in the catchment and is resistant to stress factors that disintegrate the system [7]. Protection activity taken in the lake by the administration of the national park is significant here.

5.4 Recent ecological state of the Wigry Lake

So far, various forms of active and passive protection of the lake and the immediate catchment area of the lake ecosystem have been carried out, which were included in the statutory tasks of the Wigry National Park.

Thanks to this, the current ecological status of Wigry Lake can be considered good, despite the unfavorable catchment conditions.

This is confirmed by the assessments carried out using the provisions and indicators of the EU Water Directive [28]. During the state monitoring of rivers and waters,

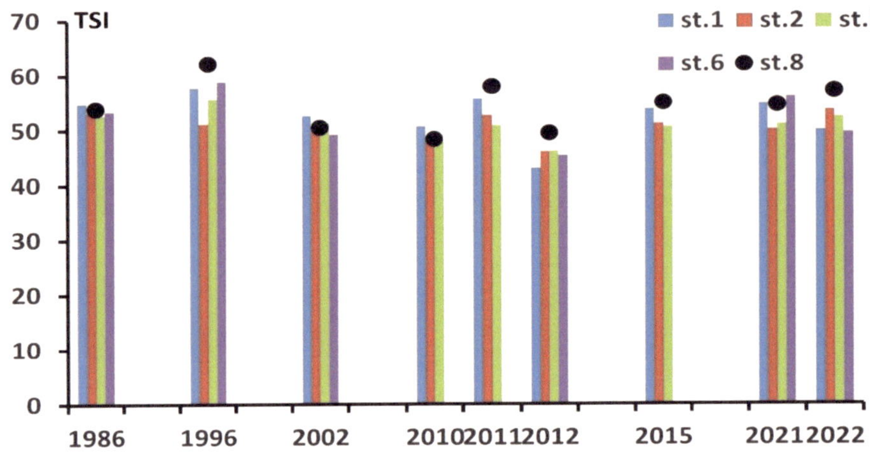

Figure 14.
Long-term and spatial changes of trophic state index (TSI) for Wigry Lake in years 1986–2022.

periodic presence of several hazardous substances specified in the list of the EU Water Directive is found in the water of lakes. In a shallow bay (stage 8 in **Figure 1**), substances were identified in the bottom sediments, formerly deposited during the period of strong eutrophication, and considered a dangerous environment. They are chemically inactive with the natural deposit of lacustrine chalk constantly present at the bottom. There is a high habitat potential in the lake, ensuring the presence of a variety of crustacean zooplankton described at the beginning of the twentieth century [29, 30]. On over 125 hectares of the Wigry Lake, there are charophyte underwater meadows with 10 species of the genera *Chara* and *Nitell*a. Four fish species and two species of mollusks from Natura 2000 list are present in the lake. The existing natural values of the lake enable the presence of 175 species of birds (found in the WNP), i.e., over one third of the list of Polish birds, of which 90 species can be found in Wigry Lake [31].

6. Conclusion

The Hydrobiological Station on Wigry, founded by Dr. A. Lityński in 1920, thanks to the exceptional commitment of the entire staff, played a significant role in learning about the aquatic environment of northern Poland and contributed to the development of limnology in Poland and in the world. Wigry Lake belongs to the flow-through, deep, polymictic, with a long time of water exchange, with underground supply constituting more than one third of the amount of water in annual circulation, with a general tendency to decrease. This trend has been disturbed in the last decade by the effects of climatic changes, and in particular by the increase in the intensity of vertical water exchange in the lake. Despite climate changes, no significant long-term changes in the annual amount of river water supply to the lake have been recorded. On the other hand, the slowly increasing amount of treated sewage, especially in dry years and periods, may account for over 12% of the total river flow, which is almost twice as much as at the beginning of the second half of the twentieth century. A potential threat to the functioning of the lake and its ecological status are the effects of more frequently heavy rainfall. The investment activities undertaken in the catchment area with increasing urbanization reduced the load of phosphorus and nitrogen to the level acceptable according to the Vollenweider criteria. Reduction of catchment eutrophication load in the lake lasted much longer for phosphorus than for nitrogen. Recovery of lakes after nutrient loading reduction may be confounded by concomitant global warming effects.

Conflict of interest

The authors declare no conflict of interest.

Thanks

We would like to thank Ms. Agnieszka Zega from the Suwałki Department of Environmental Monitoring of the Chief Inspectorate of Environmental Protection and Ms. Aleksandra Mackiewicz from the Wigry National Park for their help in archival hydrochemical and hydrological data collection. Many thanks to Professor B. Zdanowski for providing access to archival data from the years 1991 to 1994 and valuable methodological comments. We would like to thank Dr. Maciej Ambroziewicz, the Director of the Wigry Lake Museum, WNP, in Stary Folwark, for providing us with archival photographs. A.G. would like to express his sincere thanks to Dr. M. Kaminski, who over 35 years ago was the first to show him hydrobiological secrets of the Wigry Lake.

Author details

Andrzej Górniak* and Adam Więcko
Department of Freshwater Ecology, University of Białystok, Białystok, Poland

*Address all correspondence to: hydra@uwb.edu.pl

References

[1] Lityński A. Jezioro Wigry jako zbiorowisko fauny planktonowej. Prace Stacji Hydrobiologicznej na Wigrach. Inst im. N. Nenckiego. 1922;**1**(1):42 (in polish)

[2] Lityński A. Studja limnologiczne na Wigrach. Architecture Hydrobiologia Ryb. 1926;**1**(1-2):1-78 (in polish)

[3] Zdanowski B, editor. Lakes of Wigry National Park. Trophic Status and Protection Direction. Vol. 3. Ossolineum, Wrocław: Zesz. Nauk. PAN, "Człowiek i środowisko"; 1992. p. 249 (in polish)

[4] Solon J, Borzyszkowski J, Bidłasik M, Richling A, Badora K, Balon J, et al. Physico-geographical mesoregions of Poland: Verification and adjustment of boundaries on the basis of contemporary spatial data. Geographia Polonica. 2018;**91**(2):143-170. DOI: 10.7163/GPol.0115

[5] Ber A. The origin of Lake Wigry in connection to the deep basement structures. Prace i Studia Geograficzne. 2009;**4**(1):37-51 (in polish)

[6] Weckwerth P, Wysota W, Piotrowski JA, Adamczyk A, Krawiec A, Dąbrowski M. Late Weichselian glacier outburst floods in north-eastern Poland: Landform evidence and palaeohydraulic significance. Earth-Science Review. 2019;**194**:216-233. DOI: 10.1016/j.earscirev.2019.05.006

[7] Bajkiewicz-Grabowska E. Circulation of Matter in the River-Lake Systems. Warsaw: Warsaw Univ. Press; 2002. p. 274 (in polish)

[8] APHA. Standard Methods. Standard Methods for the Examination of Water and Wastewater. 21st ed. Washington, DC, USA: American Public Health Association; 2005

[9] Carlson RE. A trophic state index for lakes. Limnology and Oceanography. 1977;**22**:361-369

[10] Brzęk G. Stacja Hydrobiologiczna na Wigrach. Wyd. Lubelskie, Lublin; 1988. p. 476

[11] Lityński A. Hydrobiologia ogólna. Warsaw: PWN; 1952. p. 545 (in polish)

[12] Kamiński M. Lake Wigry – The lake "adopted" by the International Association of Theoretical and Applied Limnology (SIL) "Lake adoption" project. Polish Journal of Ecology. 1999;**57**(2):215-224

[13] Górniak A. Klimat województwa podlaskiego w czasie globalnego ocieplenia. Białystok: Univ. Białystok Press; 2021. p. 223 (in polish)

[14] Górniak A. Current climatic conditions of Lake regions in Poland, and impacts on their functioning. In: Korzeniewska E, Harnisz M, editors. Polish Rivers Basins and Lakes - Part I, 2020. Springer Nature Switzerland AG: Hydrology and Hydrochemistry 86; 2020. pp. 1-25

[15] Górniak A, Piekarski K. Seasonal and multiannual changes of water levels in lakes of northeastern Poland. Polish Journal of Environmental Studies. 2002;**11**(4):349-354

[16] Choiński A, Jańczak J, Ptak M. Wahania poziomów wody jezior w Polsce w latach 1956-2015. Przegl. Geography. 2020;**92**(1):41-54. DOI: 10.7163/PrzG.2020.1.3 (in polish)

[17] Stangenberg M. Skład chemiczny osadów głębinowych jezior Suwalszczyzny. Rozpr i spraw. Inst. Bad. Lasów Państw. 1938;**31**:44 (in polish)

[18] Rutkowski J, Król K, Szczepańska J. Lithology of the profundal sediments in Słupiańska Bay (Wigry Lake, NE Poland) – Introduction to interdisciplinary study. Geochronometria. 2007;**27**:47-52

[19] Stangenberg M. Chemische untersuchungen am Wigrysee. Arch. Hydrob. Ryb. 1935;**9**(3—4):185-220

[20] Korycka A. Characteristics of the chemical composition of water in lakes of North Poland. Poland Agriculture Annals Series H. 1991;**102**(3):112 (in polish)

[21] Górniak A, Kajak Z. Hydrobiologia. Limnologia. Warsaw: PWN; 2019. p. 452 (in polish)

[22] Zdanowski B. Precipitation of phosphorous in the zone of river and Lake water mixing: R. Hańcza and lake Wigry (north-East Poland). Polish Journal of Ecology. 2003;**51**(2):143-154

[23] Jeppesen E, Søndergaard M, Jensen JP, et al. Lake responses to reduced nutrient loading – An analysis of contemporary long-term data from 35 case studies. Freshwater Biology. 2005;**50**:1747-1771. DOI: 10.1111/j.1365-2427.2005.01415.x

[24] Wetzel RG. Limnology: Lake and River Ecosystems. Third ed. New York: Academic Press; 2001

[25] Zdanowski B, Karpiński A, Prusik S. Environmental conditions of waters in Wigry National Park. In: Zdanowski B, editor. Lakes of Wigry National Park. Trophic Status and Protection Direction. Vol. 3. Ossolineum, Wrocław: Zesz. Nauk. PAN, "Człowiek i środowisko"; 1992. pp. 35-62 (in polish)

[26] Schröder B. Schwebepflanzen aus dem Wigrysee bei Suwalki in Polen. Botanica acta By. Deutsche Botanische Gesellschaft. 1917;**50**(3):256-266

[27] Hutorowicz A. Morphologic variability of Dinophyceae from the genus *Ceratium* in a mesotrophic Lake Wigry (Poland). Polish Journal of Ecology. 2000;**48**(1):89-95

[28] Hołdyński Cz, Zawadzka D. Natural Habitats and Species of the Natura 2000 Network in the Wigry National Park and Ostoja Wigierska. Characteristic, Recognition and Management. Krzywe: WNP Press; 2014. p. 243 (in polish)

[29] Karpowicz M, Górniak A. Crustacean zooplankton of harmonious lakes in Wigry National Park and relationship with the trophic state. Monitor. Środ. Przyr. 2013;**14**:97-101 (in polish)

[30] Karpowicz M, Ejsmont-Karabin J. Diversity and structure of pelagic zooplankton (Crustacea, Rotifera) in NE Poland. Water. 2021;**13**:456. DOI: 10.3390/w13040456

[31] Zawadzka D, Zawadzki G. Birds Od Wigry Lake. Kielce: Paśny Buriat Press; 2021. p. 284 (in polish)

Chapter 4

The Impact of Lake-Level Fluctuation on Earthquake Recurrence Interval over Historical and Prehistorical Timescales: The Case of the Dead Sea

Mariana Belferman and Amotz Agnon

Abstract

We review the impact of large historical lake water-level changes on seismicity *via* the stress field of the shallow crust where devastating earthquakes nucleate. A novel backward earthquake simulation presented in this chapter can be used to investigate the geological record for the past ten millennia (presented in this study) and even more. The simulation is based on a theoretical model, which explains the variability in the recurrence interval of strong earthquakes. We suggest that the water-level changes in ancient lakes located in tectonic depressions along the Dead Sea transform could contribute to the observed differences. It is found that the increase in the water level moderates the seismic recurrence interval. Based on this empirical correlation together with mechanical considerations, an additional indication is established regarding the water-level reconstruction and location of earthquakes in the Dead Sea area. This indication is based on simulated earthquakes, by superimposing the effective normal stress change due to the reconstructed water-level change on the estimated tectonic shear stress accumulated since the preceding seismic event.

Keywords: earthquake recurrence interval, water-level change, lake, Dead Sea, poro-elasto-plasticity, strike-slip tectonic deformation, induced seismicity

1. Introduction

Earthquakes are one of the primary geological hazards, with historical and modern accounts serving as vivid testimony of the threats they pose. With catastrophic losses of human lives and health [1, 2], earthquakes may cause major economic losses and property damage [3, 4]. Earthquakes also play a significant role in sedimentary systems evolution, liquefying water-saturated clastic sediments by cyclic loading [5–7].

Despite the global, social, and scientific impact of earthquakes, in many cases, their triggering mechanisms remain poorly defined. Specifically, tectonic motion is the primary cause and consequence of earthquakes: For example., large earthquakes along transform faults are typically associated with slip rate and stress buildup and release (e.g., Dead Sea transform (DST)—[8–13], San-Andreas fault—[14, 15] or North Anatolian fault—[16–18]).

However, earthquake triggers may have no direct relation to the local tectonic dynamics. Observations supported by theoretical calculations demonstrate that surface processes (e.g., glaciations, sedimentation, and water loading) are able to change significantly the stress field in the Earth's crust [19, 20]. Field data from northern Europe and America suggest that melting of the ice sheets was connected to the paleoseismologic evidence and to the development of late Pleistocene/early Holocene fault scarps [21–24], whereas the seismicity in the tectonically stable Greenland and Antarctica may be solely due to the current ice unloading [25].

Modeling the slip-rate evolution of localized normal faults located below the glaciated Earth crust, Hetzel and Hampel [26], Hampel and Hetzel [27], and Hampel et al. [28] demonstrate that variations in the slip-rate in these faults depend on the changes in the differential stress in the crust, which are in turn connected to changes in the loading on the Earth surface. The glaciations induce a decrease in the differential stress in the crust and, thus, a decrease in the faulting rate. In contrast, during postglacial rebound and isostatic adjustment, the differential stress in the crust grows, resulting in increased rates of faulting, inducing in turn temporal clustering of the earthquakes.

Additionally, it has long been recognized that fluids can influence the occurrence of earthquakes. Injection of liquid into (or removal from) the subsurface is capable to induce seismicity [29–32] even outside the region subjected to the direct fluid perturbations [33–35].

Moreover, seismic response in the field was observed to be connected to reservoir impoundments and their seasonal water level variations [36–40]. An increase in pore pressure may lead to two different kinds of seismic response: swarm-like rapid seismicity induced by immediate pore pressure buildup, and delayed seismicity associated with pore pressure diffusion from the reservoir to hypocentral depths [37, 41]. Delayed response, as observed in the field, is often associated with large magnitude earthquakes [37, 42]; it may extend significantly beyond the confines of the reservoir and may not show an immediate correlation with major changes in reservoir level.

Seismic activity change was associated with historical water-level changes in the Dead Sea lake [43, 44] and Salton Sea lake [45–47] at much longer timescales. The Salton Sea, a remnant of ancient Lake Cahuilla, lies on the southern San-Andreas Fault (SSAF) [48]. Luttrell et al. [47] conducted a study to examine how Coulomb stress perturbations influenced local faults in relation to changes in the level of Lake Cahuilla. Their objective was to assess whether the lake could potentially impact the timing of fault rupture. The researchers proposed that the loading resulting from the ancient Lake Cahuilla impeded fault failure along most of the SSAF, except for a potential small area within the lake boundary.

Subsequently, Brothers et al. [46] modeled changes in Coulomb stress due to the combined effects of lake loading caused by periodic flooding of Lake Cahuilla, movement of stepover faults, and increased pore pressure. They determined that these factors could elevate stress levels along the southern SSAF to the extent of triggering an earthquake. Over the past 1100 years, the periodic flooding of Lake Cahuilla has exhibited fluctuations of up to 100 meters. These variations have shown throughout this time period a significant correlation with the incidence of earthquakes along the southern, the least active segment, of the SAF.

Belferman et al. [44] took advantage of the high temporal resolution of the historical earthquake records for the Dead Sea Lake during the last two millennia. The relatively small water-level hikes of 15 m, characteristic for time intervals of centuries to millennia, were analyzed and shown to be able to exacerbate the seismicity pattern at the Dead Sea fault. Belferman et al. [43] demonstrated for the first time that plausible scenarios for lake-level history show striking correlations with the historical record of earthquake recurrence intervals. Additionally, the plausible correlation observed between these phenomena suggests a complementary indicator that could be used to constrain either the locations of historical earthquakes or lake-level fluctuations during periods of uncertainty in one of these variables.

More recently, Hill et al. [45] present a direct relationship between changes in Lake Cahuilla water level and earthquakes based on computed time-dependent Coulomb stress changes for the last 1100 years. The authors present coupled 3D finite element model that takes into account time-varying surface water loads, fault geometry, crustal poroelasticity, and viscoelastic relaxation in the ductile substrate. The simulation results indicate that the past six major earthquakes on the southern San-Andreas Fault probably occurred during high stands of Lake Cahuilla. It was shown that increases in lake-level result in positive Coulomb stress changes on most of the southern San-Andreas Fault, with stressing rates as high as two to three times the tectonic loading rate.

In this chapter, we summarize the results and methods to identify the correlation between the historic water-level reconstructions at the Dead Sea and seismicity patterns in the area over the past two millennia. We also offer an additional analytical model to cope with dependence on initial conditions, which can be used in the future for interpreting these phenomena for earlier extended periods.

2. The study area: Dead Sea

2.1 Tectonics

The Dead Sea Lake fills tectonic depressions along the Arabian and African plate boundaries. The Arabian Plate is rifting away from the African and Nubian plates along the active spreading axis of the Red Sea, with the Sinai-Levant Block lagging behind, sheared along the Dead Sea fault (**Figure 1**) [49–51].

The Dead Sea fault is the major fault system along the plate boundary, and one of the best studied continental transform faults in the world [8, 11, 12, 49, 52–55]. The tectonic motion at the DST is characterized predominantly by a left-lateral strike-slip regime with a velocity of up to ~5 mm/yr. along various segments [51, 56–58]. Motion is transferred from the opening Red Sea to the escaping Anatolian Block *via* sinistral shear motion along the Dead Sea transform (e.g., [49, 51, 59]). This sinistral strike-slip plate boundary comprises the basins of the Gulf of Aqaba/Elat, Araba/Arava, the Dead Sea, the Jordan, Hula, Beqa'a, Ghab, and Karasu, a distance of about 1000 km [60].

Many factors and processes operating on different scales may have a profound influence on the dynamics of the deformations in the Dead Sea region. One of them—the influence of the water bodies deep underneath the Jordan-Arava valley—is explored in depth in this study.

2.2 Seismicity

The Dead Sea fault system is a major source of prehistoric, historical, and contemporary earthquakes [8, 11, 61–65]. Tectonic motion on the DST has always been

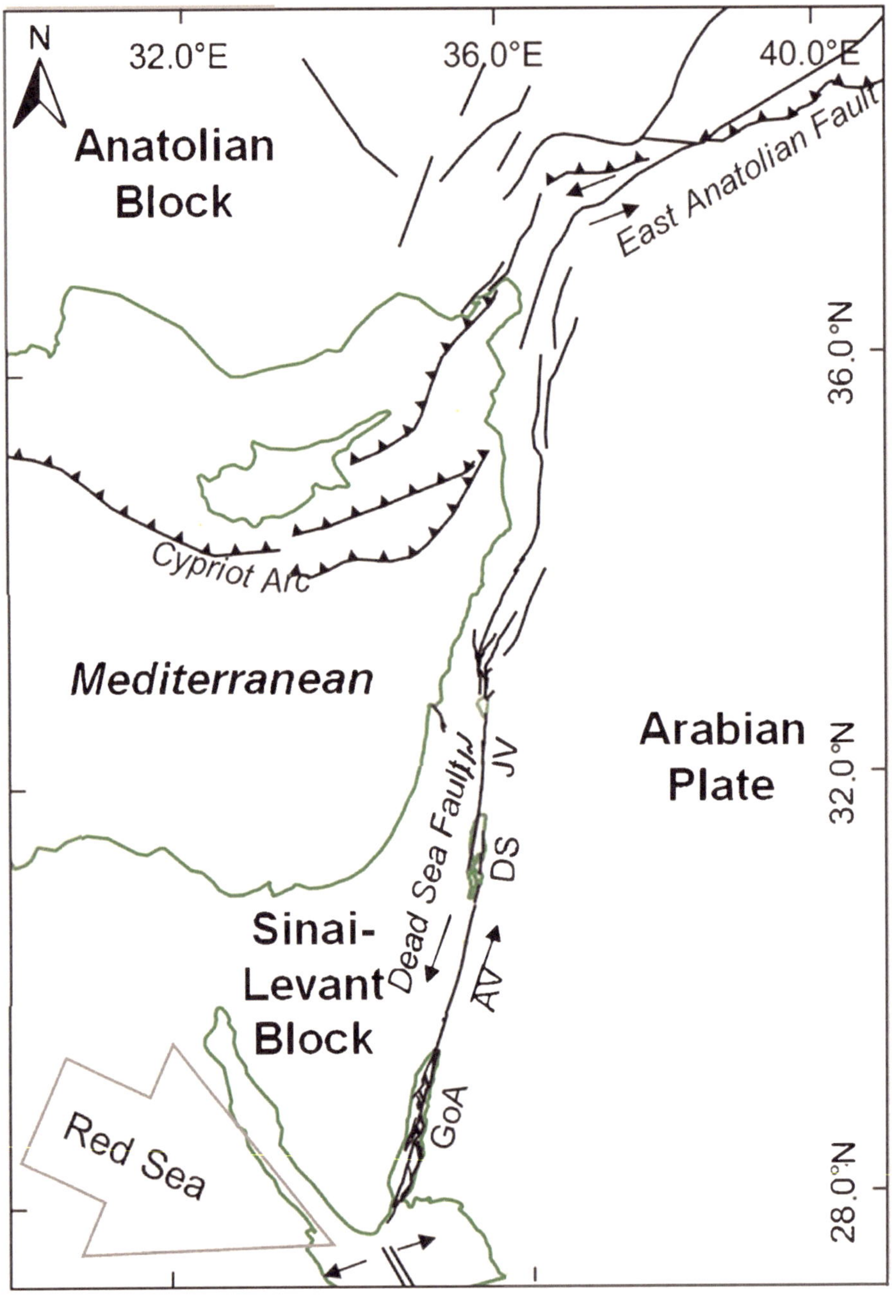

Figure 1.
Tectonic plates in the Middle East. The Dead Sea fault is one of the Dead Sea transform (DST) fault system, which transfers the opening at the Red Sea to the east Anatolian fault. Jordan Valley (JV), Dead Sea (DS), Arava/Araba Valley, and Gulf of Aqaba (GoA) located along the DST.

addressed as a primary cause of earthquakes in the region. Large earthquakes (M > 5.5) are typically associated with slip rate and stress buildup at the strike-slip faults of the transform [9–11, 13, 66–69]. Yet, normal faults might release M > 6 earthquakes [64].

Due to paleo-seismological research, we have access to 220 kyr of earthquake records, obtained based on different dating methods (e.g., uranium-thorium dating, radiocarbon dating, infrared stimulated luminescence dating) of deformation or changes that occur in rock structures due to earthquake shaking (namely seismites) [65]. A promising environment for paleoseismological investigations is provided by caves because they serve well as repositories for geological information. Kagan et al. [70] constrained dates of 38 seismites from the Soreq and Har-Tuv caves, 40 km of the DST. This was based on interpretation and amalgamation of individual seismite ages for bracketing the chronology of the strongest (M7.5-8) earthquakes covering the time interval since 185 ka.

Kagan et al. [70] found in their study a correlation between the paleoseismic events that occurred during the past 75 kyr identified with speleothems with other independent paleoseismic records in the region. Some of them were recorded in Dead Sea cores [71] and correlate with damage at archeological sites [8, 72]. Others correlate with seismites in the Lisan Formation [73, 74] or in the pre-Lisan Formations [65, 75].

Earthquake imprints formed in laminated sediments of the Lisan Formation [76] such as intraclast breccias [7], represent an uninterrupted geological record of the earthquake in the Dead Sea region for the past 14 to 70 kyr. Porat et al. [77] inferred that injection of clastic dikes is one of the most impressive liquefaction features generated by strong, $M \geq 6.5$ earthquakes. The authors constrained the liquefaction date to shortly after the deposition of the Lisan Formation (15 kyr) by using optically stimulated luminescence ages. The dikes ages vary between 15 and 10 ka, thus suggesting that the period following the draw-down of Lake Lisan period was accompanied by strong earthquakes.

Migowski et al. [71] reconstructed an earthquake record using the dated Ein Gedi and Ze'elim cores seismites from the Dead Sea shores for the past 10 kyr. Later, for the last 4000 years, significantly more seismites were detected from the Ein Feshkha sedimentary sections in the Dead Sea area [64].

Recently, Liu et al. [65] reconstructed the longest archive of earthquakes from a 250 m core from the Deed Sea bottom, covering 220 kyr. Many of the seismites in the studies mentioned above, dated for the last 2 millennia, are matched by independent historical records [8, 68, 73, 78–84]. Using geological, archeological, biblical, historical, and seismological data, researchers infer part of the historical records for the Dead Sea area. Ben-Menahem [85] surmised that about 110 major earthquakes affected the area during the past 2500. Of these, 42 originated along the Dead Sea fault system itself, while 68 were imported from the Helenic-Cyprian arcs and the Anatolian-Elburz-Zagros fault systems.

Numerous publications list earthquakes that hit the Dead Sea and its surroundings during the last 2 millennia [8, 61–63, 79, 80]. Belferman et al. [44] filtered from the scores of listed events only the most destructive ones in the Dead Sea basin, typically causing local intensities of VII or higher in Jerusalem. For a minimal epicentral distance of 30 km, this would translate to a magnitude $M > 5.7$, according to the attenuation relations of Hough and Avni [86]. Considering site effects in the rocky terrain typical of the Jerusalem vicinity, where the near-surface shear velocity is relatively high, the modeled magnitudes exceed M6 (see Eq. (6) and Table 2: Beit HaKerem and Motza, which can be found in the reference [87]). In Belferman et al. [43], most earthquakes from the earlier list were amended. All earthquakes in Belferman et al. [43] were selected by simultaneously satisfying two criteria: (1) The acceptance regularizes the relation between recurrence intervals and lake level, and

(2) they are corroborated by evidence from reassessed historical archives and/or seismites in the northern basin of the Dead Sea (Ein Feshkha and Ein Gedi sites).

2.3 Lake-level fluctuation

Several water bodies occupied the tectonic depression along the Dead Sea transform during the Neogene-Quaternary. The size and composition of these water bodies have changed through time, reflecting the changing climatic [88–95] and physiographic [90, 96] conditions in the region. Layers of sediments accumulated in the tectonic depression reflecting several lake phases: the Neogene Sedom lagoon [97–101], the mid- to late-Pleistocene Lake Amora [93, 101–103], the last interglacial Lake Samra ~130 kyr [104], the last glacial Lake Lisan 18 ~ 70 kyr [101, 105, 106], and the Holocene to modern Dead Sea ~15 kyr [92, 107].

The history of saline water bodies in the Dead Sea basin starts with the Sedom lagoon which has been related to the Neogene ingression of seawater into the Jordan-Arava valley. The reasons for the ingression and the exact age of the lagoon are not clear, and its disconnection from the open sea may be associated with various events that include eustatic changes in the Sea, the tectonic uplift of the Judea-Samaria Anticline, the formation of morphological sill in the Jezreel Valley, etc. [93].

Mid- to late-Pleistocene Lake Amora probably extended over a large part of the Dead Sea basin-Jordan Valley [93, 95]. Zak [101] suggested that the age of the Pleistocene sedimentary sequence at Mt. Sedom deposited by the Amora Formation lies between ~1 Ma and ~ 100 ka. The transition between Amora and Samra Formations corresponds to ~130–140 ka, when the lake stood lower than ~380 m below mean sea level (bmsl). The lithology and chronology of the Samra Formation suggest that the lake fluctuated mostly between ~310 and ~ 350 m bmsl [104], being significantly lower than the last glacial Lake Lisan (mostly ~280 ± 40 m bmsl) [95, 96, 108, 109]) but higher than the Holocene Dead Sea (mostly ~400 ± 20 m bmsl) [92, 107, 110, 111]. The lake-level curve for the past 150 kyr, presented in **Figure 2**, was digitized from Waldmann et al. [104].

The most widespread sediments in the rift valley are those deposited at the bottom of Lake Lisan (**Figure 2**) that occupied the Dead Sea basin and adjacent basins between ~70 and 18 ka., during the last glacial period in the northern latitudes

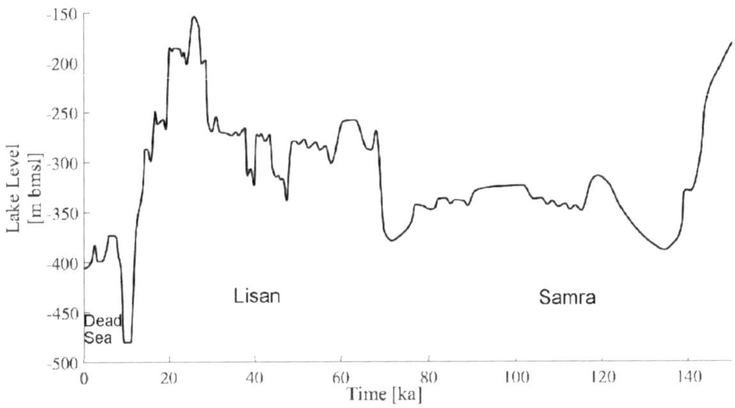

Figure 2.
Water-level change in the last 150 kyr, for Samra, Lisan and the Dead Sea (DS) Lake. Derived from the work of Waldman et al. [104].

[105, 112–114]. The age of the highest lake level of Lake Lisan ~27–23 ka, [93, 95, 96, 108] or ~ 26–24 ka [115, 116] corresponds to the time of a cold episode in the global climate, while the relatively low lake level of ~55–30 ka. Refs. [95, 108, 115] corresponds to the warmer climatic conditions.

The curve of Holocene Dead Sea level fluctuations presented for past 10 kyr is based on the recognition and dating of shoreline deposits in the fan delta outcrops [88, 111, 117] or facies of drilled sediments [92, 107, 118, 119]. The Dead Sea fluctuated within the range of 370 to 430 mbsl during the past 10 kyr [92]. For the last centuries, a precise reconstruction of the Dead Sea level has been available [90].

During the past 4000 yr., the levels of the Dead Sea fluctuated within the range of 390 to 415 mbsl [111, 120]. Most of the time, the lake level was below 402 mbsl, and the Dead Sea was confined to its deep northern basin [111].

The drop from the late nineteenth century high stand of 390 mbsl to the present (2023) elevation of 428 m bmsl (www.water.gov.il) resulted mainly due to artificial water diversion of runoff water for the basin, superimposed on the climatic trend [111]. The current rate of lake-level decline is significantly high. However, drops of similar magnitude are not unusual in the late Holocene record, which yields evidence of long periods of droughts in the region [111].

3. Poroelasticity and its role in induced seismicity

Water-level-induced seismicity was addressed in several studies of water reservoirs [41, 121, 122] and was also analyzed using the theory of linear poroelasticity [123, 124]. For the first time, a relationship of this kind was considered in the pull-apart basin of tectonic depressions (Dead Sea lake) for earthquake intervals of time-scale 100–200 years in Belferman et al. [43, 44]. The increase in water level contributes to an immediate increase in vertical stress due to the weight of the water mass (the loading effect) and subsequently to the increase in horizontal stress due to elastic coupling controlled by Poisson's ratio [44]. Resulting stress changes are then superimposed on the background tectonic stress field [44]. Further, generation or absence of post-impounding seismicity depends on several additional factors, presented in **Figure 3**, and discussed in Belferman et al. [44].

Based on the real-time observation of 30 cases of post-impounding seismicity it was indicated that the complex interaction of the water-induced stress with the state of pre-existing stress near the reservoir, together with geological and hydrologic conditions at the site, determines whether these stress changes are sufficient to generate the seismic activity [125]. It was indicated that in areas of high strain accumulation and high levels of natural seismicity, stress changes induced by the reservoir may be small compared to natural variations in contrast to regions of moderate strain accumulation (low to moderate natural seismicity). Thus, in the low to moderate natural seismicity regions, the post-impounding can usually be accompanied by earthquakes. In aseismic areas with slow strain accumulation, the reservoir-induced stresses may be insufficient to raise the stress level to a state of failure [125].

Seismic activity induced by surface water-level fluctuations is also affected by the faulting regime (**Figure 3**), determined in turn by the respective orientations of the three principal stresses [126]. In regions where the vertical compressive stress is not minimal (normal and strike-slip faulting), seismic activity is more sensitive to the change in the effective stress due to water-level change than in regions where it is (thrust faulting) [123, 125, 127–129].

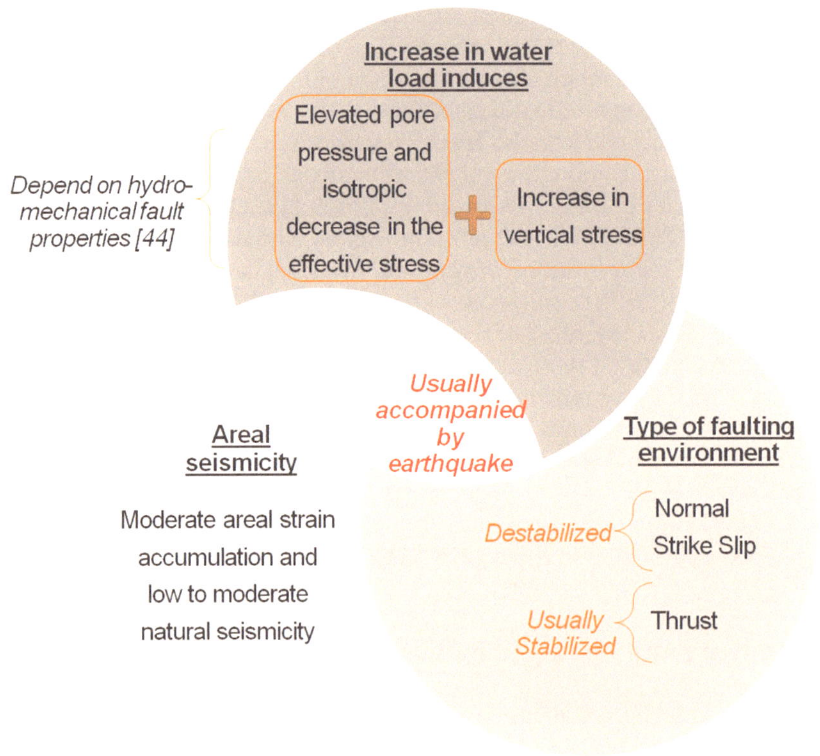

Figure 3.
Factors that have the potential to influence the seismic activity prompted by alterations in water levels. Please refer to the text for more information.

In many cases (for instance, the Koyna, Aswan, and Oroville reservoirs), the hypocenter of earthquakes, which are associated with changes in water levels, does not occur directly under the reservoir. In these cases, the faults passing through the reservoir serve as a conduit for pore pressure diffusion [41, 121, 130]. This allows the influence of the reservoir to extend to greater distances and depths, where larger magnitude earthquakes take place.

In addition to the conditions mentioned above, based on observations that some reservoirs continue to be active, whereas others show seismic quiescence under similar water-level fluctuations [37], it was indicated in Belferman et al. [44] that seismicity can strongly depend on the hydromechanical properties of faults (**Figure 3**).

4. Methods

4.1 Modeling setup

We consider a vertical fault underneath the lake (or reservoir) bed, extending along the outplane axis y (**Figure 4**), embedded in 2D (plain strain) geometry of the upper crust. We assume that fault failure will result from a change in the effective stress normal to the fault [123, 124, 131–134]. A step change in the water load, σ_w,

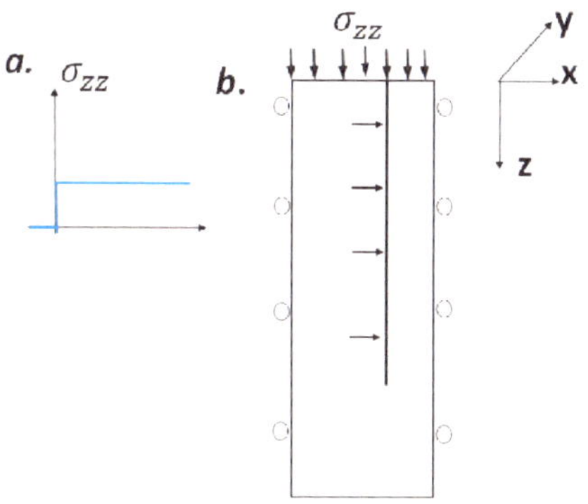

Figure 4.
This figure, derived from the work of Belferman et al. [44], illustrates the model used in this study. (a) A step-function water load, applied on a top boundary of the structure, representing a poro-elastic seabed; (b) schematic of the structure below the lake with a vertical fault oriented along the outplane axis y. the calculations are based on uniaxial strain conditions.

(**Figure 4a**) is applied on the entire top boundary of the half-infinite structure modeled, representing a valley filled by the lake, where uniaxial strain conditions apply.

4.2 The impact of Lake-level fluctuation on pore pressure

As a result of the water load on the seabed during water-level (WL) change, considering the diffusion, the pore pressure (p) is a function of time (t) and sub-bottom depth (z). Belferman et al. [44] consider a step function load on the half-space, for which, following Roeloffs [123], the analytical solution is:

$$p(z,t) = \gamma p_s + (1-\gamma)p_s erfc\left(z/\sqrt{4ct}\right) \tag{1}$$

where p_s is the pore pressure on the top boundary of the half-space, γ is the loading efficiency, and c is the hydraulic diffusivity. For the post-diffusion stage $erfc\left(\frac{z}{\sqrt{4ct}}\right) \to 1$ at $t \to \infty$ (where t is the diffusion time). For any earthquake hypocentral depth, z, the assumption of the post-diffusion stage is valid, if the diffusion time scale, indicating the travel time of excess pore pressure signal from seabed to hypocentral depth, $t_{diff} = z^2/c$, is small compared to the earthquake's recurrence interval (RI) ($t_{diff} \ll t_{RI}$) [44]. For an average earthquake hypocentral depth underneath the DSF, $z \cong 20\ km$ [135–137] and $c = 4\ m^2/s$ justified for the faulted rock, the diffusion time scale indicating a travel time of excess pore pressure from lake bed to the hypocentral depth $t_{diff} \cong 3\ yr$. It is negligible compared to the RI of moderate-to-large earthquakes (M > 5.5) in this area, estimated as $t_{RI} = 1600\ yr$ for Lake Lisan and as $t_{RI} = 100\ yr$ for the Dead Sea, based on seismite analyses from the Peratzim Creek and Ein Gedi cores [7, 8, 71, 74].

The water-level variation in Lake Cahuilla has a different tendency compared to the Dead Sea. Subsequently, Hill et al. [45] consider the periodic loading on the poroelastic half-space for which the analytical solution following Roeloffs [123] is:

$$\tilde{p}(z) = \gamma p_s + (1-\gamma)p_s e^{-z\sqrt{\omega/2c}} e^{-iz\sqrt{\omega/2c}} \tag{2}$$

were ω is the angular frequency of water loading ($\omega = \frac{2\pi}{T}$, where T is period time). Similar to a step function load for the post-diffusion stage, for periodic loading, we obtain the long-term limit: $e^{-z\sqrt{\frac{\omega}{2c}}}e^{-iz\sqrt{\frac{\omega}{2c}}} = e^{-z\sqrt{\frac{\pi}{Tc}}}e^{-iz\sqrt{\frac{\pi}{Tc}}} \rightarrow 1$ at $T \rightarrow \infty$.

Here, T is defined by the periodic change of water level. Hence, for a long period of time, the diffusion effect is negligible because the pore pressure at the depth reached the value at the top boundary of the half-space. This is expressed in Hill et al. [45] in the "memory" effect of pore pressure at depth, whereby subsequent lakes can contribute to higher pore pressure owing to the diffusive time lag of a previous lake superimposing on the next. This memory effect is pronounced only in one case where the time interval (T) between maximal lake stands is about 41 years (Lake F and Lake E in **Figure 2**, [45]).

Dead Sea lake WL change is distinctly different from the periodic trend that is characteristic of Lake Cahuilla. This may be due to the pulverization of fault rock on the southern San-Andreas Fault [138].

4.3 Analytical formulation

A poroelastic model with brittle yielding was formulated by Belferman et al. [44], utilizing the Mohr-Coulomb failure criterion. This model was used to calculate the changes in earthquake recurrence interval in response to the increase of water level at basins overlying strike-slip faults.

Within the poroelastic component of the model, produced by the WL change, horizontal stress changes normal to the strike-slip fault calculated under a uniaxial (vertical) strain condition (Eq. (10b) in [44]). This applies to a post-diffusion stage, i.e., when pore pressure at hypocentral depth equilibrates with the lake's bed ($P_f = \sigma_w$). In the case discussed, it is appropriate to skip the diffusion stage since the diffusion time is much shorter than the interseismic period. In this case, the effective vertical stress change becomes zero, while the effective horizontal stress ($\Delta\sigma'$) change becomes:

$$\Delta\sigma' = \frac{1-2\nu}{1-\nu}(\beta-1)\sigma_w \tag{3}$$

where β is Biot's coefficient and ν is Poisson's ratio, $\sigma_w = \rho g \Delta h$, where ρ is density of water and g is the acceleration of gravity.

Further, under the general loading conditions, the critical value of shear stress (τ_c) responsible for yielding at the fault is connected to the effective stress normal to the fault (σ'_n), according to the Mohr-Coulomb failure criterion (e.g., [139]):

$$\tau_c = C + tan\,(\varphi)\sigma'_n \tag{4}$$

where $\sigma'_n = \sigma_n - P_f$ is effective normal stress affected by pore pressure (P_f), C is cohesion, and φ is the angle of internal friction at the fault.

In order to calculate the change of recurrence interval (Δt) induced by the water level change, we use the Coulomb failure envelope and Mohr circle [140] with the following initial characteristics:

$$\sigma_0 = \frac{\sigma_1+\sigma_3}{2}, \tau_0 = 0, R_0 = \frac{\sigma_1-\sigma_3}{2} \tag{5}$$

where σ_1 and σ_3 is the minimal and maximal principal stress, which is horizontal for the strike-slip faults. In this case, if during the interseismic period, the water level increased by Δh, the horizontal stress change can be specified by $\Delta\sigma' = \Delta\sigma'_{xx} = \Delta\sigma'_{yy}$, following the definitions of principal stresses in 2D [140]. This change manifested in the shift of the Mohr circle toward the origin by $\Delta\sigma'$ while for the pre-seismic stress state, the Mohr circle center is: $\sigma'_0 = \sigma_0 + \Delta\sigma'$ (where $\Delta\sigma'$ is negative at the post-diffusion stage, see **Figure 5b**, [44]).

We assume that the far-field tectonic strain rate, $\dot{\varepsilon}_{xy}$, stays constant over any interseismic time period, Δt, [44] while the elastic strain, $\Delta\varepsilon$, accumulated during this period can be presented by: $\Delta\varepsilon = \dot{\varepsilon}_{xy}\Delta t$. Based on Hooke's law and considering only horizontal shear strain, shear stress accumulation at the fault during Δt is:

$$\Delta\tau_{xy} = 2G\dot{\varepsilon}_{xy}\Delta t = \frac{Ccos(\varphi)}{t_{RI}}\Delta t \tag{6}$$

where t_{RI} is a tectonic recurrence interval and C $\neq 0$ at the healed fault (e.g., [141]). Hence, a permanent tectonic strike-slip stress rate can be defined as $\frac{Ccos(\varphi)}{t_{RI}}$ [44]. The change of the shear stress during the interseismic period increases the radius of the Mohr circle while for the pre-seismic stress state, the radius is: $R'_0 = R_0 + \Delta\tau_{xy}$.

Byerlee's law envelope [142] was used to define the strength of a seismogenic zone at the fault immediately after the earthquake. An initial stress state, σ_{init}, is defined as a Mohr circle with a radius R_0 centered at the point $(\sigma_0, 0)$ and restricted by the Byerlee's law envelope. The larger hypocentral depth is associated with a larger σ_0. Based on the results of Byerlee [142] and on additional laboratory experiments (for a review, see [143]), we adopt, for simplicity's sake, the friction angle $\varphi = 0.54$ and initial cohesion $C = 0$.

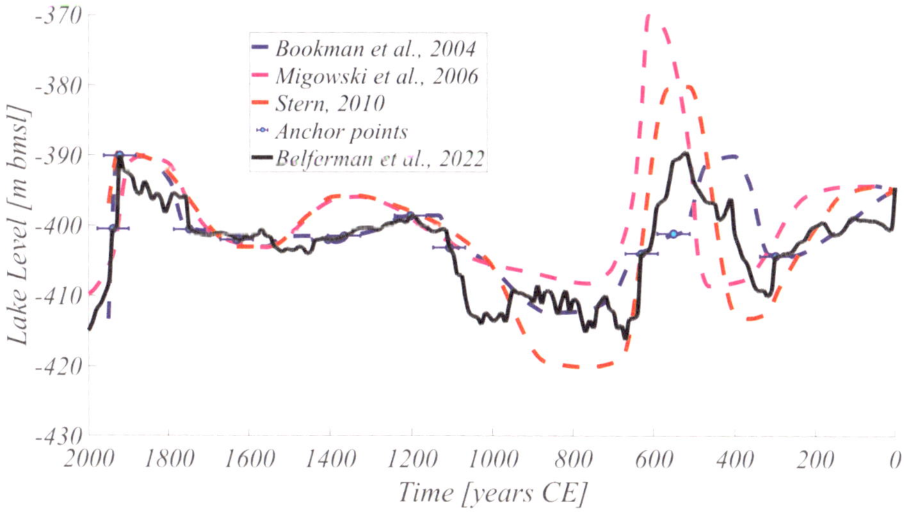

Figure 5.
The Dead Sea WL reconstructions for the last two millennia. The dashed curves are suggested by the literature sources. Turquoise anchor points follow Bookman et al. [111] used in WL interpretation, while one point (in dark blue) is shifted to left in error interval of ±45 yr. solid, black line water curve is a plausible scenario suggested by this study.

The pre-seismic Coulomb failure envelope is defined by the nonzero cohesion coefficient, $C \neq 0$, specific for the healed fault zone (e.g., [141] and references therein) and friction angle $\varphi = 0.54$. When the circle reaches the failure envelope, the rock fails at the fault oriented most favorably for sliding (in our case, it is the pre-defined strike-slip fault). The stress will then drop again to σ_{init}.

After stress release, the time to the next earthquake, Δt, is calculated from the solution of the Mohr-Coulomb failure criterion for a strike-slip tectonic regime and a water level change, Δh, characteristic to the Dead Sea lake [44]:

$$\begin{cases} (\tau - \tau_0)^2 + (\sigma - (\sigma_0 + \Delta\sigma'))^2 = \left(R_0 + \frac{Ccos(\varphi)}{t_{RI}} \Delta t \right)^2 \\ \tau = C + tan\,(\varphi)\sigma \end{cases} \tag{7}$$

Solving this problem for single solution, we get Δt us a linear function of the water level change Δh:

$$\Delta t = (C + tan\,(\varphi)\rho g \Delta h) \frac{t_{RI}}{C} \tag{8}$$

4.4 Best fit random method of WL curve prediction

In the interpretation of water level over the past two millennia [92, 111, 144], there are several uncertainty. Specifically, the water-level dating method (Radiocarbon dating) could have an error of about ±45 yr., as estimated from the radiocarbon dating of shoreline deposits in fan delta outcrops [111]. The entire past bi-millennial Dead Sea level record is constrained by less than twenty "anchor points" (the data obtained by the dating collected from surveyed paleo-shorelines, [111]). However, historical water level records are quite precise elevation-wise, as they are obtained from different points around the lake [92, 111].

Consequently, the interpretations of the curves (**Figure 5**) are not identical and are not unambiguous, except for some limitations [43]. Accounting for these limitations, which include the "anchor points" (the data obtained by the dating collected from surveyed paleo-shorelines, [111]) of water level, the 10 million WL curves were generated [44] for the last bi-millennial interval, using a uniformly distributed random number generator. For the simulation, we are setting a ten-year time step.

The linear correlation between RI of widely recorded medium to large historical earthquakes (M > 5.5) available from the literature and WL interpolations was tested on the basis of an estimate of the value of the Pearson product-moment correlation coefficient, R. These statistics were used to assess the suitability of each randomly interpolated WL curve and for identifying out-of-correlation earthquakes.

4.5 Earthquake simulation

The analytical model presented in this section has been discretized with the time step of 1 year and applied to the sequence of WL change samples [44]. From the starting point (AD 33, see Table 1 which can be found in the Appendix of the reference [44]), the simulator moves forward with time, along the WL curve. For each step, the WL change, Δh_i, and the amount of accumulated tectonic stress are calculated. After each stress release, the time to the next earthquake, Δt, is calculated from the solution of the Mohr–Coulomb failure criterion for a strike-slip tectonic regime. Eq. (9) takes the following discrete form:

$$\begin{cases} (\tau_i - \tau_0)^2 + \left(\sigma - \left(\sigma_0 + \frac{1-2\nu}{1-\nu}(\beta-1)\rho g \Delta h_i\right)\right)^2 = \left(R_0 + \frac{C cos(\varphi)}{t_{RI}}\Delta t\right)^2 \\ = \tau_i = C + tan\,(\varphi)\sigma_i \end{cases} \tag{9}$$

For each time step, the algorithm determines whether there is a single solution, or two, or nil. A case of no solutions means that the Mohr circle is yet to reach the failure envelope, as the accumulating tectonic stress and the WL increase are still insufficient. At this case, the Δt increase with the time steps. The system of Eq. (7) may have a single solution when the failure criterion is met at the end of some timestep or two solutions when it is met before the end of the timestep. A case of two solutions is rounded down to a case of a single solution if a time step (1 year) is small compared to the earthquake RI (about hundreds of years), where the Mohr circle reaches the failure envelope and the earthquake accrues. In this case, the calculated Δt saved and resets to zero. At the final stage, we get an array of RI.

4.6 The backward earthquake simulation approach

This formulation is based on the same basic assumptions that were proposed in Belferman et al. [43, 44], as well as set out in the "Analytical formulation of the direct model" section. Similar to the previous simulation, we assume that after an earthquake, the Mohr's circle returns to the initial state of stress. But in contrast to the previous case, our starting point of reference is the last earthquake, as it is much better constrained. Considering that at the moment before the specific earthquake occurred, Mohr's circle reaches the failure envelope, the set of Eq. (7) gets one solution. The horizontal effective stress change ($\Delta\sigma'$) calculated from Eq. (3). While $\sigma_w = \rho g \Delta h$, calculated for the WL change Δh at the moment of this specific earthquake.

The recurrence interval, Δt, calculated from Eq. (8), and based on this value, we calculate the time of the previous earthquake $t_p = t_c - \Delta t$. Running this algorithm on the WL change during the period 2–10 kyr (based on the level history of [92]), combined with the WL change received from the best fit random method [44] including the anchor points [111] for the past two millennia, we get a new catalog of simulated earthquakes presented in **Figure 6**.

This formulation, like the previous one, is based on the solution of a system of equations (Eq. (7)) in the domain determined by positive definite values of Δt. This is only possible when the WL change (Δh) does not exceed the value $\frac{C}{tan\,(\varphi)\rho g}$. When the WL change exceeds the defined value, the model jumps to the next point where the WL change is within the definition domain.

5. Results and discussion

Even though historical water-level records exhibit a significant degree of elevation accuracy, given that they are derived from diverse surveyed locations around the lake [92, 111], the dating of water levels might carry an approximate error of ±45 years and even more, as deduced from radiocarbon dating of sediment deposits in fan delta outcrops [111] or sediment cores [92]. The substantial differences in potential interpolations arise from the uncertainties in dating and resolution aspects of WL reconstructions. This was shown in Belferman et al. [43], using best fit random method WL curve generation. Ten million different curves were generated, constrained with anchor points established by field surveys and radiocarbon dating [111].

Dating historical earthquakes can exhibit a high level of precision, and the validation of accuracy occurs when various historical references align in agreement [61–63]. But in many cases, the epicenter of even a historical earthquake can be imprecise or not known.

This chapter extends our simulation to prehistoric time (10 ka), for which further challenges include identifying and dating shoreline and earthquake markers [71, 92]. The water-level curve is combined from the one indicated in Belferman et al. [43] (for 2 ka) with that provided by Migowski et al. [92] (for 2–10 ka). Our model for that level curve generates 50 earthquakes.

The recurrence intervals are simulated using the reverse earthquake simulation algorithm, while the starting point of reference is a simulated earthquake that occurred in 1907 CE. We start from this point because it is the last earthquake received from the earthquake simulation presented in Belferman et al. [43], which correlates with the documented earthquake from the field of 1927 CE. The recurrence intervals are depicted in **Figure 6** as black points plotted against the water level (orange curve). The calculated years of simulated earthquakes are presented along the time axis by blue stars (**Figure 6**).

Two types of spatial segments have been identified (gray bars in **Figure 5**). Light gray bar (~3490–4020, 4370–4730, and 9380–10,000 years) indicates the time segments where the WL change exceeds the definition domain specified in our model. The dark gray bars indicate the time segments of earthquake sequence (4024–4065, 4724–5068, and 9141–9196) received for periods of height WL stand, with recurrence interval smaller than 35 yrs. The domain excludes the time segments where the WL change almost reaches the critical value of $\frac{C}{tan\,(\varphi)\rho g}$ (marked in dark gray in **Figure 6**).

Simulated earthquakes are presented vs. earthquakes from the literature in **Figure 7**. The blue points (**Figure 7**) indicate the simulated events supported by

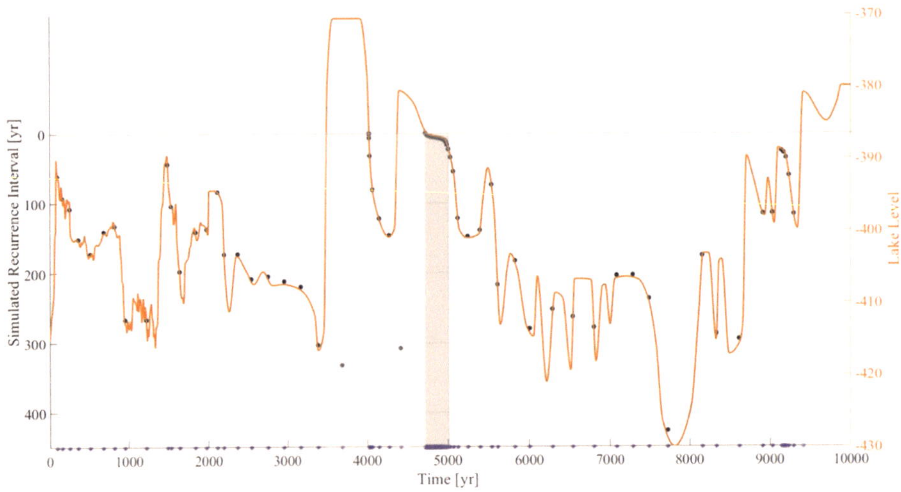

Figure 6.
Orange curve represents the best fit random water level [43] combined with Migowski et al. [92] vs. simulated and historic RIs, correspondingly. The blue dots mark the dates of the seismic events, while the black dots indicate the recurrence interval between these events.

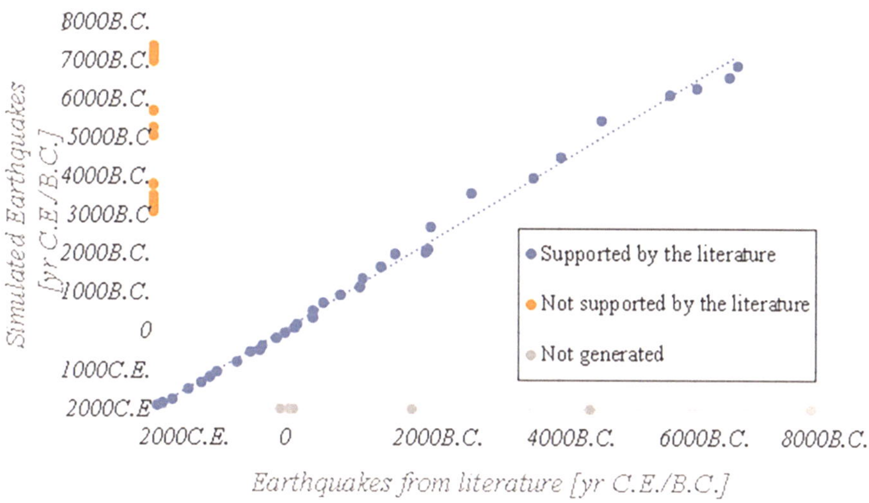

Figure 7.
A comparison dates of historic vs. simulated earthquakes based on the suggested best-fit WL curve [43] combined with WL curve presented by Migowski et al. [71].

literature. Orange points (**Figure** 7) indicate simulated events, for which there is no evidence in the literature. A cluster of earthquakes was generated in these segments. In **Figure** 7, we left only the first and the last earthquake of the cluster.

In the segments marked in light gray, our model did not perform the calculation. The earthquakes received in these segments were calculated in the previous step. Therefore, our model may not have generated earthquakes that occurred in these segments, such as the one indicated by Ben-Menahem [85] to 1560 B.C. or the seismite estimated by Migowski et al. [71] to have occurred around 7700 B.C.

Of the 50 simulated earthquakes during a period of 10,000 years, only 15 events were not confirmed in the literature (**Figure** 7). All these 15 events occurred before 3000 BCE. In other words, for all simulated earthquakes by our model within a span of 5000 years, there is evidence in the literature of a significant earthquake occurring in the Dead Sea area. Five earthquakes not confirmed by literature generated at 3068–3533 BCE. In Ben-Menachem's 1991 study, the archeological excavations at Tleilat el-Rassul of Kenyon were referenced, indicating the possibility of destruction caused by earthquakes that occurred between 3300 and 3600 BCE. In addition, Migowski et al. [71] dated the possible earthquake to 3300–3367 BCE from Ein Gedi core. This evidence can satisfy the simulated events 339 3 BCE, 3533 BCE. Another prehistoric cluster (four events) of earthquakes was simulated in the period 7196–7410 BCE due to a relatively high water level. This period is not well represented by the archeological record due to scarcity of masonry structures.

The advantage of using the backward simulation method, presented for the first time in this chapter, is that it starts from the last earthquake along the Dead Sea fault, for which ample information is available. By contrast, the previous simulation [43] starts from the last earthquake of the studied period. Therefore, it is suitable for a period of 2000 years where earthquakes have historical evidence. The backward method enables us to simulate earthquakes that occurred in prehistoric times, including during the Lake Lisan period. For this period, the water level was significantly higher, and the data regarding both phenomena are subject to even greater uncertainty. Nevertheless, the paleoseismicity studies [8, 12] indicate that despite such high water levels, the seismic activity may have been subdued compared with activity for the Holocene Dead Sea.

However, it is worth noting that our model is limited by the maximum change in water level, and the water level during the Lisan period exceeds this limit. On this occasion, in future studies, it is worth integrating additional physical properties of Earth's crust. Field data from Scandinavia and North America suggest that the melting of Pleistocene ice sheets was connected to paleoseismic activity [19, 21–24, 145, 146]. Conversely, the low level of seismicity in Greenland and Antarctica may be due to the current ice load [25], which in turn induces changes in the differential stress in the Earth crust [26, 28].

Hampel et al. [20], based on numerical models that include one or several fault planes embedded in a rheologically layered lithosphere, indicate how loading and flexure of the lithosphere by ice or water decreases the slip rates of nearby faults, whereas unloading and rebound of the lithosphere accelerates the slip (see also [47]). Slip rate variations are caused by changes in the differential stress, which is the difference between the maximum (σ_1) and minimum (σ_3) principal stresses. When a fault is under a load, the differential stress decreases during loading and flexure of the lithosphere and increases during unloading and rebound. This leads to a decrease and subsequent increase in the fault slip rate.

This configuration needs to be considered when developing the earthquake simulation model for the Lisan period. Additional considerations for further studies are the mechanical condition of the recorder (namely the sediment bed) and the effect of water column height on the dynamics of seismite generation [147, 148].

6. Conclusions

The correlation model, developed in [44] and supported by the comparability of a simulated earthquake with a historical catalog for 2,000 years [43] imitates reaction of pore-elasto-plastic crust. This model is validated by the results obtained in this chapter for 10 millennia of geological records of earthquakes and shows a good correlation for the past five millennia.

Based on the findings presented in the Results-Discussion chapter, we indicate that flexure of the lithosphere, suggested by Hampel et al. [20], may not be significant over two or five millennia, but it should be considered when analyzing longer time periods as Lake Lisan.

Considering this in our future model, we expect to receive an explanation of the low seismicity pattern during the Lisan high stand and solve the model limitations where the WL change exceeds the definition domain, specified in our model.

Nevertheless, the new back-counting model presented in this work enables us to extend the relationship between WL changes and seismicity over a period span of ten millennia.

Acknowledgements

This project was supported by grants from the Ministry of Energy (grant no. 213-17-002) and the German–Israeli Foundation for Scientific Research and Development (GIF; grant no. I-1280-301.8). We would like to express our gratitude to Dr. Regina Katsman and Prof. Zvi Ben-Avraham for their valuable contributions in initiating this chapter.

Author details

Mariana Belferman[1*] and Amotz Agnon[2]

1 Physics Department, Shamoon College of Engineering, Ashdod, Israel

2 The Fredy and Nadine Herrmann Institute of Earth Sciences, The Hebrew University of Jerusalem, Jerusalem, Israel

*Address all correspondence to: mkukuliev@gmail.com

References

[1] Daito H, Suzuki M, Shiihara J, Kilgore PE, Ohtomo H, Morimoto K, et al. Impact of the Tohoku earthquake and tsunami on pneumonia hospitalisations and mortality among adults in northern Miyagi, Japan: A multicentre observational study. Thorax. 2013;**68**:544-550

[2] Xu Y, Huang J, Zhou J, Zeng Y. Patterns of abdominal injury in 37 387 disaster patients from the Wenchuan earthquake. Emergency Medicine Journal. 2013;**30**:538-542

[3] Mileti D. Disasters by Design: A Reassessment of Natural Hazards in the United States. Washington, DC: Joseph Henry Press; 1999. DOI: 10.17226/5782

[4] Noriega GR, Ludwig LG. Social vulnerability assessment for mitigation of local earthquake risk in Los Angeles County. Natural Hazards. 2012;**64**: 1341-1355. DOI: 10.1007/s11069-012-0301-7

[5] Allen JRL. Sedimentary Structures: Their Character and Physical Basis. Vol. 30. Amsterdam, The Netherlands: Elsevier; 1982. (Developments in Sedimentology)

[6] Youd T, Perkins D. Mapping of liquefaction severity index. Journal of Geotechnical Engineering. 1987;**113**: 1374-1392

[7] Agnon A, Migowski C, Marco S. Intraclast breccias in laminated sequences reviewed: Recorders of paleo-earthquakes. In: Enzel Y, Agnon A, Stein M, editors. New Frontiers in Dead Sea Paleoenvironmental Research. Vol. 401. Colorado, US: Geological Society of America; 2006. p. 195. DOI: 10.1130/2006.2401(13)

[8] Agnon A. Pre-instrumental earthquakes along the Dead Sea rift. In: Garfunkel Z, Ben-Avraham Z, Kagan E, editors. Dead Sea Transform Fault System: Reviews. Dordrecht: Springer; 2014. pp. 207-261. DOI: 10.1007/978-94-017-8872-48

[9] Ben-Menahem A, Nur A, Vered M. Tectonics, seismicity and structure of the Afro-Eurasian junction—The breaking of an incoherent plate. Physics of the Earth and Planetary Interiors. 1976; **12**(1):1-50. DOI: 10.1016/0031-9201(76) 90005-4

[10] Ferry M, Meghraoui M, Karaki NA, Al-Taj M, Amoush H, Al-Dhaisat S, et al. A 48-kyr-long slip rate history for the Jordan Valley segment of the Dead Sea fault. Earth and Planetary Science Letters. 2007;**260**(3):394-406. DOI: 10.1016/j.epsl.2007.05.049

[11] Garfunkel Z. Internal structure of the Dead Sea leaky transform (rift) in relation to plate kinematics. Tectonophysics. 1981;**80**(1–4):81-108. DOI: 10.1016/0040-1951(81)90143-8

[12] Marco S, Klinger Y. Review of on-fault Palaeoseismic studies along the Dead Sea fault. In: Garfunkel Z, Ben-Avraham Z, Kagan E, editors. Dead Sea Transform Fault System: Reviews. Modern Approaches in Solid Earth Sciences. Vol. 6. Dordrecht: Springer; 2014. DOI: 10.1007/978-94-017-8872-4_7

[13] Niemi T, Zhang H, Atallah M, Harrison JB. Late Pleistocene and Holocene slip rate of the northern Wadi araba fault, Dead Sea transform, Jordan. Journal of Seismology. 2001;**5**:449-474. DOI: 10.1023/A:1011487912054

[14] Schwartz DP, Coppersmith KJ. Fault behavior and characteristic earthquakes:

Examples from the Wasatch and San Andreas fault zones. Journal of Geophysical Research: Solid Earth. 1984; **89**(B7):5681-5698. DOI: 10.1029/JB089iB07p05681

[15] Sieh KE. Slip along the San Andreas fault associated with the great 1857 earthquake. Bulletin of the Seismological Society of America. 1978;**68**(5):1421-1448. Available from: https://authors.library.caltech.edu/49783/1/1421.full.pdf

[16] Barka AA. The north Anatolian fault zone. In: Annales Tectonicae. Vol. 6. Bristol, England: Bristol University Press; 1992. pp. 164-195. Available from: ftp://ftp.ecn.purdue.edu/Barka.PDF

[17] Stein RS, Barka AA, Dieterich JH. Progressive failure on the North Anatolian fault since 1939 by earthquake stress triggering. Geophysical Journal International. 1997;**128**:594-604

[18] Toksöz MN, Shakal AF, Michael AJ. Space-time migration of earthquakes along the north Anatolian fault zone and seismic gaps. Pure and Applied Geophysics. 1979;**117**:1258-1270. DOI: 10.1007/BF00876218

[19] Muir-Wood R. Deglaciation seismotectonics: A principal influence on intraplate seismogenesis at high latitudes. Quaternary Science Reviews. 2000;**19**:1399-1411. DOI: 10.1016/S0277-3791(00)00069-X

[20] Hampel A, Lüke J, Krause T, Hetzel R. Finite-element modelling of glacial isostatic adjustment (GIA): Use of elastic foundations at material boundaries versus the geometrically non-linear formulation. Computers & Geosciences. 2019;**122**:1-14. DOI: 10.1016/j.cageo.2018.08.002

[21] Arvidsson R. Fennoscandian earthquakes: Whole crustal rupturing related to postglacial rebound. Science. 1996;**274**:744. DOI: 10.1126/science.274.5288.744

[22] Davenport CA, Ringrose PS. Deformation of Scottish quaternary sediment sequences by strong earthquake motions. In: Jones ME, Preston RM, editors. Deformation of Sediments and Sedimentary Rocks. Vol. 29. London, UK: Geological Society of London; 1987. pp. 299-314. DOI: 10.1144/GSL.SP.1987.029.01.23

[23] Fjeldskaar W. The amplitude and decay of the glacial forebulge in Fennoscandia. Norsk Geologisk Tidsskrift. 1994;**74**(1):2-8. Available from: http://njg.geologi.no/images/NJG_articles/NGT_74_1_002-008.pdf

[24] Stewart IS, Sauber J, Rose J. Glacio-seismotectonics: Ice sheets, crustal deformation and seismicity. Quaternary Science Reviews. 2000;**19**:1367-1389. DOI: 10.1016/S0277-3791(00)00094-9

[25] Turpeinen H, Hampel A, Karow T, Maniatis G. Effect of ice sheet growth and melting on the slip evolution of thrust faults. Earth and Planetary Science Letters. 2008;**269**:230-241. DOI: 10.1016/j.epsl.2008.02.017

[26] Hetzel R, Hampel A. Slip rate variations on normal faults during glacial–interglacial changes in surface loads. Nature. 2005;**435**:81-84. DOI: 10.1038/nature03562

[27] Hampel A, Hetzel R. Response of normal faults to glacial-interglacial fluctuations of ice and water masses on Earth's surface. Journal of Geophysical Research: Solid Earth. 2006;**111**:B06406. DOI: 10.1029/2005JB004124

[28] Hampel A, Hetzel R, Densmore AL. Postglacial slip-rate increase on the Teton normal fault, northern basin and range province, caused by melting of the

Yellowstone ice cap and deglaciation of the Teton range? Geology. 2007;**35**: 1107-1110. DOI: 10.1130/G24093A.1

[29] Ellsworth WL. Injection-induced earthquakes. Science. 2013;**341**(6142): 1225942. DOI: 10.1126/science.1225942

[30] Suckale J. Induced seismicity in hydrocarbon fields. In: Advances in Geophysics. Vol. 51. Cambridge UK: Elsevier; 2009. pp. 55-106. DOI: 10.1016/S0065-2687(09)05107-3

[31] Majer EL, Baria R, Stark M, Oates S, Bommer J, Smith B, et al. Induced seismicity associated with enhanced geothermal systems. Geothermics. 2007; **36**(3):185-222. DOI: 10.1016/j.geothermics.2007.03.003

[32] Zoback MD, Gorelick SM. To prevent earthquake triggering, pressure changes due to CO_2 injection need to be limited. Proceedings of the National Academy of Sciences. 2015;**112**(33):E4510-E4510. DOI: 10.1073/pnas.1508533112

[33] Megies T, Wassermann J. Microseismicity observed at a non-pressure-stimulated geothermal power plant. Geothermics. 2014;**52**:36-49. DOI: 10.1016/j.geothermics.2014.01.002

[34] Stark MA, Davis SD. Remotely triggered microearthquakes at the geysers geothermal field, California. Geophysical Research Letters. 1996;**23**(9):945-948. DOI: 10.1029/96GL00011

[35] Yeck WL, Hayes GP, McNamara DE, Rubinstein JL, Barnhart WD, Earle PS, et al. Oklahoma experiences largest earthquake during ongoing regional wastewater injection hazard mitigation efforts. Geophysical Research Letters. 2017;**44**(2):711-717. DOI: 10.1002/2016GL071685

[36] Gupta HK. Reservoir induced earthquakes. In: Gupta HK, editor. Development in Geotechnical Engineering. Vol. 71. Amsterdam: Elsevier; 1992. pp. 1-364. DOI: 10.1016/B978-0-444-88906-5.50022-3

[37] Talwani P. On the nature of reservoir-induced seismicity. Pure and Applied Geophysics. 1997;**150**:473-492. DOI: 10.1007/978-3-0348-8814-1_8

[38] Talwani P. Seismotectonics of the Koyna-Warna area, India. Pure and Applied Geophysics. 1997;**150**:511-550. DOI: 10.1007/s000240050091

[39] Tomic J, Abercrombie R, Do Nascimento A. Source parameters and rupture velocity of small M≤2.1 reservoir induced earthquakes. Geophysical Journal International. 2009;**179**:1013-1023. DOI: 10.1111/j.1365-246X.2009.04233.x

[40] Hua W, Chen Z, Zheng S. Source parameters and scaling relations for reservoir induced seismicity in the Longtan reservoir area. Pure and Applied Geophysics. 2013;**170**:767-783. DOI: 10.1007/s00024-012-0459-7

[41] Hua W, Chen Z, Zheng S, Yan C. Reservoir-induced seismicity in the Longtan reservoir, southwestern China. Journal of Seismology. 2013;**17**(2): 667-681. DOI: 10.1007/s10950-012-9345-0

[42] Simpson DW, Leith W, Scholz C. Two types of reservoir-induced seismicity. Bulletin of the Seismological Society of America. 1988;**78**:2025-2040. Link to full.pdf

[43] Belferman M, Agnon A, Katsman R, Ben-Avraham Z. Identifying plausible historical scenarios for coupled lake level and seismicity rate changes: The case for the Dead Sea during the last 2 millennia. Natural Hazards and Earth System Sciences. 2022;**22**(8):2553-2565. DOI: 10.5194/nhess-22-2553-2022

[44] Belferman M, Katsman R, Agnon A. Effect of large-scale surface water level fluctuations on earthquake recurrence interval under strike-slip faulting. Tectonophysics. 2018;**744**:390-402. DOI: 10.1016/j.tecto.2018.06.004

[45] Hill RG, Weingarten M, Rockwell TK, et al. Major southern San Andreas earthquakes modulated by lake-filling events. Nature. 2023;**618**:761-766. DOI: 10.1038/s41586-023-06058-9

[46] Brothers D, Kilb D, Luttrell K, et al. Loading of the San Andreas fault by flood-induced rupture of faults beneath the Salton Sea. Nature Geoscience. 2011; **4**:486-492. DOI: 10.1038/ngeo1184

[47] Luttrell K, Sandwell D, Smith-Konter B, Bills B, Bock Y. Modulation of the earthquake cycle at the southern San Andreas fault by lake loading. Journal of Geophysical Research. 2007;**112**:B08411. DOI: 10.1029/2006JB004752

[48] Tostrud MB. The Salton Sea, 1906-1996, computed and measured salinities and water levels. In: Draft Report. California, US: Colorado River Board of California; 1997

[49] Freund R, Garfunkel Z, Zak I, Goldberg M, Weissbrod T, Derin B, et al. The shear along the Dead Sea rift. Philosophical Transactions for the Royal Society of London. Series A, Mathematical and Physical Sciences. 1970;**267**(1181):107-130. Available from: https://www.jstor.org/stable/73612

[50] Le Beon M, Klinger Y, Amrat AQ, Agnon A, Dorbath L, Baer G, et al. Slip rate and locking depth from GPS profiles across the southern Dead Sea transform. Journal of Geophysical Research: Solid Earth. 2008;**113**(B11):B11403. DOI: 10.1029/2007JB005280

[51] McClusky S, Reilinger R, Mahmoud S, Ben Sari D, Tealeb A. GPS constraints on Africa (Nubia) and Arabia plate motions. Geophysical Journal International. 2003;**155**(1):126-138. DOI: 10.1046/j.1365-246X.2003.02023.x

[52] Quennel AM. Tectonics of the Dead Sea Rift, Paper Presented at the 20th International Geological Congress. Africanos, Mexico: Assoc. Serv. Geol; 1956. p. 1959

[53] Wilson JT. A new class of faults and their bearing on continental drift. Nature. 1965;**207**:343-347

[54] Ben-Avraham Z, Garfunkel Z, Lazar M. Geology and evolution of the southern Dead Sea fault with emphasis on subsurface structure. Annual Review of Earth and Planetary Sciences. 2008; **36**:357-387

[55] Weber FA, Voegelin A, Kretzschmar R. Multi-metal contaminant dynamics in temporarily flooded soil under sulfate limitation. Geochimica et Cosmochimica Acta. 2009;**73**(19): 5513-5527. DOI: 10.1016/j.gca.2009.06.011

[56] Hamiel Y, Amit R, Begin ZB, Marco S, Katz O, Salamon A, et al. The seismicity along the Dead Sea fault during the last 60,000 years. Bulletin of the Seismological Society of America. 2009;**99**:2020-2026. DOI: 10.1785/0120080218

[57] Masson F, Hamiel Y, Agnon A, Klinger Y, Deprez A. Variable behavior of the Dead Sea fault along the southern Arava segment from GPS measurements. Comptes Rendus Geoscience. 2015;**347** (4):161-169. DOI: 10.1016/j.crte.2014.11.001

[58] Sadeh M, Hamiel Y, Ziv A, Bock Y, Fang P, Wdowinski S. Crustal deformation along the Dead Sea transform and the Carmel fault inferred from 12 years of GPS measurements.

Journal of Geophysical Research: Solid Earth. 2012;**117**:B08410. DOI: 10.1029/2012jB009241

[59] Joffe S, Garfunkel Z. Plate kinematics of the circum Red Sea—A re-evaluation. Tectonophysics. 1987;**141**: 5-22. DOI: 10.1016/0040-1951(87)90171-5

[60] Bartov Y, Steinitz G, Eyal M, Eyal Y. Sinistral movement along the Gulf of Aqaba—Its age and relation to the opening of the Red Sea. Nature. 1980; **285**(5762):220-222

[61] Ambraseys N. Earthquakes in the Mediterranean and Middle East: A Multidisciplinary Study of Seismicity Up to 1900. Cambridge, UK: Cambridge University Press; 2009

[62] Guidoboni E, Comastri A. Catalogue of Earthquakes and Tsunamis in the Mediterranean Area from the 11th to the 15th Century. Rome, Italy: Istituto nazionale di geofisica e Vulcanologia; 2005. DOI: 10.1515/BYZS.2008.854

[63] Guidoboni E, Comastri A, Traina G. Catalogue of Ancient Earthquakes in the Mediterranean Area Up to the 10th Century. Rome, Italy: Istituto nazionale di geofisica e Vulcanologia; 1994. DOI: 10.1163/182539185X01377

[64] Kagan E, Stein M, Agnon A, Neumann F. Intrabasin paleoearthquake and quiescence correlation of the late Holocene Dead Sea. Journal of Geophysical Research: Solid Earth. 2011; **116**(B4):B04311. DOI: 10.1029/2010JB007452

[65] Liu Y, Rubin AM. Role of fault gouge dilatancy on aseismic deformation transients. Journal of Geophysical Research - Solid Earth. 2010;**115**(B10): B10414. DOI: 10.1029/2010JB007522

[66] Ellenblum R, Marco S, Agnon A, Rockwell T, Boas A. Crusader castle torn apart by earthquake at dawn, 20 may 1202. Geology. 1998;**26**:303-306. DOI: 10.1130/0091-7613(1998)026<0303:CCTABE>2.3.CO;2

[67] Matmon A, Shaked Y, Porat N, Enzel Y, Finkel R, Lifton N, et al. Landscape development in an hyperarid sandstone environment along the margins of the Dead Sea fault: Implications from dated rock falls. Earth and Planetary Science Letters. 2005;**240**:803-817. DOI: 10.1016/j.epsl.2005.06.059

[68] Reches ZE, Hoexter DF. Holocene seismic and tectonic activity in the Dead Sea area. Tectonophysics. 1981;**80**: 235-254. DOI: 10.1016/0040-1951(81)90151-7

[69] Shaked Y, Agnon A, Lazar B, Marco S, Avner U, Stein M. Large earthquakes kill coral reefs at the north-West Gulf of Aqaba. Terra Nova. 2004; **16**:133-138. DOI: 10.1111/j.1365-3121.2004.00541.x

[70] Kagan E, Agnon A, Bar-Matthews M, Ayalon A. Dating large infrequent earthquakes by damaged cave deposits. Geology. 2005;**33**:261-264. DOI: 10.1130/G21193.1

[71] Migowski C, Agnon A, Bookman R, Negendank JF, Stein M. Recurrence pattern of Holocene earthquakes along the Dead Sea transform revealed by varve-counting and radiocarbon dating of lacustrine sediments. Earth and Planetary Science Letters. 2004;**222**: 301-314. DOI: 10.1016/j.epsl.2004.02.015

[72] Karcz I, Kafri U, Meshel Z. Archaeological evidence for subrecent seismic activity along the Dead Sea-Jordan rift. Nature. 1977;**269**:234-235

[73] Haase-Schramm A, Goldstein SL, Stein M. U-Th dating of Lake Lisan (late Pleistocene dead sea) aragonite and implications for glacial East Mediterranean climate change. Geochimica et Cosmochimica Acta. 2004;**68**:985-1005. DOI: 10.1016/j.gca.2003.07.016

[74] Marco S, Stein M, Agnon A, Ron H. Long-term earthquake clustering: A 50,000-year paleoseismic record in the Dead Sea Graben. Journal of Geophysical Research: Solid Earth. 1996; **101**(B3):6179-6191. DOI: 10.1029/95JB01587

[75] Waldmann N. The geology of the Samra Formation in the Dead Sea Basin. Unpublished [MSc thesis]. Jerusalem, Israel: The Hebrew University (in English, Hebrew summary). 2002

[76] Marco S. Recognition of earthquake-related damage in archaeological sites: Examples from the Dead Sea fault zone. Tectonophysics. 2008;**453**(1–4):148-156. DOI: 10.1016/j.tecto.2007.04.011

[77] Marco S, Rockwell TK, Agnon A, Heimann A, Frieslander U. Late Holocene slip of the Dead Sea transform revealed in 3D palaeoseismic trenches on the Jordan gorge fault. Earth and Planetary Science Letters. 2005;**234**(1–2):189-205. DOI: 10.1016/j.epsl.2005.01.017

[78] Ambraseys NN, Jackson JA. Faulting associated with historical and recent earthquakes in the eastern Mediterranean region. Geophysical Journal International. 1998;**133**(2): 390-406. DOI: 10.1046/j.1365-246X.1998.00508.x

[79] Ambraseys NN, Melville CP, Adams RD. The Seismicity of Egypt, Arabia and the Red Sea: A Historical Review. Cambridge: Cambridge Univ; 1994. DOI: 10.1017/S1356186300007240

[80] Amiran DH, Arieh E, Turcotte T. Earthquakes in Israel and adjacent areas: Macroscopic observations since 100 B.C. E. Israel Exploration Journal. 1994;**44**: 260-305 Available from: http://www.jstor.org/stable/27926357

[81] Enzel Y, Kadan G, Eyal Y. Holocene earthquakes inferred from a fan-delta sequence in the Dead Sea graben. Quaternary Research. 2000;**53**(1):34-48. DOI: 10.1006/qres.1999.2096

[82] Ken-Tor R, Agnon A, Enzel Y, Stein M, Marco S, Negendank JF. High-resolution geological record of historic earthquakes in the Dead Sea basin. Journal of Geophysical Research-Solid Earth. 2001;**106**:2221-2234. DOI: 10.1029/2000JB900313

[83] Niemi TM, Ben-Avraham Z. Evidence for the Jericho earthquakes from slumped sediments of the Jordan River delta in the Dead Sea. Geology. 1994;**22**:395-398. DOI: 10.1130/0091-7613(1994)022<0395:EFJEFS>2.3.CO;2

[84] Russell KW. The Earthquake Chronology of Palestine and Northwest Arabia from the Second through the Mid-8th Century AD. Cambridge, Mass: Bulletin of the American Schools of Oriental Research; 1985. pp. 37-59. DOI: 10.2307/1356863

[85] Ben-Menahem A. Four thousand years of seismicity along the Dead Sea rift. Journal of Geophysical Research. 1991;**96**:20195-20216. DOI: 10.1029/91JB01936

[86] Hough SE, Avni R. The 1170 and 1202 CE Dead Sea rift earthquakes and long-term magnitude distribution of the Dead Sea fault zone. Israel Journal of Earth Sciences. 2011;**58**:295-308. DOI: 10.1560/IJES.58.3-4.295

[87] Darvasi Y, Agnon A. Calibrating a new attenuation curve for the Dead Sea

region using surface wave dispersion surveys in sites damaged by the 1927 Jericho earthquake. Solid Earth. 2019; **10**(2):379-390. DOI: 10.5194/se-10-379-2019

[88] Enzel Y, Bookman(Ken-Tor) R, Sharon D, Gvirtzman H, Dayan U, Ziv B, et al. Late Holocene climates of the near east deduced from Dead Sea level variations and modern regional winter rainfall. Quaternary Research. 2003;**60**: 263-273. DOI: 10.1016/j.yqres.2003.07.011

[89] Frumkin A, Magaritz M, Carmi I, Zak I. The Holocene climatic record of the salt caves of Mount Sedom Israel. The Holocene. 1991;**1**:191-200. DOI: 10.1177/09596836910010

[90] Klein C, Flohn H. Contributions to the knowledge of the fluctuations of the Dead Sea level. Theoretical and Applied Climatology. 1987;**38**:151-156. DOI: 10.1007/BF00868099

[91] Lisker S, Vaks A, Bar-Matthews M, Porat R, Frumkin A. Stromatolites in caves of the Dead Sea fault escarpment: Implications to latest Pleistocene lake levels and tectonic subsidence. Quaternary Science Reviews. 2009;**28**: 80-92. DOI: 10.1016/j.quascirev.2008.10.015

[92] Migowski C, Stein M, Prasad S, Negendank J, Agnon A. Holocene climate variability and cultural evolution in the near east from the Dead Sea sedimentary record. Quaternary Research. 2006;**66**(3):421-431. DOI: 10.1016/j.yqres.2006.06.010

[93] Stein M. The sedimentary and geochemical record of Neogene-quaternary water bodies in the Dead Sea Basin – Inferences for the regional paleoclimatic history. Journal of Paleolimnology. 2001;**26**:271-282

[94] Stein M, Starinsky A, Katz A, Goldstein SL, Machlus M, Schramm A. Strontium isotopic, chemical, and sedimentological evidence for the evolution of Lake Lisan and the Dead Sea. Geochimica et Cosmochimica Acta. 1997;**61**:3975-3992

[95] Torfstein A, Goldstein SL, Stein M, Enzel Y. Impacts of abrupt climate changes in the Levant from last glacial Dead Sea levels. Quaternary Science Reviews. 2013;**60**:1-7. DOI: 10.1016/j.quascirev.2013.02.015

[96] Bartov Y, Stein M, Enzel Y, Agnon A, Reches Z. Lake levels and sequence stratigraphy of Lake Lisan, the late Pleistocene precursor of the Dead Sea. Quaternary Research. 2002;**57**:9-21. DOI: 10.1006/qres.2001.2284

[97] Belmaker R, Lazar B, Beer J, Christl M, Tepelyakov N, Stein M. ^{10}Be dating of Neogene halite. Geochimica et Cosmochimica Acta. 2013;**122**:418-429. DOI: 10.1016/j.gca.2013.08.033

[98] Starinsky A. Relationship between Ca-chloride brines and sedimentary rocks in Israel [PhD thesis]. The Hebrew Univ. of Jerusalem; 1974

[99] Stein M, Starinsky A, Agnon A, Katz A, Raab M, Spiro B, et al. The impact of brine-rock interaction during marine evaporite formation on the isotopic Sr record in the oceans: Evidence from Mt. Sedom, Israel. Geochimica et Cosmochimica Acta. 2000;**64**:2039-2053. DOI: 10.1016/S0016-7037(00)00370-7

[100] Steinitz G, Bartov Y. The Miocene-Pliocene history of the Dead Sea segment of the rift in light of K–Ar ages of basalts. Israel Journal of Earth Sciences. 1991;**40**: 199-208

[101] Zak I. The Geology of Mount Sedom [PhD thesis]. Jerusalem: The Hebrew University; 1967

[102] Neev D, Emery KO. The Dead Sea; Depositional Processes and Environments of Evaporites. Jerusalem: Geological Survey; 1967

[103] Torfstein A, Haase-Schramm A, Waldmann N, Kolodny Y, Stein M. U-series and oxygen isotope chronology of the mid-Pleistocene Lake Amora (Dead Sea basin). Geochimica et Cosmochimica Acta. 2009;**73**:2603-2630. DOI: 10.1016/j.gca.2009.02.010

[104] Waldmann N, Stein M, Ariztegui D, Starinsky A. Stratigraphy, depositional environments and level reconstruction of the last interglacial Lake Samra in the Dead Sea basin. Quaternary Research. 2009;**72**:1-15

[105] Bartov Y. The geology of the Lisan Formation at Massada plain and Lisan Peninsula [MSc thesis]. The Hebrew University of Jerusalem; 1999

[106] Begin ZB, Broecker W, Buchbinder B, Druckman Y, Kaufman A, Magaritz M, et al. Dead Sea and lake Lisan levels in the last 30,000 years. Geological Survey of Israel. 1985;**85**:1-18

[107] Yechieli Y, Magaritz M, Levy Y, Weber U, Kafri U, Woelfli W, et al. Late quaternary geological history of the Dead Sea area, Israel. Quaternary Research. 1993;**39**:59-67. DOI: 10.1006/qres.1993.1007

[108] Bartov Y, Goldstein SL, Stein M, Enzel Y. Catastrophic arid episodes in the eastern Mediterranean linked with the North Atlantic Heinrich events. Geology. 2003;**31**(5):439-442

[109] Torfstein A, Goldstein SL, Kagan EJ, Stein M. Integrated multi-site U–Th chronology of the last glacial Lake Lisan. Geochimica et Cosmochimica Acta. 2013a;**104**:210-231. DOI: 10.1016/j.gca.2012.11.003

[110] Bookman R, Bartov Y, Enzel Y, Stein M. Quaternary lake levels in the Dead Sea basin: Two centuries of research. In: Enzel Y, Agnon A, Stein M, editors. New Frontiers in Dead Sea Paleoenvironmental Research. Geological Society of America Special Paper. Vol. 401. Colorado, US. 2006. pp. 155-170. DOI: 10.1130/2006.2401(10)

[111] Bookman R, Enzel Y, Agnon A, Stein M. Late Holocene lake levels of the Dead Sea. Geological Society of America Bulletin. 2004;**116**:555-571. DOI: 10.1130/B25286.1

[112] Kaufman A, Yechieli Y, Gardosh M. Reevaluation of Lake-sediment chronology in the Dead Sea basin, Israel, based on new 230Th/U dates. Quaternary Research. 1992;**38**:292-304. DOI: 10.1016/0033-5894(92)90039-L

[113] Schramm A, Stein M, Goldstein SL. Constancy of 234u/Z38u ratios in late Pleistocene aragonitic sediments—Lake Lisan, paleo-Dead Sea. Terra Nova. 1995; 7:325

[114] Stein M, Goldstein SL, Ron H, Marco S. Precise TIMS 23 Th-231U ages and magnetostratigraphy of Lake Lisan sediments (paleo-Dead Sea.). EOS Trans. AGU 73, (Spring Meet. Suppl. 154). 1992

[115] Bartov Y, Agnon A, Enzel Y, Stein M. Late quaternary faulting and subsidence in the Central Dead Sea basin. Israel Journal of Earth Sciences. 2006;**55**:17-31. DOI: 10.1560/E3RK-0RDY-3CUP-LN5A

[116] Waldmann N, Starinsky A, Stein M. Primary carbonates and Ca-chloride

brines as monitors of a paleo-hydrological regime in the Dead Sea basin. Quaternary Science Reviews. 2007;**26**:2219-2228

[117] Kadan G. Evidence of Dead-Sea level fluctuations and neotectonic events in the Holocene fan-delta of Nahal Darga. M.Sc., Ben Gurion University of the Negev. 1997. p. 54

[118] Ben-Avraham Z, Niemi T, Heim C, Negendank J, Nur A. Holocene stratigraphy of the Dead Sea: Correlation of high-resolution seismic reflection profiles to sediment cores. Journal of Geophysical Research. 1999;**104**(B8): 617-625. DOI: 10.1029/1999JB900084

[119] Heim C, Leroy S, Negendank J. Near east desertification—Evidence from the Dead Sea. Naturwissenschaften. 1997;**84**:398-401

[120] Torfstein A, Gavrieli I, Katz A, Kolodny Y, Stein M. Gypsum as a monitor of the paleo-limnological–hydrological conditions in Lake Lisan and the Dead Sea. Geochimica et Cosmochimica Acta. 2008;**72**:2491-2509. DOI: 10.1016/j.gca.2008.02.015

[121] Pandey AP, Chadha RK. Surface loading and triggered earthquakes in the Koyna–Warna region, western India. Physics of the Earth and Planetary Interiors. 2003;**139**:207-223. DOI: 10.1016/j.pepi.2003.08.003

[122] Durá-Gómez I, Talwani P. Reservoir-induced seismicity associated with the Itoiz reservoir, Spain: A case study. Geophysical Journal International. 2010;**181**(1):343-356. DOI: 10.1111/j.1365-246X.2009.04462.x

[123] Roeloffs EA. Fault stability changes induced beneath a reservoir with cyclic variations in water level. Journal of Geophysical Research - Solid Earth. 1988;**93**:2107-2124. DOI: 10.1029/JB093iB03p02107

[124] Wang H. Theory of Linear Poroelasticity with Applications to Geomechanics and Hydrogeology. Princeton: University Press; 2000

[125] Simpson DW. Seismicity changes associated with reservoir loading. Engineering Geology. 1976;**10**(2–4): 123-150. DOI: 10.1016/0013-7952(76)90016-8

[126] Anderson EM. The Dynamics of Faulting and Dyke Formation with Applications to Britain. Edinburgh: Oliver and Boyd; 1951

[127] Bell ML, Nur A. Strength changes due to reservoir-induced pore pressure and stresses and application to Lake Oroville. Journal of Geophysical Research - Solid Earth. 1978;**83**(B9):4469-4483. DOI: 10.1029/JB083iB09p04469

[128] Snow DT. Hydrogeology of induced seismicity and tectonism: Case histories of Kariba and Koyna. In: Narasimhan TN, editor. Recent Trends in Hydrogeology. Geological Society of America Special Paper. Vol. 189. Colorado, US. 1982. pp. 317-360. DOI: 10.1130/SPE189-p317

[129] Rao NP, Shashidhar D. Periodic variation of stress field in the Koyna–Warna reservoir triggered seismic zone inferred from focal mechanism studies. Tectonophysics. 2016;**679**:29-40. DOI: 10.1016/j.tecto.2016.04.036

[130] Rajendran K, Thulasiraman N, Sreekumari K. Microearthquake activity near the Idukki reservoir, South India: A rare example of renewed triggered seismicity. Engineering Geology. 2013; **153**:45-52. DOI: 10.1016/j.enggeo.2012.11.004

[131] Crone TJ, Wilcock WS. Modeling the effects of tidal loading on mid-ocean ridge hydrothermal systems. Geochemistry, Geophysics, Geosystems. 2005;**6**(7): Q07010. DOI: 10.1029/2004GC000905

[132] Fletcher JB, Sykes LR. Earthquakes related to hydraulic mining and natural seismic activity in western New York state. Journal of Geophysical Research. 1977;**82**:3767-3780. DOI: 10.1029/JB082i026p03767

[133] Healy JH, Rubey WW, Griggs DT, Raleigh CB. The Denver earthquake. Science. 1968;**161**:1301-1310. DOI: 10.1126/science.161.3848.1301

[134] Raleigh CB, Healy JH, Bredehoeft JD. An experiment in earthquake control at Rangely, Colorado. Science. 1976;**191**:1230-1237. DOI: 10.1126/science.191.4233.1230

[135] Aldersons F, Ben-Avraham Z, Hofstetter A, Kissling E, Al-Yazjeen T. Lowercrustal strength under the Dead Sea basin from local earthquake data and rheological modeling. Earth and Planetary Science Letters. 2003;**214**:129-142. DOI: 10.1016/S0012-821X(03)00381-9

[136] Godey S, Bossu R, Guilbert J, Mazet-Roux G. The Euro-Mediterranean Bulletin: A comprehensive seismological Bulletin at regional scale. Seismological Research Letters. 2006;**77**:460-474. DOI: 10.1785/gssrl.77.4.460

[137] Shamir G. The active structure of the Dead Sea depression. In: Enzel Y, Agnon A, Stein M, editors. New Frontiers in Dead Sea Paleoenvironmental Research. Geological Society of America Special Paper. Vol. 401. Colorado, US. 2006. pp. 15-32. DOI: 10.1130/2006.2401(02)

[138] Dor O, Ben-Zion Y, Rockwell T, Brune J. Pulverized rocks in the Mojave section of the San Andreas fault zone. Earth and Planetary Science Letters. 2006;**245**:642-654. DOI: 10.1016/j.epsl.2006.03.034

[139] Terzaghi K. Theoretical Soil Mechanics. New York: Wiley; 1942. DOI: 10.1002/9780470172766

[140] Jaeger JC, Cook NG, Zimmerman R. Fundamentals of Rock Mechanics. Oxford: Blackwell; 2009

[141] Weiss J, Pellissier V, Marsan D, Arnaud L, Renard F. Cohesion versus friction in controlling the long-term strength of a self-healing experimental fault. Journal of Geophysical Research - Solid Earth. 2016;**121**(12):8523-8547. DOI: 10.1002/2016JB013110

[142] Byerlee J. Friction of rocks. In: Rock Friction and Earthquake Prediction. Basel, Switzerland: Birkhäuser; 1978. pp. 615-626. DOI: 10.1007/978-3-0348-7182-2_4

[143] Agnon A, Reches Z. Frictional rheology: Hardening by rotation of active normal faults. Tectonophysics. 1995;**247**:239-254. DOI: 10.1016/0040-1951(94)00178-C

[144] Stern O. Geochemistry, Hydrology and Paleo-Hydrology of Ein Qedem Spring System, Geological Survey of Israel, Jerusalem, Israel, Report GSI/17/2010. 2010. Available from: https://www.gov.il/he/Departments/publications/reports/stern-report-201 [Accessed: August 2, 2022] (in Hebrew)

[145] Mörner NA. Paleoseismicity and geodynamics in Sweden. Tectonophysics. 1985;**117**:139-153. DOI: 10.1016/0040-1951(85)90242-2

[146] Calais E, Freed AM, Van Arsdale R, Stein S. Triggering of New Madrid seismicity by late-Pleistocene erosion.

Nature. 2010;**466**:608-611. DOI: 10.1038/nature09258

[147] Heifetz E, Agnon A, Marco S. Soft sediment deformation by kelvin Helmholtz instability: A case from Dead Sea earthquakes. Earth and Planetary Science Letters. 2005;**236**(1–2):497-504. DOI: 10.1016/j.epsl.2005.04.019

[148] Wetzler N, Marco S, Heifetz E. Quantitative analysis of seismogenic shear-induced turbulence in lake sediments. Geology. 2010;**38**(4): 303-306. DOI: 10.1130/G30685.1

Section 2

The Interactions between Anthropogenic Activities and Lake Ecosystems

Chapter 5

Anthropogenic Impact on Lake Ecosystem

Lukman Lukman

Abstract

The world's population growth in various ways impacts the waters environment, and these impacts have been observed since the twentieth century. However, paleolimnological data indicates that anthropogenic activities have been affecting the aquatic ecosystem for a long time ago. The primary determinant of the lake ecosystem damage is the change and utilization of the catchment area landscapes, which contributes to siltation as well as nutrient supply. The increased activities of agriculture and domestic work are the main causes of eutrophication due to nutrient input. Additionally, the cage aquaculture in the lake waters has led to oxygen depletion in the lower water column as an impact of organic loading input. Furthermore, habitat modification, including disturbance to the shore zone, has led to changes in riparian areas. Ultimately, these processes impact the biota population structure and degrade the lake ecosystem. Therefore, understanding the anthropogenic factors and their impact on the lake ecosystem will enable humans to control their activities and manage their impact on the ecosystem.

Keywords: paleolimnology, sedimentation, siltation, eutrophication, nutrient, organic loading, habitat modification

1. Introduction

All human efforts in exploiting and modifying the environment are anthropogenic activities that significantly impact the environment and tend to have negative consequences. To manage these impacts, it is important to distinguish between environmental and anthropogenic factors' influence and understand how they affect human survival and well-being. This understanding is crucial for controlling the threats that humans face. Unfortunately, humans are often their own worst enemy when they are the main cause of pollution and environmental damage. While the cumulative effect from natural events on the environment is much smaller than the problems caused by human activities [1].

Over time, it has become increasingly apparent that human activities in various fields have caused a significant imbalance in the ecological system, leading to environmental disruption, and affecting human life. The negative impacts on our environment resulting from continued population growth accompanied by economic development cannot be overlooked. Therefore, it is crucial to analyze environmental conditions such as water, air, and soil, and observe the primary sources of pollution.

The hope is that this analysis will minimize the impacts, regardless of where and when it occurs. Sustainable development is becoming more important as the "blue economy" concept raises awareness of the adverse effects of industrial and agricultural practices on the environment. Consideration of ecological factors gets more attention, not merely economic ones. Thus, anticipatory measures will likely be taken to address the influence of anthropogenic activities and their impact on the environment [2].

Human activities significantly impacted the environment. **Figure 1** illustrates that the impact curve continues to increase with time in line with the anthropogenic activity rate, and time a peak curve pattern will reach, indicating when anthropogenic activity's impact begins to be controlled. The timing of this peak varies significantly for each country, highlighting the need for tailored solutions to address environmental issues on a local level [3]. While it is widely acknowledged that environmental damage was severe in the twentieth century, paleolimnological records indicate that anthropogenic activity and its impact on lake waters occurred much earlier than previously thought [4].

Despite various mitigation efforts, especially wastewater management, other environmental impacts are still visible. One such impact is the alteration of waters morphology, increased turbidity caused by eutrophication, and the loss of biodiversity. This is a major problem in many parts of the world. Inland waters, such as lakes, are susceptible to anthropogenic impacts. As human populations grow and agricultural and industrial activities expand, aquatic ecosystems are increasingly threatened. This puts pressure on these ecosystems, and urgent action is needed to address the negative impacts on our water resources.

As economic growth has continued, humans have altered Earth's biodiversity and ecosystems, decreasing the use value of ecosystem services. This biodiversity loss is driven by the need for various resources for human survival, which is likely to continue [5]. With a population of over 7 billion people and a rapid increase in per capita consumption of goods and services, the visible ecological footprint of human growth is changing land cover, rivers, climate systems, biogeochemical cycles, and ecosystem functions. However, historically, human attention to environmental health has been more focused on the quantitative response of the relationship between pollutant exposure and human health. More recent research has shed light on how changes in

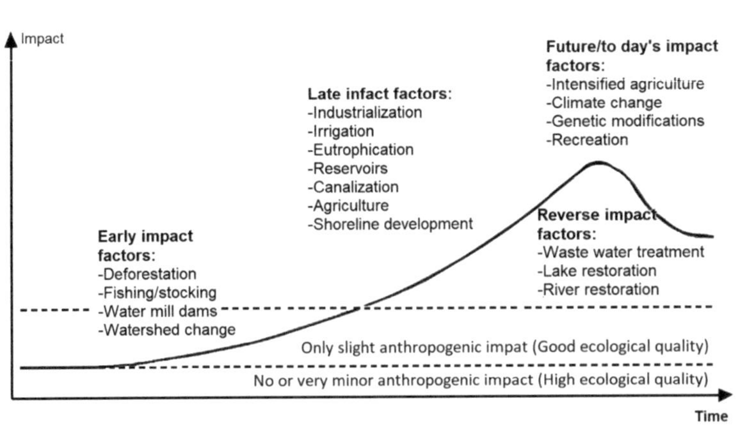

Figure 1.
General development stages of human disturbance factors to the environment in industrialized countries [3].

the structure and function of natural systems can also impact human health in various ways. This perspective is becoming increasingly important as the rate and extent of these changes continue to accelerate [6].

To achieve a harmonious balance between human life and nature and prevent the negative impacts of human activities, it is crucial to disclose the hidden potential of basic science. This will help us transition towards a better environment for all [7]. Furthermore, it will enable us to maximize the socioeconomic and bio-cultural benefits to improve human quality of life [8].

2. Paleolimnological review

The sediments store the lake's history, the lake waters state, the catchment area environment, and climatic conditions. Interpretation of past situations is recorded in sedimentary structures and mineralogy, their organic and inorganic chemical components, and the morphological remains of organisms stored in them. The history of the lake can show the change in ecosystem condition, both their saprobic state and productivity. Interpreting critical records is needed to see the variable factors of the drainage basin, redistribution of sediments, and variations in the preservation of the interpreted components [9].

Paleolimnological analysis, as a study of the history of lakes in terms of physical, chemical, and biological information preserved in sediments, is an approach to understanding both natural and anthropogenic changes in the past. The palaeoecological approach is a crucial tool for understanding the long-term ecological condition of lakes, including how human activities and climate change affect these ecosystems. While natural forces such as volcanic eruptions and climate change can impact lake ecosystems, humans are the drivers of change [10, 11].

Studies from various regions of the world have traced human influences on lake ecosystems in the past, with landscape disturbance often representing an early signal from the paleolimnological record. The impacts of human activities on aquatic systems can vary widely over space and time, with evidence of human influence on inland water ecosystems worldwide dating back to pre-1850 CE (Common Era; for the last 2000-year period) based on various paleolimnological studies of lakes and wetlands at low latitudes. The impacts of human activities on lake ecosystems in the past was recorded including pollution, eutrophication, sedimentation, acidification, and salinization [10].

The study on the anthropogenic impacts on lake waters in Lake Funda, a deep crater lake located on Flores Island, part of the Azores Islands in the middle of the North Atlantic, has been conducted. The lake, which has not been affected by volcanic activity in the last 1000 years, provides a unique opportunity to investigate the effects of human activity and climate change on its condition. The evolution process of Lake Funda is divided into three distinct phases (as shown in **Figure 2**). The first phase (A), which lasted until 1335 CE, was driven primarily by climate and lake catchment processes. The second phase (B), between 1335 and 1560, was marked by a sudden change in the composition and diversity of diatoms and chironomids, indicating a shift in the trophic status of lake waters from mesotrophic to eutrophic conditions. This change was caused by a synergistic effect of high climate variability (Medieval Climate Anomaly) and human disturbances in the catchment, such as the introduction of livestock. In the last phase (1560 CE to present) (C), the lake has maintained a eutrophic condition, which is sustained through cycles and feedback between lake

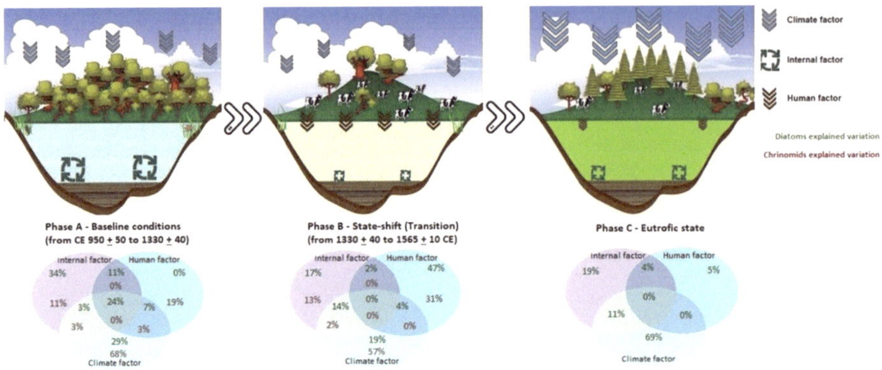

Figure 2.
The lake evolution process is divided into three main phases of conditions, namely (A) climate and lake catchment process; (B) changes in the trophic status of lake waters from mesotrophic to eutrophic; (C) the last phase with eutrophic state [12].

productivity and phosphorus remobilization in the lake. Climate variability and lake internal dynamics significantly influence lake ecosystems' variability, such as phosphorus remobilization [12].

3. Sedimentation and siltation

Land use is one of the human interventions in nature that involves altering the condition of vegetation in the catchment area. Modifying natural vegetation can significantly contribute to the increase of erosion rate and drive the sedimentation pattern in the area. Sedimentation is closely related to erosion, influenced by various factors such as geological, topography, land slope, climate, soil type, and vegetation [13].

The study conducted on global-scale sediment flux patterns suggests that, with a few exceptions, land use impacts have a greater influence on sedimentation patterns than impact of climate change, particularly in smaller catchments ($<10^3$ km^2) [14]. The authors also observed that the intensity of land use has a qualitative impact on the sedimentation rate, although there are some caveats. In the case of the catchment of Lake Egari in Papua New Guinea, land clearing and the introduction of sweet potatoes in the nineteenth century resulted in a continued increase in sediment flux until the late twentieth century. This long-term sediment supply is expected to have an impact on the siltation of Lake Egari.

The lakes in the Pacific Northwest region of Canada have seen considerable disturbance in its catchment areas due to deforestation and road-building activities, particularly in recent decades. Therefore, the lake sedimentation rate in several cases has increased significantly along with these land use changes. However, the comparison data for other lakes did not show a significant effect of logging on sedimentation rates. Road construction activity had a particularly pronounced effect, with an average increase of 137% above the background rate and a maximum of up to 307%. Additionally, mining activities in the Lake Aldrich catchment between the early 1920s and 1954 also resulted in increased sedimentation rates during the period of land use disturbance [15].

Sedimentation has seriously threatened the Lake Tana Basin in Ethiopia, which can reduce its carrying capacity. The sedimentation is the impact of soil erosion due to agricultural activities in the catchment. Severe erosion in the Lake Tana Basin catchment is supported by inappropriate land use, especially on a high slope [16]. Similarly, Lake Malawi/Nyasa/Niassa, which is shared by Malawi, Tanzania, and Mozambique country and receives water from 13 rivers, is also impacted by sediment loads from agricultural activities and deforestation in its catchment area. The watersheds in the area are steep and narrow, with forested areas having lower sediment loads than those with extensive agricultural activities [17].

The high erosion as the impact of catchment damage has seriously threatened the sedimentation of Lake Rawa Pening. Lake Rawa Pening is a small (maximum area of 2667 ha) and shallow (the most profound part was 18 m) lake in Central Java, Indonesia. An analysis based on the Sediment Delivery Ratio Sub-Model indicated that the total sediment exports from the catchment reached 501628.6 tons/year. Sediment sources mainly come from the land with high slope and the land use was dominated by dry land agriculture and horticulture [18]. Another study on the evaluation of the impact of erosion on catchments was in Lake Tondano (the area was 4800 ha; catchment area: 31,400 ha) in North Sulawesi Indonesia, using bathymetry maps in 2015 and 2020. The results indicate a decrease in water depth from 2015 (maximum depth of 31.81 m and average depth of 12.88 m) to 2020 (maximum depth of 30.22 m and average depth of 12.66 m) at a rate of 4.4 cm y^{-1}, suggesting an increase of lake sedimentation to erosion in the catchment [19].

Siltation from catchment as an impact of land use activity allows fine sediments to be efficiently transferred from hillslopes to lake basins. Fine sediment is harmed to aquatic ecosystems [20]. A comprehensive document regarding the impact of suspended material in water has been published [21] entitled Effect of Suspended Sediment on Freshwater Fish and Fish Habitat as a collection of literature reviews. The document reveals the effect of suspended matter on eggs and larvae, physiological effects, aspects of foraging and fish growth, primary production and aquatic plants, invertebrates, and effects on the habitat, abundance, and community structure of fish. One of the effects of siltation was that it affected the attachment of Walleyes *Sander vitreous* fish eggs. One experiment found that rocks covered with fine sediment could not hold eggs, whereas the clean rocks could hold 35.9 + 36.6% of eggs [22].

4. Eutrophication

Eutrophication is an excessive enrichment process of water with mineral nutrients, primarily phosphorus (P), in freshwater lakes. Eutrophication is characterized by autotroph overproduction, mainly algae and cyanobacteria, which can lead to the depletion of dissolved oxygen in the water, especially in poorly mixed bottom water areas. High bacterial populations and high respiration rates contribute to the depletion of dissolved oxygen, resulting in conditions of hypoxia and anoxia. These conditions are often more pronounced in calm and dry conditions, particularly in warm waters. The relationship between P input, primary production, dissolved oxygen, and the trophic status of water clearly explained by [23].

Phosphorus (P) is a crucial nutrient for all life forms, and its availability can significantly impact primary production in aquatic ecosystems. Ecologists have identified P as the limiting factor for primary production in freshwater lakes, and input of P into waters must be managed to prevent eutrophication. However, noted that the limiting factors for primary production vary in different ecosystems, with nitrogen (N) being the limiting factor in oceans and P in lake waters [23]. Eutrophication occurs when the availability of limiting factors for photosynthesis, such as P and N, increases, leading to an overgrowth of algae [24].

Although naturally, eutrophication can take place in lake waters and is a result of the aging process of water so that lakes are described at the level of trophic status, namely oligotrophic (poor in nutrients), mesotrophic (moderate nutrient content), eutrophic (rich in nutrients) [9, 25]. However, the process of eutrophication that began in the mid-twentieth century is called cultural eutrophication, a major environmental problem in almost all lakes worldwide [26–28].

Land use and landscape modifications in the catchment area can indeed affect sediment behavior and nutrient supply to the waters area. Catchments are indeed the primary source of nutrients for lakes, and different transport capacities in each catchment can affect water quality in receiving waters. Nutrient loads supplied from catchment to lakes are influenced by various factors such as land use variety, geological, hydrological, nutrient availability, topographical factors, and response to rainfall [29]. To determine the trophic status of a lake, Total Nitrogen (TN) and Total Phosphorus (TP) are suggested as simple indicators [30]. Various criteria for trophic status have been established based on parameters such as TN, TP, chlorophyll, and Sechi depth [27, 31–33] (**Table 1**). However, eutrophication is a multidimensional natural event, so more than a single-factor evaluation is generally required [34]. To address of this a multidimensional approach to eutrophication classification proposed. Lakes are classified as oligotrophic, mesotrophic, and eutrophic with sub-classifications within each class based on the Trophic Status Index (TSI) of Total phosphorus (TP), chlorophyll (Chl-a), and Sechi Disc Depth (SDD) [35].

Trophic state	TN (mg/L)	TP (mg/L)	Chlorophyl-a (mg/m^3)	Sechi depth (m)	Reference
Oligotrophic	—	<0.011	<2.9	>5	[31]
	<0.3	<0.015	<3	>4	[32]
	<0.4	—	—	—	[33]
	<0.35	0.01	3.5	>4	[27]
Mesotrophic		0.011–0.0217	2.9–5.6	5–3	[31]
	0.3–0.65	0.015–0.025	3–7	2.5–4	[32]
	0.4–0.6		—		[33]
	0.35–0.65	0.01–0.03	3.5–9	2–4	[27]
Eutrophic		>0.0217	>5.6	<3	[31]
	>0.6	>0.025	>7	<2.5	[32]
	>0.65				[33]
	0.65–1.2	0.03–0.1	9–25	1–2	[27]
Hypertrophic	>1.200	>0.1	>25	<1	[27]

Table 1.
Criteria for the trophic status of lake waters based on a single parameter.

It is important to note that eutrophication caused by cultural factors is a major environmental issue in almost all lakes worldwide [26–28]. Therefore, managing land use and the transport of nutrients into the water area is crucial in preventing the negative impact of eutrophication.

Aquaculture activities in water bodies that use cages have a lot on the lakes and reservoirs in Indonesia, and have contributed as an additional input of phosphorus to the aquatic system, marking a significant cultural eutrophication phenomenon [36–39]. From aquaculture activities in the cages system in Lake Maninjau, Indonesia, with a fish production of 36,219 tons, it is estimated that the phosphorus [P] load into the waters reach 387 tons y^{-1}, consisting of [P] as wasted through feces (130.5 tons y^{-1}) and wasted as dissolved (256.6 tons y^{-1}) [39]. Meanwhile, from cage aquaculture activities in Lake Toba on fish production rate 62,023 tons in 2016, the phosphorus load released into the waters was estimated at 570 tons (**Table 2**) [40]. The salmon (*Oncorynchus mykiss*; *O. salar*) culture activities in Lake Rupanco Chile, with the production of 1626 tons on August 2008 to July 2009, was estimated to supply TN and TP to the lake waters 76.4 tons y^{-1}- and 12.1tons y^{-1}, respectively, as unconsumed feed, feces, and urine [41].

Eutrophication is a well-known environmental stressor that impacts various ecosystem components, especially the phytoplankton community. The response to eutrophication can be observed at the individual, population, and community levels. At the individual level, physiologically, it can lead to increased mortality or rapid growth. Changes in species abundance or disappearance can be observed at the population level. At the community level, there are structural changes and alterations in biodiversity. At the ecosystem level, eutrophication can lead to the disruption of biochemical cycles and decreased productivity. These environmental stressors can increase in intensity and persistence over time, resulting in a greater impact on individual species and the ecosystem [42].

Based on various references, cyanobacteria have a positive response in line with the increase in phosphorus level. It is important to note that the response of one population level ultimately impacts the community level.

Production of fish	(tons)[1]	62,023.30
Estimated feed used	(tons)[a]	76,288.66
Content of P on feed	(tons)[b]	915.46
Retention of P by fish	(tons)[c]	345.13
Release of P from feces	(tons)[d]	192.25
Load of dissolved P in the form of metabolite residues	(tons)[e]	378.09
Total P release to the waters	(tons)	570.33

[1]*Maritim Affairs and Fisheries Board, North Sumatra Province, Indonesia. Fisheries Annual Report 2015 (Unpublished).*
[a]*Food Conversion Ratio (FCR) of Nile tilapia = 1.23.*
[b]*1.2% of feed.*
[c]*37.7% from feed P content.*
[d]*21.0% from P content.*
[e]*41.3% from P content of feed based on Rismeyer (1988) formulation in [40].*

Table 2.
Prediction of phosphor loading from cage aquaculture activity in Lake Toba, Indonesia.

Generally, blooming cyanobacteria strongly characterize the eutrophication of freshwater ecosystems and ecologically has a negative effect, including decreasing water transparency and causing high oxygen fluctuations. Cyanobacteria blooms, on the other hand, produce toxins that are harmful to the surrounding and may have lethal effects on many aquatic or terrestrial organisms. The toxins produced by cyanobacteria include microcystins, anatoxin-a and saxitoxins. The existence of harmful cyanobacteria blooms is related to P load and N load [43, 44].

There are some toxic phytoplankton known as toxic harmful algae (HA) although in low assemblage proportion, other groups may occur harmful algal blooms (HABs) and almost on dominant composition in algal population. In fact, many HA types can grow abundantly and increase their toxin production when nutrient concentration is not in Redfieldian balance and when the inorganic nutrient components do not predominance [45].

Harmful algal blooms have become a national concern in the United States. Their appearance has become widespread globally in recent years, on the other hand, it was also recorded in every state and causes economic losses, affects human and animal health and disrupts the condition of the aquatic ecosystem itself. When cyanobacteria grow massively, die and decompose, much oxygen will be absorbed, producing an anoxic area where other organisms cannot live [46].

In a study conducted on 12 lakes in Kosciusko County, Indiana, microcystin concentrations ranged from 0.15 to 11 μg/L during blue-green algae blooms observed between 2015 and 2017. The highest mean microcystin concentrations were observed in Big Chapman Lake (1.64 μg/L) and the lowest in Big Barbee Lake (0.17 μg/L). According to the Indiana Department of Environmental Management (IDEM), microcystin concentrations above 4 μg/L and blue-green algae abundance above 20,000 cells/L pose a risk to human health. Therefore, monitoring and managing blue-green algae blooms and their associated toxins are crucial to mitigate potential impacts on both human health and the aquatic ecosystem [47].

One eutrophication phenomenon that has received little attention is the abundance of attached filamentous algae (FABs; filamentous algal blooms) in clear lakes worldwide, which mainly grow and blooms in the littoral area of the lake. These attached algal communities generally consist of groups of green algae and/or cyanobacteria, a form of lake degradation without indicating eutrophic waters. They cannot be explained in current eutrophication models. This phenomenon is allegedly due to the pollution of groundwater nutrients. Nutrient concentrations in groundwater higher than those in the water column are thought to be a driving force for the development of FABs. In contrast, phytoplankton growth in the water column is limited [48].

The blooms of FABs in the littoral zone must be considered because it is an essential area with high biodiversity. Based on observations in Bear Lake in the USA, a large and, clear lake, *Cladophora glomerata* dominates the FABs groups [49]. *Cladophora glomerata* indicates a eutrophic aquatic flora in Windermere, English Lake District, during observations from 1992 to 1993. In the southern basin of the lake, the maximum biomass of *C. glomerata* is almost 200 g dry weight m^{-2}, up to six times the biomass in the northern basin. Observations on phosphate concentration in the southern part of the lake on average was 0.12 g P m^{-2} was higher than in the northern area was 0.01 g P m^{-2}, meanwhile, the standing stock of N in the South Basin was 1.81 g N m^{-2}, elevenfold that in the North Basin at 0.16 g N m^{-2}, on average [50].

The high biomass but low productivity of FABs can have severe consequences for lake ecosystems. The existence of periphyton assemblages, where FABs are absent,

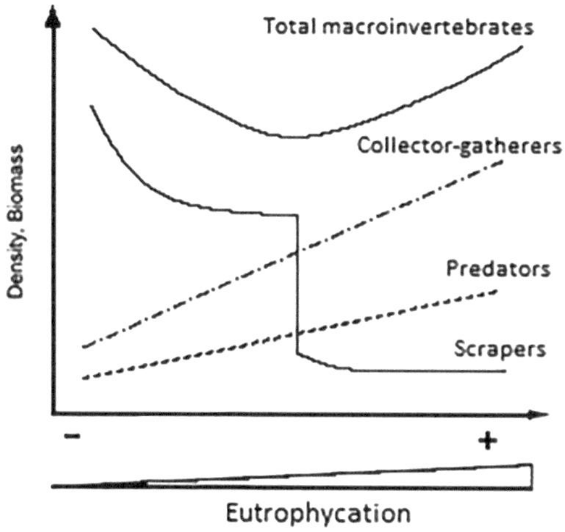

Figure 3.
Trend response macroinvertebrate groups based on functional feeding groups to eutrophication [51].

although low biomass but high productivity provide important value as food webs in lakes. The cause of the FABs phenomenon in lake waters worldwide is difficult to ascertain, mainly when various pressures occur simultaneously in a lake ecosystem. For instance, in the case discussed above, factors such as climate change, anthropogenic eutrophication, and invasive species could all contribute to the blooming of FABs [49].

The impact of eutrophication on floodplain lakes causes water conditions to change from clear to turbid and a decrease in macrophyte coverage. The decrease of macrophytes assemblages along the eutrophication gradient resulting in changes in taxonomic groups and the diversity of macrobenthic communities. These changes are attributed to the different tolerance levels of macroinvertebrate taxonomic groups to trophic states, which generally lead to decreased taxa. An increase in the abundance of phytoplankton in response to eutrophication was followed by a decrease in the euphotic depth, which inhibited macrophytes growth and impacted the number of macroinvertebrate species. Meanwhile, the response of the macroinvertebrate community to eutrophication showed a pattern of density and total macroinvertebrate biomass, where collectors-gatherers (mainly Tubificidae and Chironomidae) and predators (e.g., Tanypus) increased, scrapers (e.g., Bithyniidae) decreased, and decreased again (**Figure 3**). Macrophytes play an essential role in maintaining the integrity of macroinvertebrate diversity, as they serve as the primary source and habitat for epiphytic animals [51].

5. Organic loading

Many organic loads due to anthropogenic activities as allochthonous material have entered lake waters and impacted the water column and the sediment zone conditions. Organic load and other substances that are deposited into the water column in the bottom waters act as energy for bacteria, and throughout the process, will go through remineralization processes, respiration, oxidation of metabolites and consume much oxygen, causing a decrease in oxygen dissolved in the region [52, 53]. In the hypolimnetic column, various aerobic microbes mineralize the organic matter. Oxygen depletion is also accelerated in the hypolimnion region due to anaerobic processes, such as methane and ammonium [54, 55].

The supply of organic matter is often ignored in the environmental monitoring of lake waters. The crisis of anoxic conditions in almost all hypolimnion areas is also faced in tropical lakes, with rates of hypolimnion decline reaching 0.046–5.9 m^2y^{-1}. This is the impact of organic loads and climate change which not allows vertical mixing and oxygen-deficient water change [56].

Cage aquaculture activities in Lake Toba Indonesia contribute to the reduction of the oxic hypolimnion layer, marked by the anoxic column in the cage area (CA) being much narrower than in the non-cage area (NCA) and dissolved oxygen availability in water the column that is suitable for animal life in CA was shallower than NCA (**Figure 4**) [57].

The tolerance values of certain macroinvertebrate groups, i,e anthropod family have been classified in a family-level biotic index (FBI) as a rapid field assessment tool of organic pollution in the lotic systems. There is a relationship between FBI, water quality and the degree of organic pollution [58, 59]. Nevertheless, a biological index based on macroinvertebrates communities related to their response to environmental stressors is not easy to apply in the lake, due to the stagnant condition, and there is a diffusion factor in the lentic ecosystem. The macroinvertebrate community structure appears to be more permanent and requires high identification until species level [60].

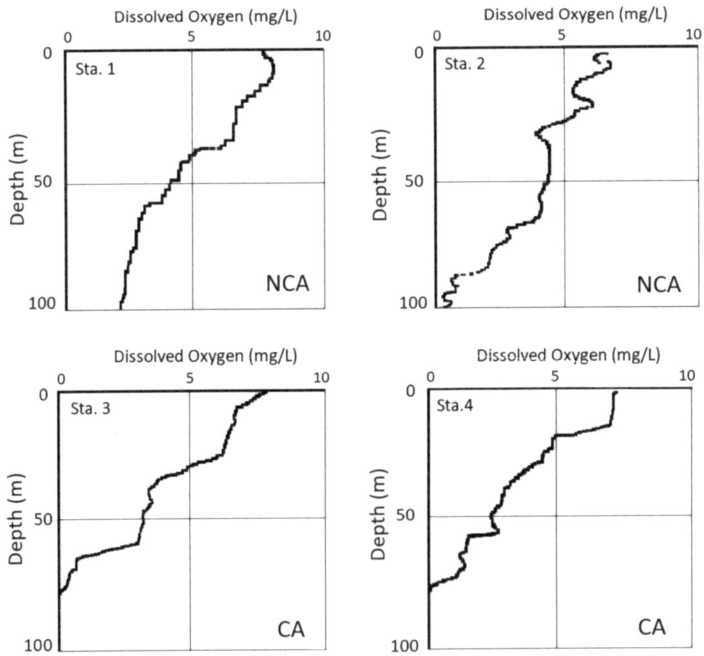

Figure 4.
Dissolved oxygen vertical stratification pattern in non-cage aquaculture (NCA) and in cage aquaculture (CA) area [57].

One postulate generally accepted that macroinvertebrates assemblages in lake benthic habitats are driven by various and integrated environmental factors including temperature, oxygen, and organic matter availability [61]. The sensitivity to organic load has formed certain of macroinvertebrates' community clusters and responses by changing their community structure. The term eutrophication is related to community of macroinvertebrates which refer to sediment conditions that show undergoing organic enrichment and are characterized by a decrease of oxygen.

The increase of organic sediment loading and other important chemical variable indicates that eutrophication can be assessed using bio indicators such as Tubificid and lumbriculid species [62]. The occurrence of lumbriculid and tubificid species is associated to the eutrophic state of the aquatic ecosystem, which assumes that organic sediment loading originates from phytoplankton production as the dominant autochthonous source. It was demonstrated in Lake Geneva, Switzerland, which sorted sediment organic loading condition based on the black layer thickness (represent organic) of sediment upper part was <10 cm (LOS; Organic sediment was low) and on black layers thickness > 10 cm (HOS; Organic sediment was high) and how the effect to macroinvertebrate community. Show that the mean abundance of oligotrophic worm species in the LOS area reached 30%, while in the HOS area, it was below 15% [63].

In observing the effect of organic pollution which validated as high-level ammonia and orthophosphate and low oxygen in Lake Lysimachia (Western Greece) showed that *Limnodrilus hoffmeisteri* in particular and species of chironomids (*Paratrichocladius rufiventris*, and adults of *Cricotopus bicinctus* dan *Rheocricotopus atripes*) characterized as polluted community. Three species that characterize clean water condition and high oxygen, namely *Psammoryctides barbatus*, *Dianella thiesseana,* and *Gammarus* sp. [60].

Organic material sourced from anthropogenic activities, includes metal and total phosphorus, exerts pressure on the lake sediment area. Based on a multivariate comparison of environment chemical variables and bio indicators species obtained six groups which characterized one level of pollution and trophic status of the sediment. The groups are as follows: (1) polluted and eutrophic (*Potamothrix hammoniensis*, *Peloscolex ferox*, *Limnodrilus claparedeanus*); (2) polluted and mesotrophic (*Psammoryctides barbatus*); (3) unpolluted and mesotrophic (*Limnodrilus hoffmeisteri*, *Limnodrilus udekemianus*); (4) polluted and oligomesotrophic (*Potamothrix vejdovskyi*); and (5) unpolluted and oligotrophic (*Stylodrilus lemani*, *Peloscolex velutinus*) [64].

Eutrophication conditions of Lake Yamanakako, one of five lakes in Fuji including lakes Kawaguchiko, Motosuko, Saiko, Shojiko, and Yamanakako in Japan, have changed the distribution of chironomid fauna density between 1994 and 2003. *Chironomus nipponensis* larvae decreased while *Propsilocerus akamusi* increased, and at the same time, the density of Tanypodia larvae had decreased. In this study, the density of *P. akamusi* larva showed a positive correlation to sediment's organic ignition loss (IL) value [65]. This show that the eutrophication of Lake Tamanakako is characterized by organic enrichment in the sediment. The decrease of Tanypodia abundance can occur due to low oxygen, as well known, Tanypodia can not adapt to low oxygen.

6. Acidification

Acidification is a process that leads to a decrease in pH in water bodies, and it can occur at both local and global scales. Local-scale acidification is often linked to soil conditions in a catchment area. On a global scale, acidification is caused by "acid rain," which results from burning fossil fuels and has a transboundary impact. On a local scale, lake acidification is the impact of catchment soil condition. Before the period of acid rain, acidification in lakes was related to changes in catchment conditions rather than acid deposition. Lake in Galloway, Scotland, undergoing acidification (pH 5.5–6.0) before peat formation based on lake acidity data from the earlier post-glacial period. However, acidification of lakes due to acid rain began to occur after AD 1850 when acid emissions increased due to the burning of fossil fuels. It is worth noting that acidification resulting from soil acidification differs from the current acidification phenomenon, which has been widely reported in Europe and North America [66].

In the 1970–1980s, acidification of lake waters caused by atmospheric sulfur deposits became a significant international concern, leading to the implementation of strong sulfur control programs in Europe and North America. However, various factors prevented their widespread adoption. While reducing industrial sector employment, the environmental impact of sulfur control programs remained unclear. Interestingly, it was discovered that nitrogen deposition is a significant contributor to lake acidification in some regions [67].

Ammonia, mainly produced from livestock, fertilizer, and industrial activities, enters the atmosphere, and neutralizes sulfuric acid with nitrogen oxides, substantially increasing pH precipitation. When ammonium compounds enter the soil and water environment, they can be oxidized to nitric acid, releasing acid [68]. Acid deposits may contain high concentrations of NO_3 and NH_4. In the Irish Lakes, for example, precipitation from ten sampling stations during 1995–1996 had a nitrogen concentration 1.8 times that of SO_4^{2+}, with NH_4^+ concentration twice that of NO_3^- [69].

Based on experiments on lake acidification (from pH 6.2 to 4.7) in Little Rock Lake (Wisconsin, USA) on the species richness and annual biomass on five tropic levels of biota (phytoplankton, herbivores, omnivore, carnivorous zooplankton, and fish) indicates an impact of this acidification. There was a marked decrease in the relative richness of fish and zooplankton compared to phytoplankton, in which the phytoplankton did not appear to be affected by acidification [70].

Acidification has an impact on macroinvertebrates community and is observed in Pennsylvania lakes; Deep (Area [A]; 3.0 ha; Maximum depth [D_{max}]: 6.8 m), Lacawac (A: 21.0 ha; D_{max}: 13.5 m), and Long Pond (A; 32.8 ha; D_{max}: 7.0 m). The lakes showed very significant differences in acidification levels. Deep was acidified (mean total alkalinity <0.0 μeq L^{-1}; the mean pH decreased from 5.5 to 4.2 between 1981 and 1983), Lacawac was moderately sensitive to acidification (mean total alkalinity: 47 μeq L^{-1}; the mean pH 6.1) and Long Pond was the least sensitive to acidification (mean total alkalinity 190 μeq L^{-1}; mean pH 6.6). The macroinvertebrate community formed on three levels of acidification show that in acid lakes dominated by Chironomidae (71.3% in number; 19.6% in wet weight), the abundance in the number of Chironomidae was dominant (43%) in moderate acidification lakes but in biomass (wet weight) dominated by Odonata (18.6%) and Mollusk (12.7%), in the least sensitive lake show that the Amphipoda (31.3%) and Chironomidae was dominant in number but in biomass (wet weight) Mollusca makes up 55.1% of total in the least sensitive lake individual abundance is dominated by Amphipoda (31.3%) and Chironomidae (27.3%) while from Mollusca biomass it makes up 55.1% of the total [71].

7. Habitat modification

Lake ecosystem habitats are uniquely characterized by paired patterns that play a crucial role in shaping nutrient cycles, predator-prey interactions, food web structure, and ecosystem stability, creating an integrated system. Habitat modification, such as anthropogenic pressures, can significantly alter the interdependence between habitats, disrupting the crucial flow of energy and nutrients in the lake ecosystem [72]. Habitat modification, including shoreline morphology and water level changes, can impact the littoral habitat area in the lake ecosystem.

The physical condition of the shore zone of the lakes can be altered, and various activities, such as recreation and development, can lead to disturbance in the riparian and littoral zone. The shore zone plays a crucial role in maintaining the biodiversity of lake ecosystems due to its complex habitats, unique plant and animal communities, and high biochemical activity. It also contributes significantly to ecological processes and structures in aquatic ecosystems. Alterations to the morphology of the shoreline and riparian habitats can significantly impact all processes within the lake ecosystem [73].

The littoral zone is an essential lake ecosystem zone supporting the highest productivity. The littoral zone and benthic habitat of the lake waters form a microhabitat for macroinvertebrate communities, providing a feeding and breeding ground for and shelter from predators and wave action [9, 74].

Various human activities can cause modification of littoral areas and lead to severe impacts on benthic habitats. These activities include forest logging in catchment areas and riparian zones, removal of debris from lake shores, macrophyte removal, wetland drainage, dredging, and water lake management. The complexity of the littoral habitat is disrupted due to these activities, and the ecological damage caused by changes in the morphology of the shore zone is much higher than that caused by eutrophication [75]. The damage caused to riparian vegetation results in a decrease in the supply of organic matter from the land to the littoral food web system, indicating the decoupling of the littoral from the riparian zone [76].

Changes in lake hydrology, particularly those resulting in water level fluctuations beyond natural limits for the purposes of hydropower, flood control, and aquatic plant management, pose a threat to the ecological integrity of the shore zone. While water level fluctuations are a common pattern in lake ecosystems, changes in hydrology beyond natural variability can harm lake ecosystems [73]. The Miorina Dam, located in Italy and Switzerland, regulates water level fluctuations in Lake Maggiore. However, this has led to significant changes in the ecology of the littoral zone, impacting the structure and function of benthic copepod groups, including the abundance of ovigerous females, opportunists, omnivores, and deposit feeders. While no missing classes were observed, variations in the composition and function of taxonomic groups can have significant implications for entire communities [77]. Water level fluctuation due to lake discharge regulation for hydroelectric power plants is expected impact the wetland area. The semi aquatic area on lake the shore zone is considered as eel (*Anguilla* spp.) habitat so in the long term will affect the eel population [78].

The hydrological balance of Lake Bracciano, the largest and deepest lake in Italy, has been disturbed due to the withdrawal of water for human needs beyond its potential, which is exacerbated by climate change and decreased rainfall. This has reduced the richness and taxonomic abundance of the invertebrate assemblage of sessile forms such as water mites, gastropods, nematodes, and naidid oligochaetes that feed on living plants and epiphytic algae, and an increase in the types of mobile detritus eaters. Lowering the water level of Lake Bracciano has led to the removal of littoral habitats, especially in areas where the bottom slopes less steeply, and the distribution of macrophytes is less frequent when the annual water level is low [79].

8. Conclusion

As the human population grows and their needs and interests become more diverse, lake conditions will likely change. The most significant impact of human activity on the ecosystem is sedimentation and habitat modification which alters the biota's habitat and physiology. Nutrient enrichment from the phosphorous loading into the water overproduces the algae and cyanobacteria, decreasing water transparency and inducing high oxygen fluctuation. High organic matter load into the water causes oxygen depletion, however, the supply of organic matter is often ignored in the environmental monitoring of lake waters. Acidification which occurs at local and global scales affects the biota community. Therefore, it is crucial to increase our knowledge and understanding of the aquatic environment, especially lakes, so that we can develop a greater appreciation for and respect towards nature.

Acknowledgements

The authors thank the Research Center for Conservation of Marine and Inland Water Resources and Research Center for Limnology and Water Resources at the National Research and Innovation Agency (BRIN), Indonesia for their support. We also would like to thank our colleagues in BRIN for constructive discussion.

Author details

Lukman Lukman
Research Center for Conservation of Marine and Inland Water Resources, National Research and Innovation Agency, Cibinong, Indonesia

*Address all correspondence to: lukm001@brin.go.id

References

[1] Lerche I. Natural and anthropogenic environmental problems. Energy & Environment. 2001;**12**(1):73-88. DOI: 10.1260/0958 3050 1150 0616

[2] Bogan E, Doina S, Varvaruc D. The impact of anthropogenic activities on components of the natural environment of the Titu Plain. GEOREVIES. 2014;**24**:54-64. DOI: 10.4316/GEOREVIEW.2014.24.1.170

[3] Sønergaard M, Jeppesen E. Anthropogenic impacts on lake and stream ecosystems, and approaches to restoration. Journal of Applied Ecology. 2007;**44**:1089-1094. DOI: 10.1111/j.1365-2664.2007.01426.x

[4] Niessen F, Wick L, Bonani G, Chondrogianni C, Siegenthaler C. Aquatic system response to climatic and human changes: Productivity, bottom water oxygen status, and sapropel formation in Lake Lugano over the last 10 000 years. Aquatic Sciences. 1992;**54**:257-276

[5] Díaz S, Fargione J, Chapin FS III, Tilman D. Biodiversity loss threatens human well-being. PLoS Biology. 2006;**4**:e277. Available from: www.plosbiology.org

[6] Myers SS, Gaffikin L, Golden CD, Ostfeld RS, Redford KH, Ricketts TH, Turner WR, Osofsky SA. Human health impacts of ecosystem alteration. Perspective 2013; 110: 18753-18760. Available from: www.pnas.org/cgi/doi/10.1073/pnas.1218656110. Downloaded from https://www/pnas.pras.org [Accessed: March 18, 2023]

[7] Dalton R. Ecologists seek sustainable future. Nature. 2002;**416**:113. DOI: 10.1038/416113b

[8] Abhilash PC. Ecology for sustainable development in the Anthropocene. Anthropocene Science. 2022;**1**:339-341. DOI: 10.1007/s44177-022-00035-z

[9] Wetzel RG. Limnology: Lake and River Ecosystems. 3rd ed. San Diego: Academic Press; 2001

[10] Dubois N, Saulnier-Talbot E, Mills M, Gell P, Battarbee R, Bennion H, et al. First human impacts and responses of aquatic systems: A review of palaeolimnological records from around the world. The Anthropocene Review. 2017;**2017**:1-41. DOI: 10.1177/2053019617740365

[11] Szabó Z, Buczkó K, Haliuc A, Pál I, Korponai JL, Begy RC, et al. Ecosystem shift of a mountain lake under climate and human pressure: A move out from the safe operating space. Science of the Total Environment. 2020;**743**:140584. DOI: 10.1016/j.scitotenv.2020.140584

[12] Ritter C, Gonçalves V, Pla-Rabes S, de Boer EJ, Bao R, Sáez A, et al. The vanishing and the establishment of a new ecosystem on an oceanic island – Anthropogenic impacts with no return ticket. Science of the Total Environment. 2022;**830**:154828. DOI: 10.1016/j.scitotenv.2022.154828

[13] Brooks KN, Ffolliott PF, Magner JA. Hydrology and the Management of Watersheds. 4th ed. Oxford: A John Wiley & Sons, Inc., Publ.; 2013

[14] Dearing JA, Jones RT. Coupling temporal and spatial dimensions of global sediment flux through lake and marine sediment records. Global and Planetary Change. 2003;**39**:147-168. DOI: 10.1016/S0921-8181(03)00022-5

[15] Schiefer E, Reid K, Burt A, Luce J. Assessing natural sedimentation patterns and impacts of land use on sediment yield: A Lake-sediment–based approach. In: Watershed Assessment in the Southern Interior of British Columbia: Workshop Proceedings. BC Ministry of Forest, Research Branch, Victoria, BC Working Paper. Issue 57. 2001. pp. 209-247. Available from: https://www.researchgate.net/publication/237389176

[16] Bogale A. Review, impact of land use/cover change on soil erosion in the Lake Tana Basin, Upper Blue Nile, Ethiopia. Applied Water Science. 2020;**10**:235. DOI: 10.1007/s13201-020-01325-w

[17] Hecky RE, Bootsma HA, Kingdon ML. Impact of land use on sediment and nutrient yields to Lake Malawi/Nyasa (Africa). International Association of Great Lakes Research. 2003;**29**(Supplement 2):139-158

[18] Nugroho NP. Sediment export estimation from the catchment area of Lake Rawapening using InVEST model. IOP Conference Series: Earth and Environmental Science. 2022;**950**:012072. DOI: 10.1088/1755-1315/950/1/012072

[19] Moningkey AT, Rampengan MMF, Tumengkol AA, Kumaat JC. Study of bathymetry and sedimentation in Tondano Lake. IOP Conference Series: Earth and Environmental Science. 2022;**986**:012038. DOI: 10.1088/1755-1315/986/1/012038 1

[20] Schiefer E, Petticrew EL, Immell R, Hassan MA, Sonderegger DL. Land use and climate change impacts on lake sedimentation rates in western Canada. Anthropocene. 2013;**3**:61-71. DOI: 10.1016/j.ancene.2014.02.006

[21] Robertson MJ, Scruton DA, Gregory RS, Clarke KD. Effect of suspended sediment on freshwater fish and fish habitat. Canadian Technical Report of Fisheries and Aquatic Sciences. 2006;**2644**:37. (PDF) Effect of suspended sediment on freshwater fish and fish habitat (researchgate.net)

[22] Crane DP, Farrell JM. Spawning substrate size, shape, and siltation influence walleye egg retention. North American Journal of Fisheries Management. 2013;**33**:329-337. DOI: 10.1080/02755947.2012.760504

[23] Correll DL. The role of phosphorus in the eutrophication of receiving waters. Journal of Environmental Quality. 1998;**27**:261-266. DOI: 10.2134/jeq1998.00472425002700020004x

[24] Schindler DW. Recent advances in the understanding and management of eutrophication. Limnology and Oceanography. 2006;**51**(2):356-363

[25] Goldman RC, Horne AJ. Limnology. Tokyo: Mc-Graw Hill International Book Company; 1983

[26] Carpenter SR, Ludwig D, Brock WA. Management of Eutrophication for lakes subject to potentially irreversible change. Ecological Applications. 1999;**9**:751-771

[27] Smith V. Eutrophication of freshwater and coastal marine ecosystems: A global problem. Environmental Science and Pollution Research. 2003;**10**:126-139. DOI: 10.1065/espr2002.12.142

[28] Nyenje PM, Foppen JW, Uhlenbrook S, Kulabako R, Muwanga A. Eutrophication and nutrient release in urban areas of sub-Saharan Africa—A review. Science of the Total Environment. 2010;**408**:447-455. DOI: 10.1016/j.scitotenv.2009.10.020

[29] Fraterrigo JM, Downing JA. The influence of land use on Lake nutrients

varies with watershed transport capacity. Ecosystems. 2008;**11**(7):1021-1034. DOI: 10.1007/s10021-008-9176-6

[30] Leach JH, Herron RC. A review of Lake habitat classification. In: Busch WDN, Sly PG, editors. The Development of an Aquatic Habitat Classification System for Lake. London: CRC Press; 1992

[31] Chapra SC, Dobson HFH. Quantification of the Lake Trophic typologies of Naumann (surface quality) and Thienemann (oxygen) with special reference to the Graet Lake. Journal of Great Lake Research. 1981;7:182-193

[32] Forsberg C, Ryding S. Eutrophication parameters and trophic state indices in 30 Swedish waste receiving Lake. Archiv für Hydrobiologie. 1980;**80**:189-207

[33] Vollenweider RA. Scientific Fundamentals of the Eutrophication of Lake and Flowing Water, with Particular Reference to Nitrogen and Phosphorus as Factor in Eutrophication. Paris: UNECD; 1968

[34] Xu FL, Tao S, Dawson RW, Li BG. A GIS-based method of lake eutrophication assessment. Ecological Model. 2001;**144**(2-3):231-244. DOI: 10.1016/S0304-3800(01)00374-X

[35] Carlson RE. A trophic state index for lakes. Limnology and Oceanography. 1977;**22**:361-369

[36] Lukman L, Syawal MS, Maghfiroh M. Sumatran major lakes: Limnological overviews. IOP Conference Series: Earth Environmental Science. 2020;**535**:012064

[37] Lukman L, Dina R, Ibrahim A, Waluyo A. Phosphorus load base carrying capacity estimation for fish culture in Lake Ranau in Indonesia. IOP Conference Series: Earth Environmental Science. 2021;**919**:012020

[38] Lukman L, Kartamihardja ES, Hariyadi S, Sunardi D, Sumiarsa D. Brief overview of allochtonous phosphorus load on cascade reservoirs in Citarum River Basin, West Java Indonesia. IOP Conference Series: Earth Environmental Science. 2022;**1062**:012021

[39] Lukman L, Setyobudiandi I, Muchsin I, Hariyadi S. Impact of cage aquaculture on water quality condition in Lake Maninjau, West Sumatera Indonesia. International Journal of Sciences: Basic and Applied Research (IJSBAR). 2015;**23**(1):120-137

[40] Lukman L, Hidayat H, Subehi L, Dina R, Mayasari N, Melati I, et al. Pollution loads and its impact on Lake Toba. IOP Conference Series: Earth Environmental Science. 2019;**299**:012051

[41] Munoz JL, Echeverria C, Marce R, Sherman WB, Iriarte JL. The combined impact of land use change and aquaculture on sediment and water quality in oligotrophic Lake Rupanco (North Patagonia, Chile, 40.8o S). Journal of Environmental Management. 2013;**128**:283-291. Available from: www.elsevier.com/locate/jenvman

[42] Salmaso N, Tolotti M. Phytoplankton and anthropogenic changes in pelagic environments. Hydrobiologia. 2021;**848**:251-284. DOI: 10.1007/s10750-020-04323-w

[43] Vasconcelos V. Eutrophication, toxic cyanobacteria and cyanotoxins: when ecosystems cry for help. Limnetica. 2006;**25**:425-432. DOI: 10.23818/limn.25.30

[44] Schindler DW. Evolution of phosphorus limitation in lakes. Science.

1977;**196**:260-262. Available from: http://www.jstor.org/stable/1743244

[45] Gilbert PM, Burkholder JM. Harmful algal blooms and eutrophication: "Strategies" for nutrient uptake and growth outside the Redfield comfort zone. Chinese Journal of Oceanology and Limnology. 2011;**29**(4):724-738. DOI: 10.1007/s00343-011-0502-z

[46] Gatz L. Freshwater harmful algal blooms: Causes, challenges, and policy considerations. Congressional Research Service. 2019;**2019**:26. Available from: https://crsreports.congress.gov

[47] Bosch NS, Bingham SA, Farwell K. Blue-green algae (cyanobacteria) patterns and prediction in 12 lakes in Kosciusko County, Indiana. Proceedings of the Indiana Academy of Science. 2019;**128**(1):1-12. Available from: https://lakes.grace.edu/wp-content/uploads/2020/01/pias-128-01-1-1.pdf

[48] Vadeboncoeur Y, Moore MV, Stewart SD, Chandra S, Atkins KS, Baron JS, et al. Blue Waters, green bottoms: Benthic filamentous algal blooms are an emerging threat to Clear Lakes worldwide. Bioscience. 2021;**71**(10):1011-1027. DOI: 10.1093/biosci/biab049

[49] Page M, Goldhammer T, Hilt S, Tolentino S, Brothers S. Filamentous algae blooms in a large, Clear-Water Lake: Potential drivers and reduced benthic primary production. Water. 2022;**14**:2136. DOI: 10.3390/w14132136

[50] Planas D, Maberly SC, Parker JE. Phosphorus and nitrogen relationships of *Cladophora glomerata* in two lake basins of different trophic status. Freshwater Biology. 1996;**35**:609-622

[51] Pan BZ, Wang HZ, Pusch MT, Wang HJ. Macroinvertebrate responses to regime shifts caused by eutrophication in subtropical shallow lakes. Freshwater Science. 2015;**34**(3):942-952. DOI: 10.1086/682077

[52] Marcé R, Moreno-Ostos E, López P, Armengol J. The role of allochthonous inputs of dissolved organic carbon on the hypolimnetic oxygen content of reservoirs. Ecosystems. 2008;**11**(7):1035-1053. DOI: 10.1007/s10021-008-9177-5

[53] Kritzberg ES, Cole JJ, Pace ML, Granéli W, Bade DL. Autochthonous versus allochthonous carbon sources of bacteria: Results from whole-lake 13C addition experiments. Limnology and Oceanography. 2004;**49**(2):588-596. DOI: 10.4319/lo.2004.49.2.0588

[54] Carstens D, Köllner KE, Bürgmann H, Wehrli B, Schubert CJ. Contribution of bacterial cells to lacustrine organic matter based on amino sugars and d-amino acids. Geochimica et Cosmochima Acta. 2012;**89**:159-172. DOI: 10.1016/j.gca.2012.04.052

[55] Steinsberger T, Schwefel R, Wüest A, Müller B. Hypolimnetic oxygen depletion rates in deep lakes: Effects of trophic state and organic matter accumulation. Limnology and Oceanography. 2020;**65**:3128-3138. DOI: 10.1002/lno.11578

[56] Fukushima T, Matsushita B, Subehi L, Setiawan F, Wibowo H. Will hypolimnetic waters become anoxic in all deep tropical lakes? Scientific Report. 2017;7(24320):1-8. DOI: 10.1038/srep45320

[57] Lukman L, Subehi L, Dina R. A preliminary study of stratification in Lake Toba: The largest tropical Lake

in Southeast Asia. European Modern Studies Journal. 2020;**4**(6):113-122

[58] Hilsenhoff WL. Rapid field assessment of organic pollution with a family-level biotic index. Journal of the North American Benthological Society. 1988;**7**(1):65-68

[59] Hilsenhoff WL. An improved biotic index of organic stream pollution. The Great Lakes Entomologist. 1987;**1987**:20. Available from: https://scholar.valpo.edu/tgl/vol120/issl/7

[60] Petridis D. Macroinvertebrate distribution along an organic pollution gradient in Lake Lysimachia (Wester Greece). Archeology Hidrobiology. 1993;**128**(3):367-384

[61] Jyvasjarvi J, Boros G, Jones RI, Hamalainen H. The importance of sedimenting organic matter, relative to oxygen and temperature, in structuring lake profundal macroinvertebrate assemblages. Hydrobiologia. 2013;**709**:55-72. DOI: 10.1007/s10750-012-1434-0

[62] Brinkhurst RO. The Benthos of Lakes. London and Basingstoke: The Macmillan Press Ltd; 1974

[63] Lang C. Oligochaetes, organic sedimentation, and trophic state: How to assess the biological recovery of sediments in lakes? Aquatic Sciences. 1997;**59**:26-33

[64] Lang C, Doble-Lang B. The chemical environment of tubificid and lumbriculid worms according to the pollution level of the sediment. Hydrobiologia. 1979;**65**(3):273-282

[65] Hirabayashi K, Yoshizawa K, Yoshida N, Kazama F. Progress of eutrophication and change of chironomid fauna in Lake Yamanakako, Japan. Limnology. 2004;**5**:47-53. DOI: 10.1007/s10201-003-0113-2

[66] Jones VJ, Stevenson AC, Battarbee RW. Lake acidification and the land-use hypothesis: A mid-post glacial analogue. Nature. 1986;**322**(10):157-158

[67] Keller W, John M, Gunn JM, Yan ND. Acid rain – Perspectives on lake recovery. Journal of Aquatic Ecosystem Stress and Recovery. 1999;**6**:207-216. DOI: 10.1023/A:1009983318502

[68] Schuurkes JAAR, Mosello R. The role of external ammonium inputs in freshwater acidification. Hydrology. 1988;**50**(1):71-86

[69] Aherne J, Curtis CJ. Critical loads of acidity for Irish lakes. Aquatic Sciences. 2003;**65**:21-35

[70] Hogsden KL, Xenopoulos MA, Rusak JA. Asymmetrical food web responses in trophic-level richness, biomass, and function following lake acidification. Aquatic Ecology. 2009;**43**:591-606. DOI: 10.1007/s10452-008-9169-8

[71] Bradt PT, Berg MB. Macrozoobenthos of three Pennsylvania lakes: Responses to acidification. Hydrobiologia. 1877;**150**:63-74

[72] Schindler DE, Scheuerell MD. Habitat coupling in Lake ecosystems. Oikos. 2002;**98**(2):177-189. http://www.jstor.org/stable/3547424

[73] Strayer DL, Findlay SEG. Ecology of freshwater shore zones. Aquatic Sciences. 2010;**72**:127-163. DOI: 10.1007/s00027-010-0128-9

[74] McGoff E, Irvine K. A test of the association between Lake habitat quality

assessment and macroinvertebrate community structure. Aquatic Conservation: Marine Freshwater Ecosystem. 2009;**19**:520-533. DOI: 10.1002/aqc.1024

[75] Pilotto F, Bazzanti M, Di Vito V, Frosali D, Livretti F, Mastrantuono L, et al. Relative impacts of morphological alteration to shorelines and eutrophication on littoral macroinvertebrates in Mediterranean lakes. Freshwater Science. 2015;**34**(2):410-422. DOI: 10.1086/680523

[76] Brauns M et al. Human lakeshore development alters the structure and trophic basis of littoral food web. Journal of Applied Ecology. 2011;**48**:916-925. DOI: 10.1111/j.1365-2664.2011.02007.x

[77] Cifoni M, Boggero A, Rogora M, Ciampittiello M, Martínez A, Galassi DMP, et al. Effects of human-induced water level fuctuations on copepod assemblages of the littoral zone of Lake Maggiore. Hydrobiologia. 2022;**849**:3545-3564. DOI: 10.1007/s10750-022-04960-3

[78] Lukman L, Triyanto T, Haryani GS, Samir O, Gogali L, Bandjolu KP. Eel (*Anguilla* spp.) fishing activity in Poso Area Central Sulawesi, Indonesia. IOP Conference Series: Earth Environmental Science. 2021;**869**:012022

[79] Mastrantuono L, Solimini AG, et al. Plant-associated invertebrates and hydrological balance in the large volcanic Lake Bracciano (Central Italy) during two years with different water levels Plant-associated invertebrates and hydrological balance in the large volcanic Lake Bracciano (Central Italy) during two years with different water levels. Hydrobiologia. 2008;**599**:143-152. DOI: 10.1007/s10750-007-9196-9

Chapter 6

Anthropogenic Impacts as Determinants of Tropical Lake Morphology: Inferences for Strategic Conservation of Lake Wetland Biodiversity

Aina O. Adeogun and Azubuike V. Chukwuka

Abstract

Lakes as essential ecosystems for diverse life forms, including humans, have suffered altered morphology with adverse effects on biodiversity including amphibians and amphibious species. Thus, it is imperative for effective conservation strategies to simultaneously consider lake morphology, landscape variables, and the role of keystone species as ecosystem engineers for biodiversity preservation. Keystone species, particularly birds and large-bodied predators, i.e., crocodylians, play a critical role in maintaining the health of lake ecosystems as ecosystem engineers, bringing about large-scale changes in lake morphology and hydrology that determine the abundance and survival of other species in the ecosystem. Conservation strategies should, therefore, prioritize the protection of these keystone species and their habitats. To balance the needs of human society with the protection of lake ecosystems and their biodiversity, conservation practices must involve stakeholder engagement, including government agencies, local communities, traditional ecological knowledge, and scientists. A multidisciplinary approach, incorporating ecological, hydrological, and social factors, is considered necessary for effective lake conservation. This approach will encompass the preservation of lake biodiversity and consider important variables such as lake morphology, landscape variables, and the role of keystone species as ecosystem engineers in providing insights for strategic conservation practices.

Keywords: lake conservation, biodiversity, anthropogenic impacts, keystone species, multidisciplinary approach

1. Introduction

Lakes are important ecosystems that support a high level of biological diversity, making them valuable resources for conservation efforts. They provide critical habitats for many plant and animal species, including numerous rare and endemic species and are essential for the survival of many migratory birds [1]. Additionally, lakes

 IntechOpen

contribute to the regulation of global biogeochemical cycles and serve as a source of drinking water for millions of people worldwide [2, 3]. Conserving the biological diversity of lakes is, therefore, crucial for maintaining the integrity of these ecosystems and ensuring their continued provision of important ecological services.

The factors that influence lake morphology and biodiversity are complex and interconnected, making conservation efforts challenging. Some of the most significant drivers of lake biodiversity loss are anthropogenic activities, including urbanization, agriculture, and mining [1, 4]. These activities alter the natural characteristics of lake ecosystems, leading to habitat degradation, water quality degradation, and loss of biodiversity [5]. Therefore, understanding the impact of these factors is critical to developing effective conservation strategies. Keystone species are a crucial component to ensure the conservation of biological diversity in natural habitats like lakes. They play an essential role in regulating the population sizes of other species, often by controlling the availability of resources such as food or habitat. In many cases, the loss of a keystone species can have cascading effects on the rest of the ecosystem, leading to the decline of other species and ultimately compromising ecosystem health [6, 7]. Hence, the conservation of keystone species is critical for maintaining the biodiversity and ecological integrity of lake ecosystems.

1.1 Lakes ecosystems and biological diversity conservation

Lakes play a critical role in the conservation of biological diversity, as they support a diverse array of aquatic and terrestrial habitats that are home to a wide range of plant and animal species. They are recognized as one of the most important ecosystems in terms of biodiversity, and their conservation is essential for maintaining ecological balance and the provision of ecosystem services. Lakes provide optimal habitats for numerous keystone species, which are important in maintaining the balance and health of lake wetland ecosystems [8].

Human activities such as deforestation, urbanization, agriculture, and mining have resulted in significant changes in lake ecosystems, leading to loss of biodiversity and degradation of ecosystem services. Climate change also poses a significant threat to the health and functioning of lake ecosystems, as rising temperatures and changes in precipitation patterns can alter water chemistry, nutrient cycling, and ecosystem processes [9]. Climate change-related changes in the inflow water volumes of lakes may also bring about morphological changes (**Figure 1**) [10]. These threats highlight the importance of conserving lake ecosystems to maintain biodiversity and ensure the continued provision of vital ecosystem services.

Conservation efforts that focus on the protection and restoration of lake ecosystems have become increasingly important in recent years [11]. These efforts involve the implementation of policies and management practices that aim to reduce the impact of human activities on lake ecosystems, while also promoting sustainable use and the conservation of biodiversity [12]. Examples of conservation strategies include the establishment of protected areas. Implementation of sustainable fishing practices, restoration of degraded habitats, and reduction of pollution and eutrophication [13, 14].

1.2 Factors that influence lake morphology and biodiversity

Several factors influence the morphology and biodiversity of lakes, including geology, climate, water chemistry, and physical characteristics such as depth,

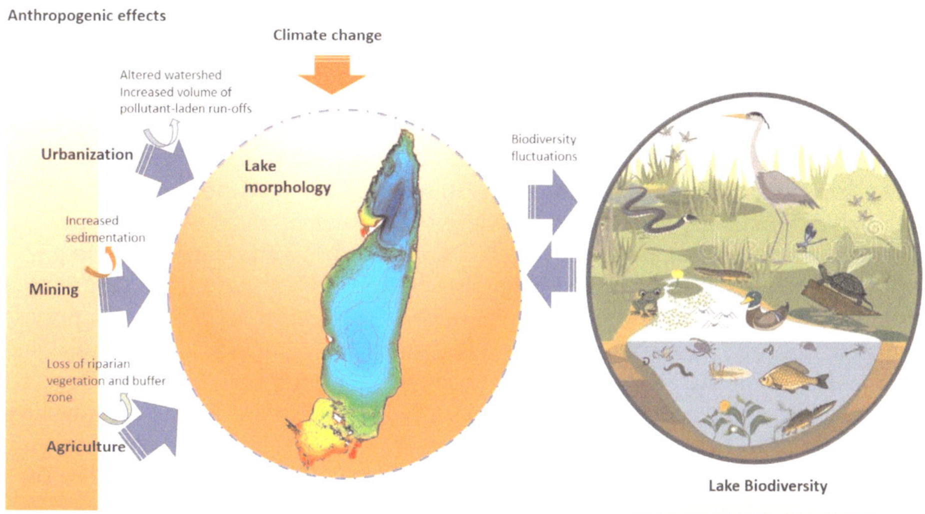

Figure 1.
Direct linkages between anthropogenic activities, lake morphology, and biodiversity occurrence.

shoreline length, and connectivity to other water bodies [15, 16]. Variations in these factors can lead to alterations in lake morphology, including changes in water depth, temperature, and nutrient availability, with significant impacts on lake biodiversity [17, 18]. For example, increased nutrient inputs from human activities such as agriculture and urbanization may lead to eutrophication and algal blooms, which can negatively affect fish populations and other aquatic organisms [19, 20]. Similarly, changes in the physical structure of lake habitats, for example, the removal of shoreline vegetation for the construction of dams, can lead to habitat loss and fragmentation, with negative impacts on the diversity and abundance of aquatic and terrestrial species [21].

In addition to the factors mentioned above, other human activities: deforestation, mining, and land use changes can also influence the morphology and biodiversity of lakes [22]. Activities like deforestation around lake watersheds may lead to increased sedimentation and nutrient runoff, altering the water chemistry of a lake and leading to decreased water clarity and dissolved oxygen levels [23]. Similarly, mining activities may introduce heavy metals and other toxic substances into lake ecosystems, which can negatively impact the health and diversity of aquatic species [24]. Land use changes, such as the conversion of wetlands to agricultural, industrial, or urban areas, can also result in the loss of crucial habitats, especially for wetland-dependent species and further contribute to biodiversity loss [25]. Thus, understanding the factors that influence lake morphology and biodiversity is crucial for effective lake conservation and management. By monitoring changes in these factors and implementing appropriate management strategies, such as nutrient reduction programs or habitat restoration efforts, it is possible to mitigate the negative impacts of human activities and promote the health and diversity of lake ecosystems [26]. Furthermore, addressing these factors can help to maintain the provision of ecosystem services that lakes offer to human society, such as drinking water, food, and recreation [27].

1.3 Keystone species and health of lake wetland ecosystems

Keystone species are essential for the maintenance of lake wetland ecosystems, and their loss can have cascading effects throughout the food web [28]. Amphibians are important in many ecosystems; however, negative large-scale effects such as climate change or massive pollution events on amphibians may have cascading effects on many other animals in the ecosystem [29]. In tropical lake wetland ecosystems, the conservation and management of keystone species, particularly amphibians and amphibious species, are vital to the regulation of food webs and nutrient cycling [30]. Amphibious species, such as mudskipper fish, birds, reptiles, and some mammals, are among the keystone species that play vital roles in nutrient cycling, predator-prey dynamics, and vegetation management [31]. Other species include amphibious fish species such as lungfish (Dipnoi), Bichir (Polypteridae), Air-breathing catfish (Claridae), and snakehead fish (Channidae). These species regulate the abundance of phytoplankton, aquatic plants, and periphyton growth in tropical lake ecosystems. While the bird, African Jacana controls the abundance of aquatic invertebrates, the mudskipper regulates the abundance of prey species, preventing overconsumption and maintaining the balance of the food web. Another example is where large-bodied apex predators like alligators undertake ecosystem roles by dam-building activities, and creating complex hydrological and ecological systems that support diverse aquatic and terrestrial species [32].

The loss of keystone species due to habitat destruction, pollution, and other threats may have significant impacts on the balance and health of lake ecosystems. For instance, the decline of birds due to hunting, habitat destruction, and climate change has been reported to have cascading effects throughout the food web. Similarly, the loss of amphibious species can negatively impact nutrient cycling and the regulation of phytoplankton and aquatic plant abundance in these ecosystems. The removal of top predators which constitute the amphibious group can lead to an increase in the abundance of their prey resulting in the overconsumption of aquatic vegetation and a decline in water quality [33, 34]. Developing effective conservation strategies for keystone species, (such as restoring bird habitats and conserving top predator species), is crucial for the long-term management and sustainability of tropical lake wetland ecosystems. Conservation efforts that focus on protecting and restoring keystone species, including amphibians and amphibious species, can have cascading effects that benefit multiple species and ecosystem processes [35].

2. Anthropogenic impacts on lake morphology and biodiversity

2.1 Overview

The current state of the world's lakes is indeed alarming, and people around the world will have to make a concerted effort to reverse the trend toward degradation. Human activities have been shown to have a significant impact on the morphology and biodiversity of lakes. For example, agricultural practices, deforestation, and urbanization can cause changes in land use and land cover, leading to alterations in the hydrological regime of lakes [36]. This may result in increased sedimentation and eutrophication, with negative effects on the diversity and abundance of aquatic plants and animals [37]. Pollution from domestic, industrial, and agricultural sources can also introduce toxins and excess nutrients into lakes, causing algal blooms and other forms of ecological disturbance [38]. Furthermore, the construction of dams, canals, and other forms of water infrastructure can alter the natural flow and connectivity of rivers and lakes, and impact biodiversity and ecosystem services [39]. The fragmentation of lake habitats can lead to the isolation of populations, reducing gene flow and increasing the risk of extinction for certain species [40, 41]. Invasive species introduction, either intentionally or unintentionally, can also have a profound impact on lake biodiversity by outcompeting native species for resources and altering food webs [42]. In essence, anthropogenic activities have a significant impact on the morphology and biodiversity of lakes, and understanding these impacts along with developing strategies to mitigate them is crucial for the sustainable management of these freshwater ecosystems.

2.2 Specific anthropogenic activities and impact on Lake morphology

2.2.1 Urbanization

Urbanization refers to the process of population growth and expansion of urban areas, resulting in the conversion of natural landscapes into built-up areas. This process is accompanied by a variety of anthropogenic activities, such as land-use change, construction of buildings, and infrastructure development, that can have significant impacts on the morphology of lakes and their associated ecosystems. One of the primary impacts of urbanization on lakes is the alteration of their hydrology. The expansion of impervious surfaces, such as roads and buildings, can increase surface runoff and reduce infiltration, leading to changes in the hydrology (volume, timing, and frequency of water inputs) of the lakes [43] and resultant negative effects on lake health and its inhabitants [44]. Additionally, urbanization can lead to the destruction of natural vegetation that provides essential ecosystem services such as water purification, nutrient cycling, and erosion control, leading to reduced water quality and increased sedimentation [45]. The input of nutrients and organic matter increases the chances of eutrophication and toxic algal blooms in receiving habitats [19]. In essence, changes due to urbanization can negatively impact the diversity and abundance of aquatic species, as well as the ecosystem processes that support them. The study by Saha et al. [46] underscored the importance of considering the impacts of anthropogenic activities on lake morphology and biodiversity and the need for effective management strategies to mitigate these impacts. They found that the water quality of oxbow lake was affected by both point and non-point sources of pollution, including domestic sewage and agricultural runoff. The authors also found that the distribution of fish species within the lake was influenced by the hydrological connectivity of the lake with adjacent habitats. Fish species that were more adapted to stagnant water conditions were found in the inner parts of the oxbow lake, while species that were more adapted to flowing waters were found in the outer parts of the lake where the water was more connected to the main river channel. Urbanization also contributes to the introduction and spread of invasive species in lakes. The construction of waterways, channels, and drainage systems for urban development can facilitate the movement of non-native species, that have been observed to outcompete native species and alter the ecological balance of lake ecosystems [47]. In essence, the complex interactions between urbanization, hydrology, water quality, and biodiversity in oxbow lake ecosystems highlight the need for sustainable urban planning and management practices that minimize the negative impacts of human activities on lake ecosystems.

2.2.2 Agriculture

Agricultural practices and associated activities such as fertilizer and pesticide use, land clearing, and irrigation can result in increased sedimentation, nutrient enrichment, and water pollution in lakes [48]. These inputs may cause eutrophication, (a process whereby excessive nutrients stimulate the growth of algae and other aquatic plants), ultimately leading to oxygen depletion and fish kills [49]. In addition, irrigation practices can reduce water levels in lakes, altering lake morphology and reducing water availability for other uses [50]. Livestock grazing, crop production, and the use of fertilizers and pesticides may also have negative impacts on lake ecosystems (as runoff from agricultural lands laden with excess nutrients such as nitrogen and phosphorus), leading to eutrophication and harmful algal blooms [51]. Pesticides are a class of endocrine disrupters with reported estrogenic effects and modulated vitellogenin production in male and female aquatic species [52, 53]. Other effects resulting from vitellogenin induction in male species include kidney failure and impairment of reproductive success, increasing the risks of declines in local populations' biodiversity [54, 55].

To mitigate the impact of agricultural practices on lake ecosystems, best management practices (BMPs) have been developed to reduce nutrient and sediment runoff from agricultural lands. These BMPs include practices such as reducing fertilizer application rates, implementing cover crops, and maintaining vegetative buffer strips along streams and lakes [56]. Additionally, the implementation of water conservation measures in agricultural practices, such as drip irrigation and precision agriculture, may reduce water use and increase water availability for other uses [57]. By implementing these practices, the negative impact of agriculture on lake morphology and biodiversity can be minimized, allowing for the sustainable use of these important ecosystems.

The agricultural sector is crucial to the economy of many African countries, and it accounts for a significant portion of their GDP, hence its widespread practice drives unregulated chemical applications, overuse of water resources for irrigation and other agricultural purposes leading to the depletion of lakes and other freshwater ecosystems. As such the negative impacts of agriculture on lake habitats may outweigh the economic benefits of agricultural production. Therefore, adopting integrated approaches that consider both the environmental and economic implications of agricultural practices are necessary to achieve sustainable agriculture and promote the protection and conservation of lake habitats.

2.2.3 Mining

Mining activities such as excavation, blasting, and sedimentation alter the physical and chemical properties of lake ecosystems [58]. The discharge of toxic chemicals and heavy metals from mining activities also has reported harmful effects on aquatic organisms and their habitats [59, 60]. For example, the gold mining industry in the Amazon basin has been linked to high levels of mercury contamination in local waterways and aquatic food webs, posing a threat to the health of human populations that rely on these resources for sustenance [61]. In addition to contaminating water bodies, mining activities can also lead to habitat destruction and fragmentation through the construction of mines and access roads, leading to the isolation of different species and populations and further reducing the biodiversity of lake ecosystems

[62]. Soil erosion caused by mining can increase sedimentation in nearby water bodies and reduce water quality, further degrading the habitat. The use of heavy machinery and chemicals like cyanide and mercury also contributes to soil and water quality degradation, leading to the loss of vegetation cover and contamination of water bodies, affecting aquatic species diversity. Mining can cause significant disturbances to the soil and subsurface habitats, altering soil structure and composition, microbial communities, and nutrient availability, resulting in a reduction in the biodiversity of lake ecosystems. The excavation of minerals leads to increased sedimentation and erosion, with negative impacts on the health of lakes and resident organisms [63]. The disposal of mining waste can also lead to the release of heavy metals and other toxic substances (organics) from subsurface sediments into the surface waters of lakes [64].

Efforts to mitigate the impacts of mining on lake ecosystems include the implementation of best management practices, such as the use of sediment traps and the reduction of chemical usage in mining operations [65]. However, the effectiveness of these measures is often limited by weak regulatory frameworks and insufficient enforcement mechanisms [66], especially in developing countries. In order to maintain the long-term health and sustainability of lake ecosystems affected by mining activities, it is important to adopt a comprehensive approach that combines sustainable mining practices with effective governance and monitoring. The negative impacts of mining activities on lake ecosystems can lead to a reduction in the diversity and abundance of aquatic species, as well as changes in ecosystem processes that support them. Therefore, management practices that minimize the impacts of these activities on lake ecosystems should be implemented to maintain their health and sustainability.

Anthropogenic activities, such as urbanization, agriculture, and mining, have been found to have significant impacts on lake morphology and biodiversity [5]. These activities can lead to changes in water quality, habitat degradation, and increased sedimentation, resulting in decreased biodiversity and harm to aquatic organisms. Additionally, climate change exacerbates the impacts of anthropogenic activities by altering water cycles, lake chemistry, and increasing water temperatures [67] (**Box 1**). Therefore, to preserve lake biodiversity, conservation efforts must consider the impacts of anthropogenic activities and their effects on habitat and food sources for aquatic organisms and other species dependent on lentic water systems, taking into account recommendations from previous studies.

1. Eutrophication: excessive nutrient enrichment of water can lead to harmful algal blooms and oxygen depletion, ultimately causing death of fish and other aquatic organisms.
2. Sedimentation: the deposition of sediment on the lake bed, which causing turbidity, alter the nutrient cycle, and reduce light penetration, affecting aquatic plants and algae.
3. Shoreline alteration: the modification of the lake's natural shoreline, leading to habitat loss, fragmentation, and changes in vegetation, which impacts wildlife diversity and abundance
4. Habitat destruction and fragmentation- the clearing of natural vegetation around the lake, impacting the distribution and abundance of species.

5. Introduction of non-native species: outcompete native species and disrupt the lake's ecological balance

6. Climate change: alterations in temperature, and precipitation impacts hydrological regimes around the lake

7. Overfishing: impacts the structure and function of the food web within the lake

Box 1.
Threats to biodiversity through altered lake morphology.

2.3 Lake morphology as a diagnostic for habitat degradation

Pressures on lakes are unevenly distributed around the world, with the most severe lake health problems across the entire continent of Africa and other densely-populated low-income countries. There is also a need for better diagnostics, in particular in low-income countries. Diagnostics include the implementation of a standardized classification system for lake health conditions. To do this we suggest focusing on a few key variables that directly reflect lake health, with empirical support and validation from The Ramsar Convention on Wetlands, an intergovernmental treaty that provides a framework for national action and international cooperation for the conservation and wise use of wetlands and their resources (**Box 2**). One of its objectives is to promote the conservation of wetlands, including lakes, through wise use and management practices that take into account ecological, social, cultural, and economic considerations. The Convention emphasizes the importance of monitoring, assessing, and reporting on the status and trends of wetlands, including their ecological character,

Lake Habitat Degradation	Altered Morphology	Landscape Variables
1 (minimal)	• No significant changes in lake morphology. • Natural shoreline with no or minimal human-made structures.	• High percentage of undisturbed natural vegetation in the surrounding landscape. • Presence of a continuous riparian buffer of at least 30 meters from the shoreline, consisting of native vegetation.
2 (slight)	• Slight alteration of lake morphology, such as minor sedimentation or nutrient enrichment. • Shoreline may have minor human-made structures, such as docks or small retaining walls.	• Moderate percentage of undisturbed natural vegetation in the surrounding landscape. • Presence of a riparian buffer of at least 20 meters from the shoreline, consisting of native vegetation.
3 (moderate)	• Moderate alteration of lake morphology, such as moderate sedimentation or nutrient enrichment.	• Low percentage of undisturbed natural vegetation in the surrounding landscape. • Presence of a riparian buffer of at least 10 meters from the shoreline, consisting of native vegetation.

4 (severe)	• Shoreline may have significant human-made structures, such as bulkheads or seawalls. • Severe alteration of lake morphology, such as significant sedimentation or nutrient enrichment. • Shoreline may have extensive human-made structures, such as large marinas or commercial development.	• Minimal percentage of undisturbed natural vegetation in the surrounding landscape. • Presence of a narrow or intermittent riparian buffer, consisting of non-native or disturbed vegetation.
5 (critical)	• Critical alteration of lake morphology, such as complete loss of natural shoreline or severe pollution.	• Absence of natural vegetation in the surrounding landscape. • Absence of a riparian buffer, or presence of a degraded buffer consisting of non-native or disturbed vegetation.

Box 2.
Lake health diagnostics.

functions, and services. The Convention also guides the development of management plans and the implementation of conservation measures for wetlands, including lakes.

In addition, the United Nations Sustainable Development Goals (SDGs), particularly SDG 6 on clean water and sanitation, provided additional empirical support and validation for the suggested diagnostics. SDG 6 aims to ensure the availability and sustainable management of water and sanitation for all, including the protection and restoration of water-related ecosystems, such as lakes. The SDGs provide a framework for countries to set targets and indicators for measuring progress toward sustainable development, including the conservation and management of lakes and their biodiversity.

2.4 Lake morphology change and biodiversity loss: Focus on bird diversity

The link between changes in lake morphology and biodiversity loss is well established, with numerous studies highlighting the negative impacts of anthropogenic activities on biodiversity in lake ecosystems. For example, Chukwuka et al. [1] observed that urbanization, agriculture, and mining activities all have significant impacts on the morphology of tropical lakes and wetlands, which may have cascading effects on the biodiversity of these ecosystems. Bird diversity is one aspect of biodiversity that has been particularly affected by changes in lake morphology. Birds are important keystone species in lake ecosystems and play a crucial role in maintaining ecosystem processes such as nutrient cycling and food webs. However, urbanization and other anthropogenic activities have led to habitat loss and fragmentation, as well as changes in water quality and hydrology, with negative impacts on bird populations [68].

For example, the construction of dams and reservoirs for hydroelectric power generation and irrigation has resulted in the fragmentation of bird habitats and reduced the availability of suitable nesting sites, leading to declines in bird populations in some areas [69, 70]. Similarly, the conversion of wetlands to agricultural land or urban areas could result in the loss of important feeding and breeding grounds for birds, which also contribute to declines in bird diversity [71]. In addition to habitat loss and fragmentation, changes in water quality and hydrology also have negative impacts on bird populations. For example, the discharge of pollutants such as nutrients, pesticides, and heavy metals into lakes and wetlands can lead to eutrophication and toxic algal blooms, which can directly or indirectly impact bird populations [72]. Changes in hydrology such as altered water flow and water level fluctuations can also impact bird populations by altering the availability of suitable foraging and breeding habitats [73]. Therefore, it is important to recognize the link between changes in lake morphology and bird diversity loss and to develop effective conservation strategies that focus on preserving and restoring bird habitats and populations in lake ecosystems.

3. Lake morphology and landscape co-factors affecting biodiversity

The morphology of a lake is a key factor in determining the biodiversity of the aquatic ecosystem. Physical characteristics such as water depth, shoreline complexity, and substrate composition have a significant impact on the distribution, abundance, and diversity of species in lakes [74]. Changes in lake morphology, particularly those caused by anthropogenic activities, can have significant impacts on biodiversity. One of the most significant impacts of changes in lake morphology is the loss of shoreline habitats such as wetlands and riparian zones. These habitats are important for supporting a diverse array of aquatic and terrestrial species, including birds [75]. Wetlands offer critical nesting, foraging, and roosting areas for various bird species, particularly those that rely on shallow water habitats. Riparian zones are equally essential for bird diversity, serving as significant sources of food and habitats for various bird species, including migratory species that use them as stopover sites [76, 77].

Lake morphology can have a significant impact on the biodiversity of aquatic and terrestrial species. For example, shallow lakes may support higher plant and fish diversity than deep lakes [21]. Furthermore, surrounding landscape variables such as land use, vegetation cover, and water quality can influence biodiversity in lakes. For example, agricultural land use surrounding lakes has been linked to decreased amphibian diversity, while forested land use has been associated with increased amphibian diversity [78].

Changes in water depth and substrate composition can also have significant impacts on bird diversity in lakes. For example, shallow water habitats such as marshes and wetlands are important for supporting many bird species, particularly waterfowl and wading birds. Changes in water depth due to dredging or filling can result in the loss of these important habitats, leading to declines in bird diversity [79]. Similarly, changes in substrate composition due to sedimentation or erosion can impact bird diversity by altering the availability of food and nesting materials. Overall, changes in lake morphology can have significant impacts on bird diversity and the broader aquatic ecosystem. Conservation efforts that focus on maintaining and restoring important shoreline habitats, preserving water depth and substrate composition, and reducing the impacts of anthropogenic activities can help to maintain the health and resilience of lakes and their associated ecosystems.

3.1 Loss of riparian vegetation and lake morphology

Loss of riparian vegetation and alterations in lake morphology due to human activities may have significant impacts on lake ecosystems and their biodiversity. Riparian vegetation plays a crucial role in maintaining the physical, chemical, and biological integrity of lakes by reducing erosion, filtering pollutants, and providing habitat and food for aquatic and terrestrial organisms [80]. However, land use changes such as deforestation, urbanization, and agriculture have resulted in the loss and degradation of riparian vegetation around many lakes, leading to reduced water quality, altered hydrological regimes, and changes in the structure and function of lake ecosystems [81]. Anthropogenic alterations to lake morphology, such as shoreline modification, dredging, and damming, can also affect the physical and chemical characteristics of lakes and alter available habitats for aquatic organisms [82]. These changes can ultimately lead to a decline in lake biodiversity and ecosystem services.

Studies have demonstrated the negative impacts of riparian vegetation loss and alterations in lake morphology on lake biodiversity and ecosystem functioning [83]. Lewin et al. [84] documented that the development of residential areas along lake shores often leads to the conversion of natural littoral habitats to various structures such as riprap, sheet piles, beaches, parks, or marinas, driving the loss of littoral vegetation, which negatively affects the structural diversity and fish communities in the littoral zone. Litterfall from riparian vegetation is a significant source of organic matter for benthic and pelagic habitats in lakes [85]. The impact of this input depends on various factors such as the riparian habitat's characteristics, shoreline complexity, and the overall productivity of the aquatic ecosystem. While most organic inputs come from litterfall, terrestrial insects sustained by riparian areas can also provide a substantial source of prey for aquatic predators and contribute tothe lake nutrient cycle in some cases [86, 87]. Furthermore, Saha et al. [46] and Chukwuka et al. [1] demonstrated the role of hydrological connectivity, water quality, and landscape structure in shaping the distribution and abundance of fish and bird communities in urban oxbow lakes and wetlands. In this light, it is imperative that conservation and restoration plans for lake ecosystems should acknowledge the importance of riparian vegetation and lake morphology and account for the complex interplay between physical, chemical, and biological factors affecting lake biodiversity and ecosystem services.

3.2 Adjacent landscape variables and lake biodiversity

The surrounding landscape variables of a lake can also have a significant impact on the biodiversity of the lake ecosystem. The composition and structure of surrounding vegetation, as well as the presence of nearby habitats such as forests, wetlands, and agricultural lands, can influence the diversity and abundance of species in the lake [21]. For example, studies have shown that lakes surrounded by diverse and extensive vegetation have higher species diversity than lakes surrounded by sparse vegetation [88, 89]. This is attributable to surrounding vegetation that provides food and shelter for lake species and contributes to the regulation of water quality and temperature. Similarly, adjacent wetlands and forests also sustain bird species, providing nesting sites and foraging areas [90]. The loss or degradation of these habitats due to land-use changes portends negative effects on the diversity and abundance of bird species within lake ecosystems [91].

3.3 Keystone species as ecological engineers

Like other animals, amphibians (salamanders, frogs, toads, and caecilians) are affected by numerous environmental stressors that often act in complex ways [92]. However, in at least some regions, amphibian losses appear to be more severe than losses in other vertebrate taxa [93, 94]. Amphibious species, on the other hand, include lungfishes, mudskippers, newts, otters, herons, crocodilians, and alligators. It is important to note that the large-bodied predators among amphibious species are highly recognized as both keystone species (which contribute to nutrient and energy translocation across ecosystems) and highlighted as ecosystem engineers. Keystone species that function as ecosystem engineers are relevant to the concept of lake morphology dynamics. This is because, these organisms have the ability to create, modify, and maintain physical attributes of habitats in ways that can affect the distribution, life histories, behaviors, and abundance of other species within that ecosystem [95]. Crocodilians (alligators and crocodiles) perform the keystone function of maintaining ecosystem balance by controlling the population growth of prey species, maintaining residual waterholes during dry periods, and inhibiting encroachment of aquatic plants. However, they can modify habitats in ways that can influence the distribution and abundance of other species. They excavate open holes, dens, and tunnels that serve as refuges from environmental extremes and predation and can also store fresh water. Mound-nesting species can build elevated structures or construct nests on floating vegetation, creating elevated "islands" [32]. These modifications can have a significant impact on lake morphology, such as shaping shorelines and affecting water flow patterns. The excavation of large holes, tunnels, and the construction of mound nests by crocodylians can result in significant modifications to the landscape. The holes can be more than 20 meters in diameter and have depths greater than 1 meter, while tunnels can reach over 50 meters in length [32]. Mound nests can be as large as 7 meters in diameter and 1 meter in height. The pathways used by crocodylians between these features can also create depressions that can retain water during dry periods. These modifications have the potential to strongly influence topographic heterogeneity within the landscape due to the substantial size of the habitat features and the total area of physical modifications created by crocodylians [32].

Considering the role of crocodylians in habitat modification, we can infer that population dynamics for fluctuations of nesting crocodylians for instance can amplify their effects on the landscape and lake morphology by extension. While both amphibians and amphibious vertebrates play a crucial role as keystone species in the biodiversity of lakes and wetlands, [96] the morphological characteristics of lakes, such as size, depth, and shoreline complexity, as well as surrounding landscape variables, are driven and largely determined by the large-bodied amphibious species [97]. This brings to bear the concept of "keystone structures" which refers to the significance of habitat heterogeneity in creating areas of high species richness. This concept has been extensively studied in terrestrial environments but has received less attention in freshwater systems. Keystone structures can concentrate species activity and thus influence density and biodiversity. For instance, the presence of complex shorelines, which provide a variety of niches and microhabitats for amphibians, is positively correlated with their abundance and diversity [98]. In addition, larger and deeper lakes are associated with higher amphibian species richness [99]. The importance of habitat complexity is further seen in the presence of other keystone structures in freshwater systems including weed beds, patches of gravel or rock outcrops, and deep pools in lotic systems, which provide refuges to different sizes of crayfish and their predators [100].

While the loss of amphibious keystone species in lakes and wetlands may have far reaching ecological consequences, the decline of amphibian populations could also impact the structure and function of food webs in these ecosystems via an unregulated abundance of invertebrate species [101]. In addition, the loss of amphibians can have negative impacts on nutrient cycling, as they play a critical role in the regulation of organic matter decomposition and nutrient release [102]. Thus, incorporating measures to protect and enhance these species can contribute to the sustainability of these ecosystems and the provision of essential ecosystem services.

4. Lake conservation and multidisciplinary approaches

4.1 Importance of a multidisciplinary approach to lake conservation

Managing a lake as an environmental indicator poses significant challenges due to the need to address complex technological, financial, and institutional issues. It requires support from both the public and industry, which may prioritize short-term economic gains over long-term environmental sustainability. Effective lake management requires careful consideration of various factors, including monitoring and assessment of water quality, management of surrounding land use, and regulation of pollutant discharge into the lake. In addition, stakeholder involvement and collaboration are crucial for developing effective management plans that balance economic and environmental concerns. A multidisciplinary approach to lake conservation is critical to addressing the complex challenges facing these ecosystems. The management of lakes requires collaboration between various fields, including ecology, hydrology, limnology, geography, and social sciences. The integration of these disciplines allows for a comprehensive understanding of the environmental, social, and economic factors that affect lake ecosystems. Studies have highlighted the importance of considering the social context of lake management and found that the attitudes and perceptions of residents toward lake management influenced the success of conservation efforts [103, 104]. Aside from aiding in the development of sustainable and adaptive management strategies, integrating ecological and social knowledge, lake management can be tailored to address the unique challenges and goals of each ecosystem [105]. This approach has accounted for the interconnectedness of social and ecological systems and identified trade-offs between conservation and human well-being [106]. Furthermore, a multidisciplinary approach can help address the uncertainties and complexities associated with lake ecosystems. The integration of multiple perspectives and data sources can also improve the accuracy and robustness of predictions and models used in management decisions. By working together, scientists, managers, and stakeholders can develop effective and sustainable strategies that balance the conservation of biodiversity with the needs of local communities.

4.2 Ecological, hydrological, and social factors in conservation strategies

Effective management of lake health requires a holistic approach, recognizing the interconnectedness of ecological, hydrological, and social factors. Ecological factors, such as the physical and biological characteristics of a lake and its surrounding landscape, and hydrological factors (referring to the water cycle and human impacts on water quality and quantity) must be considered alongside social factors, including the cultural, economic, and political context in which the lake exists. To address the complex interactions that affect lake ecosystems, a multidisciplinary approach is necessary.

A multidisciplinary approach recognizes that lake ecosystems are complex and interconnected systems that require a holistic understanding of the interactions between biotic and abiotic factors, human activities, and their impacts on lake ecosystems [107]. Therefore, ecological studies that assess the physical and biological features of a lake should be complemented by hydrological studies that examine water flow dynamics and quality, as well as social studies that identify the attitudes, perceptions, and behaviors of local communities toward a lake and its resources [108]. Understanding the hydrological aspects including water availability, flow, and quality, and how these are affected by changes in land use and climate are crucial factors for developing strategies to mitigate the negative impact of human activities on lakes. These may include reducing the use of synthetic fertilizers in agriculture to prevent nutrient pollution or designing stormwater management systems to minimize runoff into lakes. On the other hand, ecological research can help identify the key drivers of ecological change and determine the appropriate conservation interventions needed to maintain or restore healthy lake ecosystems.

According to de Bisthoven et al. [109] study on the socio-ecological assessment of the Lake Manyara basin in Tanzania, they observed that improved water governance through a multi-actor approach (with a focus on distributing benefits and rights and assigning specific roles to water authorities) should be a priority for future integrated management strategies. Additionally, they highlighted the need to raise awareness among decision-makers, scientists, and local communities about the advantages of an integrated approach. Elsewhere, a lake management strategy based on both bio-physical and socio-economic aspects which also adopts a watershed/ecosystem approach at the policy level, integrates income generation in conservation activities, sharing responsibility and benefits among local stakeholders, and enhancing management capacity development and institutional strengthening for sustainable development was recommended [110]. In essence, a participatory approach involving multidisciplinary stakeholders could assist in ensuring successful lake conservation and management.

4.3 The importance of keystone species in lake conservation efforts

Lake conservation efforts have traditionally focused on preserving individual species and maintaining biodiversity. However, an approach that considers the role of keystone species in maintaining ecosystem health is becoming increasingly important [111]. Keystone species have a disproportionately large impact on the structure and function of an ecosystem relative to their abundance. They often play a critical role in maintaining ecosystem processes such as nutrient cycling, pollination, and predation [112]. In the context of lake ecosystems, amphibians and amphibious vertebrates (including birds, turtles, and crocodylia) control the flow of trophic resources including phytoplankton and aquatic plants, and also stimulates the growth of periphyton [113].

Conserving keystone species is essential for maintaining the health and resilience of lake ecosystems. Loss of keystone species can have cascading effects throughout the ecosystem, leading to declines in biodiversity and the disruption of ecosystem services [111]. For example, the loss of fish species in lakes can lead to an increase in the abundance of phytoplankton and algae, which can result in eutrophication and oxygen depletion, ultimately leading to a decline in biodiversity [114]. In addition, amphibians such as frogs and salamanders are sensitive indicators of environmental health and can provide early warning signals of changes in ecosystem function [115]. To effectively conserve keystone species in lakes, a multidisciplinary approach is necessary. This approach involves collaboration among ecologists, hydrologists, social scientists, and policymakers to develop conservation strategies that consider the ecological, hydrological, and social factors that influence the health and resilience of lake ecosystems [111]. For example, conservation efforts may involve regulating land use practices such as urbanization, agriculture, and mining that can negatively impact lake ecosystems and their keystone species. Additionally, monitoring and research programs to identify and track changes in keystone species populations and their roles in ecosystem function can inform conservation planning and management toward enhancing the resilience and sustainability of lake ecosystems [115].

5. Conclusion

Lakes are crucial ecosystems that are pivotal in conserving biological diversity. A plethora of factors influence their morphology and biodiversity, including human activities like urbanization, agriculture, and mining, as well as lake morphology and landscape features. These activities lead to changes in water quality, vegetation cover, and habitat structure, all of which affect the survival of various species in the lake. Moreover, changes in lake morphology affect the distribution and abundance of keystone species, such as fish and amphibians, which are essential for maintaining the ecosystem's functioning and health. Such species also play an important role in maintaining the balance of lake wetland ecosystems. To ensure the effective protection of lake ecosystems and their biodiversity, a multidisciplinary approach that integrates ecological, hydrological, and social factors in conservation strategies is essential. In addition, keystone species must be prioritized in conservation strategies since their decline can have cascading effects throughout the ecosystem. The overall health of lake ecosystems is critical for human well-being, as these ecosystems provide numerous services, such as water supply, fisheries, and recreation. Therefore, sustainable management practices that prioritize and strike a balance between biodiversity conservation in lake ecosystems while meeting human needs should be implemented.

Given the complex interactions between lake morphology, landscape variables, and anthropogenic impacts on biodiversity, a multidisciplinary approach is crucial for effective lake conservation. Ecological, hydrological, and social factors should all be considered when designing conservation strategies, while anthropogenic activities such as urbanization, agriculture, and mining should also be taken into account. Keystone species, such as amphibians and amphibious vertebrates, are important for maintaining lake ecosystem health and should be a top priority for conservation efforts, which must involve not only ecologists and biologists but also social scientists and policymakers. To ensure the success of conservation practices, it is also essential to engage local communities and incorporate indigenous knowledge and perspectives. Strategic conservation practices that prioritize the protection of keystone species and their habitats while balancing human needs and environmental sustainability are necessary to ensure the long-term health and sustainability of lakes as vital resources.

Author details

Aina O. Adeogun[1*] and Azubuike V. Chukwuka[2]

1 Department of Zoology, University of Ibadan, Ibadan, Nigeria

2 Department of Environmental Quality Control, National Environmental Standards and Regulations Enforcement Agency (NESREA), Nigeria

*Address all correspondence to: ao.adeogun@ui.edu.ng; ainaadeogun@yahoo.com

References

[1] Chukwuka AV, Egware TU, Okali KD, Fadahunsi AA, Oluwakotanmi PG, Emasoga P, et al. The influence of Lake morphology, landscape structure, and urbanization factors on Bird community composition in wetlands of four Tropical Lakes. Wetlands. 2022;**42**:1-15

[2] Adeogun AO, Ibor OR, Khan EA, Chukwuka AV, Omogbemi ED, Arukwe A. Detection and occurrence of microplastics in the stomach of commercial fish species from a municipal water supply lake in southwestern Nigeria. Environmental Science and Pollution Research. 2020;**27**: 31035-31045

[3] Zhang Y, Deng J, Qin B, Zhu G, Zhang Y, Jeppesen E, et al. Importance and vulnerability of lakes and reservoirs supporting drinking water in China. Fundamental Research. 2023;**3**(2):265-273

[4] Biggs J, Von Fumetti S, Kelly-Quinn M. The importance of small waterbodies for biodiversity and ecosystem services: Implications for policy makers. Hydrobiologia. 2017;**793**:3-39

[5] Jeppesen E, Meerhoff M, Davidson T, Trolle D, Sondergaar M, Lauridsen TL, et al. Climate change impacts on lakes: an integrated ecological perspective based on a multi-faceted approach, with special focus on shallow lakes. Journal of Limnology. 2014;**73**(1s):88-111

[6] Power ME, Tilman D, Estes JA, Menge BA, Bond WJ, Mills LS, et al. Challenges in the quest for keystones: Identifying keystone species is difficult—But essential to understanding how loss of species will affect ecosystems. Bioscience. 1996;**46**: 609-620

[7] Zavaleta E, Pasari J, Moore J, Hernandez D, Suttle KB, Wilmers CC. Ecosystem responses to community disassembly. Annals of the New York Academy of Sciences. 2009;**1162**: 311-333

[8] Dobson A, Lodge D, Alder J, Cumming GS, Keymer J, McGlade J, et al. Habitat loss, trophic collapse, and the decline of ecosystem services. Ecology. 2006;**87**:1915-1924

[9] Mujere N, Moyce W. Climate change impacts on surface water quality. In: Hydrology and Water Resource Management: Breakthroughs in Research and Practice. Hershey, PA: IGI Global; 2018. pp. 97-115

[10] Liu J, Kattel G, Arp HPH, Yang H. Towards threshold-based management of freshwater ecosystems in the context of climate change. Ecological Modelling. 2015;**318**:265-274

[11] Sayer CD. Conservation of aquatic landscapes: Ponds, lakes, and rivers as integrated systems. Wiley Interdisciplinary Reviews: Water. 2014; **1**:573-585

[12] Darwall W, Bremerich V, De Wever A, Dell AI, Freyhof J, Gessner MO, et al. The Alliance for freshwater life: A global call to unite efforts for freshwater biodiversity science and conservation. Aquatic Conservation: Marine and Freshwater Ecosystems. 2018;**28**:1015-1022

[13] Arthington AH, Dulvy NK, Gladstone W, Winfield IJ. Fish conservation in freshwater and marine realms: Status, threats and management. Aquatic Conservation: Marine and Freshwater Ecosystems. 2016;**26**: 838-857

[14] Ho LT, Goethals PL. Opportunities and challenges for the sustainability of lakes and reservoirs in relation to the sustainable development goals (SDGs). Water. 2019;**11**:1462

[15] Scheffer M, van Nes EH. Shallow lakes theory revisited: Various alternative regimes driven by climate, nutrients, depth and lake size. In: Shallow Lakes in a Changing World: Proceedings of the 5th International Symposium on Shallow Lakes, Held at Dalfsen, the Netherlands, 5-9 June 2005. Springer; 2007. pp. 455-466

[16] Stefanidis K, Papastergiadou E. Relationships between lake morphometry, water quality, and aquatic macrophytes, in greek lakes. Fresenius Environmental Bulletin. 2012; **21**:3018-3026

[17] Donohue I, Garcia Molinos J. Impacts of increased sediment loads on the ecology of lakes. Biological Reviews. 2009;**84**:517-531

[18] Leira M, Cantonati M. Effects of water-level fluctuations on lakes: An annotated bibliography. In: Wantzen KM, Rothhaupt K-O, Mörtl M, Cantonati M, Tóth LG, Fischer P, editors. Ecological Effects of Water-Level Fluctuations in Lakes. Springer; 2008. pp. 171-184

[19] Paerl HW, Gardner WS, Havens KE, Joyner AR, McCarthy MJ, Newell SE, et al. Mitigating cyanobacterial harmful algal blooms in aquatic ecosystems impacted by climate change and anthropogenic nutrients. Harmful Algae. 2016;**54**:213-222

[20] Wurtsbaugh WA, Paerl HW, Dodds WK. Nutrients, eutrophication and harmful algal blooms along the freshwater to marine continuum. Wiley Interdisciplinary Reviews: Water. 2019; **6**:e1373

[21] Meerhoff M, de los Ángeles González-Sagrario M. Habitat complexity in shallow lakes and ponds: Importance, threats, and potential for restoration. Hydrobiologia. 2022;**849**: 3737-3760

[22] Macháček J. Typology of environmental impacts of artisanal and small-scale mining in African Great Lakes region. Sustainability. 2019;**11**: 3027

[23] Du X, Song D, Ming K, Jin X, Wang H, Wang L, et al. Response of macroinvertebrate communities to land use and water quality in Wudalianchi Lake. Ecology and Evolution. 2021;**11**: 1368-1377

[24] Cattaneo A, Couillard Y, Wunsam S, Fortin C. Littoral diatoms as indicators of recent water and sediment contamination by metals in lakes. Journal of Environmental Monitoring. 2011;**13**: 572-582

[25] Birch WS, Drescher M, Pittman J, Rooney RC. Trends and predictors of wetland conversion in urbanizing environments. Journal of Environmental Management. 2022;**310**:114723

[26] Gross C, Hagy JD III. Attributes of successful actions to restore lakes and estuaries degraded by nutrient pollution. Journal of Environmental Management. 2017;**187**:122-136

[27] Assessment ME. Ecosystems and Human Well-Being: Wetlands and Water. Washington, DC: World Resources Institute; 2005. 248 p

[28] Gunderson LH, Pritchard L. Resilience and the Behavior of Large-Scale Systems. Washington, DC: Island

Press; 2012. 240 p. DOI: 10.5822/978-1-61091-219-8. ISBN-13: 978-1-61091-219-8

[29] Woolbright LL. Disturbance influences long-term population patterns in the Puerto Rican frog, Eleutherodactylus coqui (Anura: Leptodactylidae). Biotropica. 1996;**28**(4):493-501

[30] Kennedy G, Mayer T. Natural and constructed wetlands in Canada: An overview. Water Quality Research Journal. 2002;**37**:295-325

[31] Soni HB. Biodiversity of Wetlands and Forests: A Nature Trail. Google Book Publishers; 2020. 165 p. DOI: 10.1002/jwmg.21804. E-book

[32] Somaweera R, Nifong J, Rosenblatt A, Brien ML, Combrink X, Elsey RM, et al. The ecological importance of crocodylians: Towards evidence-based justification for their conservation. Biological Reviews. 2020;**95**:936-959

[33] Carignan V, Villard M-A. Selecting indicator species to monitor ecological integrity: A review. Environmental Monitoring and Assessment. 2002;**78**:45-61

[34] Hossack BR, Gould WR, Patla DA, Muths E, Daley R, Legg K, et al. Trends in Rocky Mountain amphibians and the role of beaver as a keystone species. Biological Conservation. 2015;**187**:260-269

[35] Johnson SA, Ober HK, Adams DC. Are keystone species effective umbrellas for habitat conservation? A spatially explicit approach. Journal for Nature Conservation. 2017;**37**:47-55

[36] Badar B, Romshoo SA, Khan M. Modelling catchment hydrological responses in a Himalayan Lake as a function of changing land use and land cover. Journal of Earth System Science. 2013;**122**:433-449

[37] Carlson Mazur ML, Smith B, Bird B, McMillan S, Pyron M, Hauswald C. Hydrologic connectivity and land cover affect floodplain lake water quality, fish abundance, and fish diversity in floodplain lakes of the Wabash-White River basin. River Research and Applications. 2022;**38**:160-172

[38] Bhateria R, Jain D. Water quality assessment of lake water: A review. Sustainable Water Resources Management. 2016;**2**:161-173

[39] Reid AJ, Carlson AK, Creed IF, Eliason EJ, Gell PA, Johnson PT, et al. Emerging threats and persistent conservation challenges for freshwater biodiversity. Biological Reviews. 2019;**94**:849-873

[40] Danley PD, Markert JA, Arnegard ME, Kocher TD. Divergence with gene flow in the rock-dwelling cichlids of Lake Malawi. Evolution. 2000;**54**:1725-1737

[41] De Meester L, Gómez A, Okamura B, Schwenk K. The monopolization hypothesis and the dispersal–gene flow paradox in aquatic organisms. Acta Oecologica. 2002;**23**:121-135

[42] Strayer DL. Alien species in fresh waters: Ecological effects, interactions with other stressors, and prospects for the future. Freshwater Biology. 2010;**55**:152-174

[43] Hobbie SE, Grimm NB. Nature-based approaches to managing climate change impacts in cities. Philosophical Transactions of the Royal Society B. 2020;**375**:20190124

[44] Pham SV, Leavitt PR, McGowan S, Peres-Neto P. Spatial variability of climate and land-use effects on lakes of the northern Great Plains. Limnology and Oceanography. 2008;**53**:728-742

[45] Decocq G, Andrieu E, Brunet J, Chabrerie O, De Frenne P, De Smedt P, et al. Ecosystem services from small forest patches in agricultural landscapes. Current Forestry Reports. 2016;**2**:30-44

[46] Saha S, Chukwuka AV, Mukherjee D, Saha NC, Adeogun AO. Hydrological connectivity, surface water quality and distribution of fish species within sub-locations of an urban oxbow lake, East India. Watershed Ecology and the Environment. 2022;**4**:44-58

[47] Zhan A, Zhang L, Xia Z, Ni P, Xiong W, Chen Y, et al. Water diversions facilitate spread of non-native species. Biological Invasions. 2015;**17**: 3073-3080

[48] Board LWS. Reducing Nutrient Loading to Lake Winnipeg and its Watershed: Our Collective Responsibility and Commitment to Action. Winnipeg, MB: Lake Winnipeg Stewardship Board; 2006. 160 p

[49] Camargo JA, Alonso Á. Ecological and toxicological effects of inorganic nitrogen pollution in aquatic ecosystems: A global assessment. Environment International. 2006;**32**:831-849

[50] Council NR. Restoration of Aquatic Ecosystems: Science, Technology, and Public Policy. Washington, DC: National Academies Press; 1992. 432 p

[51] Smith VH, Schindler DW. Eutrophication science: Where do we go from here? Trends in Ecology & Evolution. 2009;**24**:201-207

[52] Adeogun AO, Ibor OR, Adeduntan SD, Arukwe A. Intersex and alterations in reproductive development of a cichlid, Tilapia guineensis, from a municipal domestic water supply lake (Eleyele) in southwestern Nigeria. Science of the Total Environment. 2016; **541**:372-382

[53] Adeogun AO, Ibor OR, Chukwuka AV, Regoli F, Arukwe A. The intersex phenomenon in Sarotherodon melanotheron from Lagos lagoon (Nigeria): Occurrence and severity in relation to contaminants burden in sediment. Environmental Pollution. 2019;**244**:747-756

[54] Chukwuka A, Ogbeide O, Uhunamure G. Gonad pathology and intersex severity in pelagic (tilapia zilli) and benthic (Neochanna diversus and Clarias gariepinus) species from a pesticide-impacted agrarian catchment, south-south Nigeria. Chemosphere. 2019;**225**:535-547

[55] Ogbeide O, Uhunamure G, Uwagboe L, Osakpamwan T, Glory M, Chukwuka A. Comparative gill and liver pathology of tilapia zilli, Clarias gariepinus and Neochanna diversus in owan river (Nigeria): Relative ecological risks of species in a pesticide-impacted river. Chemosphere. 2019;**234**:1-13

[56] Singh G, Kaur G, Williard K, Schoonover J, Nelson KA. Managing phosphorus loss from agroecosystems of the Midwestern United States: A review. Agronomy. 2020;**10**:561

[57] Bwambale E, Abagale FK, Anornu GK. Smart irrigation monitoring and control strategies for improving water use efficiency in precision agriculture: A review. Agricultural Water Management. 2022;**260**:107324

[58] Sergeant CJ, Sexton EK, Moore JW, Westwood AR, Nagorski SA, Ebersole JL, et al. Risks of mining to

salmonid-bearing watersheds. Science. Advances. 2022;**8**:eabn0929

[59] Jerome FC, Hassan AA, Chukwuka AV. Sex-specific affinity for redox-active metals influences antioxidant responses of Callinectes amnicola (blue crab) populations in littoral and open water habitats of a tropical coastal lagoon. Marine Ecology. 2017;**38**:e12437

[60] Nriagu JO. Global metal pollution: Poisoning the biosphere? Environment: Science and Policy for Sustainable Development. 1990;**32**:7-33

[61] de Bakker LB, Gasparinetti P, de Queiroz JM, de Vasconcellos ACS. Economic impacts on human health resulting from the use of mercury in the illegal gold mining in the brazilian amazon: A methodological assessment. International Journal of Environmental Research and Public Health. 2021;**18**: 11869

[62] Pilogallo A, Scorza F. Mapping regulation ecosystem services specialization in Italy. Journal of Urban Planning and Development. 2022;**148**: 04021072

[63] Haddaway NR, Cooke SJ, Lesser P, Macura B, Nilsson AE, Taylor JJ, et al. Evidence of the impacts of metal mining and the effectiveness of mining mitigation measures on social–ecological systems in Arctic and boreal regions: A systematic map protocol. Environmental Evidence. 2019;**8**:1-11

[64] Igwe EO, Ede CO, Nnabo PN, Ukpai SN. Impact of heavy metals dispersion on water supplies around Oshiri and Ishiagu mine districts of southern Benue trough, Nigeria. Modeling Earth Systems and Environment. 2021;**7**:2015-2030

[65] Farris JL, Van Hassel JH. Freshwater Bivalve Ecotoxicology. Boca Raton, FL: CRC Press; 2006. 368 p

[66] Lauwo SG, Otusanya OJ, Bakre O. Corporate social responsibility reporting in the mining sector of Tanzania: (Lack of) government regulatory controls and NGO activism. Accounting, Auditing & Accountability Journal. 2016;**29**(6): 1038-1074

[67] Moser KA, Baron JS, Brahney J, Oleksy I, Saros JE, Hundey EJ, et al. Mountain lakes: Eyes on global environmental change. Global and Planetary Change. 2019;**178**:77-95

[68] Seress G, Liker A. Habitat urbanization and its effects on birds. Acta Zoologica Academiae Scientiarum Hungaricae. 2015;**61**:373-408

[69] Austin JE. Threats to Cranes Related to Agriculture. Cranes and Agriculture: A Global Guide for Sharing the Landscape. Baraboo, Wisconsin, USA: International Crane Foundation; 2018. pp. 83-116

[70] Baldi A, Moskat C, Zagon A. Faunal mapping of birds in a riparian area of river Danube after construction of a hydroelectric power station. Folia Zoologica-Praha. 1998;**47**:173-180

[71] Wisneskie T. The Effects of Agricultural Management on Wetland Birds. Journal of Wildlife Management. 2020;**84**(2):379-390. DOI: 10.1002/ jwmg.21804

[72] Brönmark C, Hansson L-A. Environmental issues in lakes and ponds: Current state and perspectives. Environmental Conservation. 2002;**29**: 290-307

[73] Teng J, Xia S, Liu Y, Yu X, Duan H, Xiao H, et al. Assessing habitat suitability

for wintering geese by using normalized difference water index (NDWI) in a large floodplain wetland, China. Ecological Indicators. 2021;**122**:107260

[74] Lemes da Silva AL, Petrucio MM. Relationships between aquatic invertebrate communities, water-level fluctuations and different habitats in a subtropical lake. Environmental Monitoring and Assessment. 2018;**190**: 1-14

[75] Council NR. Riparian Areas: Functions and Strategies for Management. Washington, DC: National Academies Press; 1992. 432 p

[76] Bonter DN, Gauthreaux SA Jr, Donovan TM. Characteristics of important stopover locations for migrating birds: Remote sensing with radar in the Great Lakes Basin. Conservation Biology. 2009;**23**:440-448

[77] Pennington DN, Hansel J, Blair RB. The conservation value of urban riparian areas for landbirds during spring migration: Land cover, scale, and vegetation effects. Biological Conservation. 2008;**141**:1235-1248

[78] Ray N, Lehmann A, Joly P. Modeling spatial distribution of amphibian populations: A GIS approach based on habitat matrix permeability. Biodiversity and Conservation. 2002;**11**:2143-2165

[79] Brawn JD, Robinson SK, Thompson FR III. The role of disturbance in the ecology and conservation of birds. Annual Review of Ecology and Systematics. 2001;**32**:251-276

[80] Schultz R, Isenhart T, Colletti J, Simpkins W, Udawatta R, Schultz P. Riparian and upland buffer practices. In: North American Agroforestry: An Integrated Science and Practice. Madison, WI: American Society of Agronomy, Crop Science Society of America, and Soil Science Society of America; 2009. pp. 163-218

[81] Saunders DL, Meeuwig JJ, Vincent AC. Freshwater protected areas: Strategies for conservation. Conservation Biology. 2002;**16**:30-41

[82] Ostendorp W, Schmieder K, Jöhnk KD. Assessment of human pressures and their hydromorphological impacts on lakeshores in Europe. International Journal of Ecohydrology & Hydrobiology. 2004;**4**:379-395

[83] Brauns M, Gücker B, Wagner C, Garcia XF, Walz N, Pusch MT. Human lakeshore development alters the structure and trophic basis of littoral food webs. Journal of Applied Ecology. 2011;**48**:916-925

[84] Lewin W-C, Mehner T, Ritterbusch D, Brämick U. The influence of anthropogenic shoreline changes on the littoral abundance of fish species in German lowland lakes varying in depth as determined by boosted regression trees. Hydrobiologia. 2014;**724**:293-306

[85] Gasith A, Hosier A. Airborne litterfall as a source of organic matter in lakes 1. Limnology and Oceanography. 1976;**21**:253-258

[86] Cole JJ, Caraco NF, Likens GE. Short-range atmospheric transport: A significant source of phosphorus to an oligotrophic lake. Limnology and Oceanography. 1990;**35**:1230-1237

[87] Mehner T, Ihlau J, Dörner H, Hölker F. Can feeding of fish on terrestrial insects subsidize the nutrient pool of lakes? Limnology and Oceanography. 2005;**50**:2022-2031

[88] Céréghino R, Ruggiero A, Marty P, Angélibert S. Influence of vegetation

cover on the biological traits of pond invertebrate communities. In: Annales de Limnologie-International Journal of Limnology. Paris, France: EDP Sciences; 2008. pp. 267-274

[89] Mäkelä S, Huitu E, Arvola L. Spatial patterns in aquatic vegetation composition and environmental covariates along chains of lakes in the Kokemäenjoki watershed (S. Finland). Aquatic Botany. 2004;**80**:253-269

[90] Sirami C, Jacobs DS, Cumming GS. Artificial wetlands and surrounding habitats provide important foraging habitat for bats in agricultural landscapes in the Western cape, South Africa. Biological Conservation. 2013;**164**:30-38

[91] Lepczyk CA, Flather CH, Radeloff VC, Pidgeon AM, Hammer RB, Liu J. Human impacts on regional avian diversity and abundance. Conservation Biology. 2008;**22**:405-416

[92] Blaustein AR, Kiesecker JM. Complexity in conservation: Lessons from the global decline of amphibian populations. Ecology Letters. 2002;**5**: 597-608

[93] Hoffmann M, Hilton-Taylor C, Angulo A, Böhm M, Brooks TM, Butchart SH, et al. The impact of conservation on the status of the world's vertebrates. Science. 2010;**330**:1503-1509

[94] Stuart SN, Chanson JS, Cox NA, Young BE, Rodrigues AS, Fischman DL, et al. Status and trends of amphibian declines and extinctions worldwide. Science. 2004;**306**:1783-1786

[95] Jones CG, Lawton JH, Shachak M. Organisms as ecosystem engineers. Oikos. 1994;**69**:373-386

[96] Hipsey MR, Gal G, Arhonditsis GB, Carey CC, Elliott JA, Frassl MA, et al. A system of metrics for the assessment and improvement of aquatic ecosystem models. Environmental Modelling & Software. 2020;**128**:104697

[97] Vliet KA. Courtship behavior of American alligators (Alligator mississippiensis). Crocodilian Biology and Evolution. 2001;**1**:383-408

[98] Jouney AH. Environmental factors influencing amphibian community assemblage in dune wetlands on the Lake Michigan coast. Master's thesis. Grand Rapids, MI: Grand Valley State University; 2018. 83 p

[99] Nikolay K, Nadezhda P. Temporal and spatial variability of environments drive the patterns of species richness along latitudinal, elevational, and depth gradients. Biological Communications. 2018;**63**:189-201

[100] Olsson K, Nyström P. Non-interactive effects of habitat complexity and adult crayfish on survival and growth of juvenile crayfish (Pacifastacus leniusculus). Freshwater Biology. 2009; **54**:35-46

[101] Whiles MR, Lips KR, Pringle CM, Kilham SS, Bixby RJ, Brenes R, et al. The effects of amphibian population declines on the structure and function of Neotropical stream ecosystems. Frontiers in Ecology and the Environment. 2006;**4**:27-34

[102] Meunier CL, Gundale MJ, Sánchez IS, Liess A. Impact of nitrogen deposition on forest and lake food webs in nitrogen-limited environments. Global Change Biology. 2016;**22**:164-179

[103] Hirschnitz-Garbers M, Stoll-Kleemann S. Opportunities and barriers in the implementation of protected area management: A qualitative meta-analysis of case studies from European

protected areas. The Geographical Journal. 2011;**177**:321-334

[104] Lehtoranta V, Louhi P. Does conservation in Natura 2000 areas promote water quality improvement? Findings from a contingent valuation study on environmental benefits and residents' preferences. Environmental Science & Policy. 2021;**124**:226-234

[105] Noble MM. Integrating Social-Ecological Values into Marine Protected Area Spatial Planning. Canberra, Australia: The Australian National University; 2022. 162 p

[106] Maass M, Balvanera P, Bourgeron P, Equihua M, Baudry J. Anthropogenic impacts as determinants of tropical lake morphology: inferences from the morphometry and land use change of five lakes in Hubí municipality (Michoacán, Central Mexico). In: Lakes and Water Quality. IntechOpen; 2021. pp. 83-108

[107] Brewer SK, McManamay RA, Miller AD, Mollenhauer R, Worthington TA, Arsuffi T. Advancing environmental flow science: Developing frameworks for altered landscapes and integrating efforts across disciplines. Environmental Management. 2016;**58**: 175-192

[108] Brankov J, Pešić AM, Joksimović DM, Radovanović MM, Petrović MD. Water quality estimation and Population's attitudes: A multi-disciplinary perspective of environmental implications in Tara National Park (Serbia). Sustainability. 2020;**13**:241

[109] de Bisthoven LJ, Vanhove M, Rochette A-J, Hugé J, Verbesselt S, Machunda R, et al. Social-ecological assessment of Lake Manyara basin, Tanzania: A mixed method approach. Journal of Environmental Management. 2020;**267**:110594

[110] Shrestha P. Conservation and management of Phewa Lake ecosystem, Nepal. In: Proceedings of the International Conference on Great Himalayas: Climate, Health, Ecology, Management and Conservation. Citeseer; 2004. pp. 34-39

[111] Mooney H, Larigauderie A, Cesario M, Elmquist T, Hoegh-Guldberg O, Lavorel S, et al. Biodiversity, climate change, and ecosystem services. Current Opinion in Environmental Sustainability. 2009;**1**:46-54

[112] Payton IJ, Fenner M, Lee WG. Keystone Species: The Concept and its Relevance for Conservation Management in New Zealand. Wellington: Department of Conservation; 2002. pp. 1-27

[113] Davic RD, Welsh HH Jr. On the ecological roles of salamanders. Annual Review of Ecology, Evolution, and Systematics. 2004;**35**:405-434

[114] Hecky RE, Mugidde R, Ramlal P, Talbot M, Kling GW. Multiple stressors cause rapid ecosystem change in Lake Victoria. Freshwater Biology. 2010;**55**: 19-42

[115] Collins JP, Storfer A. Global amphibian declines: Sorting the hypotheses. Diversity and Distributions. 2003;**9**:89-98

Chapter 7

Water Use in the Khanka Lake Basin – Modern and Future Estimations

Jeanna A. Balonishnikova

Abstract

An assessment of the current use of water resources in the Khanka Lake basin is given. It has been established that in the Russian and Chinese territories of the Khanka Lake basin, the use of water resources is determined by the predominant development of agriculture. On the basis of actual and expertly calculated data, the volumes of water use for the entire period of economic activity in the Russian part of the lake basin have been restored. It has been established that the waters of the lake are the source for irrigation of rice crops in Russia, the use of the waters of Lake Khanka in China is possible only in dry years. From the territory of China, in high-water years, the discharge of flood waters of the Mulinghe River into Lake Malaya Khanka and, from it through the hydraulic structures (HS) into Lake Khanka, can have a significant impact on the increase in the water level in the lake. In accordance with the plans for the development of the agricultural sector of the economy of the Lake Khanka basin, scenarios for the development of water consumption in the Russian territory of the Lake Khanka basin until 2030 are considered.

Keywords: water use, irrigated lands, rice cultivation, water withdrawal, water consumption, water management system, future in water use

1. Introduction

Since 2010, the water level in Lake Khanka has been constantly rising, which leads to the flooding of coastal, including populated, territories. In recent years, many Russian scientists have been identifying the causes of the extreme increase in the water level in the lake.

The change in the water level (volumes) of Lake Khanka depends on the ratio of water inflow to the lake and its consumption, which in turn are determined by hydrometeorological factors in its basin as well as factors of anthropogenic activity. Unfortunately, studies of the hydrological regime and water balance of the lake have not been carried out since the early 1980s. The latest fundamental studies of the lake were carried out by M. G. Vaskovsky, published in his monograph *The Hydrological Regime of Lake Khanka* in 1978.

One of the anthropogenic factors affecting the change in hydrological characteristics is water use. The purpose of this work is to study the influence of anthropogenic factors on the water level associated with the use of irrigation lands and water resources of Lake Khanka. The main tasks in achieving this goal are to determine the structure of the current use of water resources and the volume of water withdrawal for economic needs in the Russian and Chinese parts of the transboundary catchment area of Lake Khanka based on official statistical information and expert assessments. The assessment of these values was made for the first time and allows us to conclude that the impact of water use on the change in the water level in the lake is minimal.

2. Methods

2.1 Description of the study site

Lake Khanka, the largest lake in the Far East, belongs to the drainage basin of the Amur River and is located on the border of the Primorsky Territory of the Russian Federation (Ozero Khanka) and the Heilongjiang Province of the People's Republic of China (Khanka Lake) (**Figure 1**) [1].

The total area of the water surface of Lake Khanka at an average long-term level of 68.90 m BS is 4070 km^2, including within Russia (3030 km^2)—excluding Lake Small Khanka, as well as lakes-lagoons Trostnikovoe (area 22.8 km^2), Protoka (5.37 km^2), and Krylovo (1.35 km^2). The total area of these lagoon lakes, connected to Lake Khanka by channels (straits), is 29.5 km^2, or only 0.7% of the lake area.

The surface area of the lake is not constant; it varies depending on the level from 3940 to 5010 km^2. The volume of the water mass of the lake at a level of 68.90 BS is 18.3 km^3, varying depending on the level from 12.7 to 22.6 km^3. In plan, the lake is

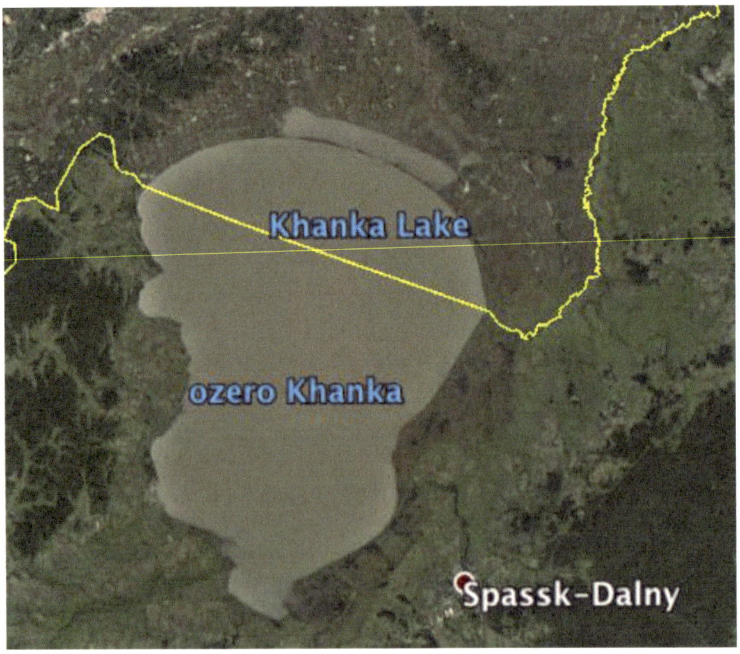

Figure 1.
Scheme of Lake Khanka.

pear-shaped, with the largest expansion in the northern part. The maximum length of the lake is 90 km, and the maximum width is 67 km [2].

There are 15 river streams flowing into the lake from the Russian territory and 9 from the Chinese side.

The Sungach River is the only natural watercourse through which the flow from Lake Khanka is carried out. The river flows out of the northeastern part of the lake and flows into the Ussuri River 450 km from its mouth through a heavily swampy lowland near Sungachen [2].

On the coast of Lake Khanka, as well as in the upper and middle reaches of the Sungach River, there is the Khanka State Natural Reserve, established in 1990. Since 2005, the Khanka State Nature Reserve has been awarded the international status of a UNESCO Biosphere Reserve. On the one hand, Lake Khanka and the surrounding wetlands of the reserve are an object of not only national but also international importance, which was the first in the Far East to be included in the list of wetlands of the Ramsar Convention as a habitat for waterfowl. On the other hand, the Khanka Lake basin is one of the main areas of agriculture in the Far East of Russia, the active economic development of which in the second half of the 20th century led to a significant transformation of landscape complexes, pollution of the aquatic environment and soils, and a change in the hydrological regime of the rivers of the basin and Lake Khanka itself. During the long economic crisis of the 1990s in the basin of Lake Khanka, there was a significant decrease in agricultural pressure. However, since the early 2000s, in both the Russian and Chinese parts of the lake basin, there has been an increase in economic activity, which indicates an increase in anthropogenic pressure in the transboundary basin as a whole.

2.2 Data sources and analysis

Given that the only anthropogenic factor affecting the state of the lake is agriculture, namely, rice cultivation, the water resources of the Khanka Lake basin are used mainly for the development of rice cultivation.

To assess the current state of water use and the long-term dynamics of water withdrawal and water consumption in the Khanka Lake basin, the data of direct accounting of water intakes and discharges in the lake basin for the period 1985–2015 were summarized and analyzed. Data of direct recording of water intakes and discharges in the lake basin for the period 1985–2015 were provided by the Amur Basin Water Administration in the framework of the project to study the hydrological features of the water regime of Lake Khanka in order to identify the causes of the abnormal rise in the water level (the author of this chapter is a participant in this project). Currently, the volume of water use for all economic needs or water withdrawal in the basin of Lake Khanka is 156.2 million m^3 per year, and the volume of discharge is 53.4 million m^3 per year. Of the total volume of water use, water withdrawal for irrigation—flooding of rice fields—makes up almost the entire volume of water used in the basin—more than 90%. At the same time, the volumes of water for industrial and communal needs (household and drinking) are approximately the same. Water withdrawal directly from the lake for irrigation is about 95%; the rest of the water is taken from rivers within the lake basin. Water losses or water consumption in the basin of Lake Khanka is 102.8 million m^3 per year, reaching up to 70% of total water withdrawal in some years.

The dynamics of the areas of irrigated land in the basin of Lake Khanka for 1995–2015 was established on the basis of officially provided data by the administration of melioration and agricultural water supply in the Primorsky and Khabarovsk territories (Primmeliovodkhoz) as part of the implementation of the above scientific project.

The availability of actual data and calculated data on irrigated areas for the entire long-term period, as well as the values of water consumption calculated from the difference between water withdrawal and discharge in those years when official accounting information was available for them, made it possible to restore the volumes of water use and water consumption for the entire period of economic activity in the lake basin. An analysis of all available data showed that irretrievable losses per hectare of irrigated land can range from 7000 to 9000 m^3 per year.

The selected time period (1960–2015) makes it possible to assess the long-term dynamics of water use and irrigated areas and assess the impact of irrigation on the water regime of Lake Khanka at various stages of the economic use of land and water resources in the basin under study.

The actual and expertly determined values of water withdrawal and water consumption in the basin of Lake Khanka for the entire period of irrigated agriculture are shown in **Figure 2** [3].

As it follows from the figure, the maximum volumes of water required for irrigation were in the mid-1980s. By 1985, the total water consumption reached more than 600 million m^3 per year.

The minimum values of water use were noted in the period 2002–2007, when rice cultivation was stopped in the lake basin, which consumes the main volume of water taken from Lake Khanka.

Restoration of rice cultivation began in 2008, which immediately affected the increase in total water withdrawal. By 2012, water use increased to 292 million m^3;

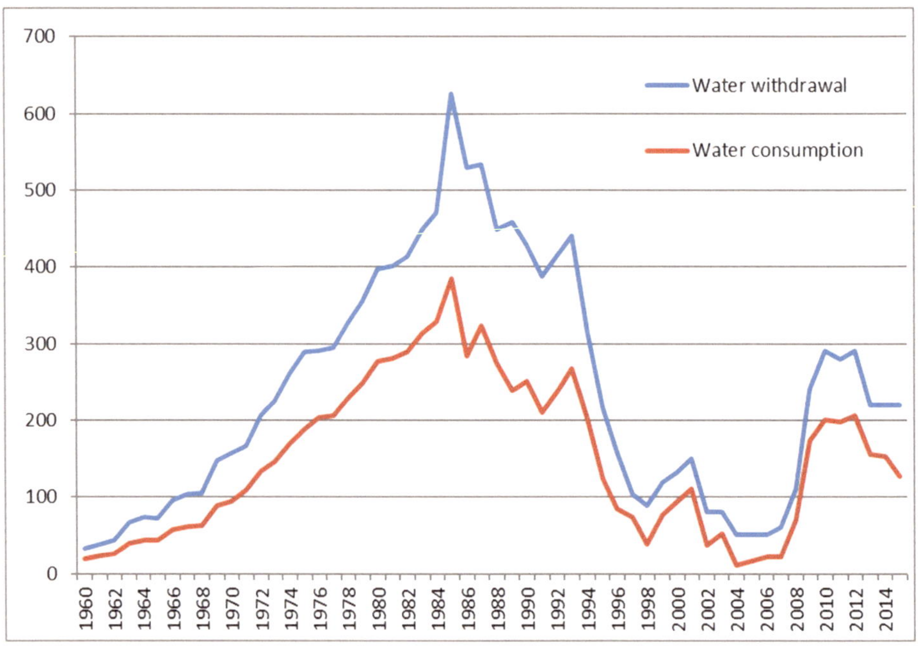

Figure 2.
Long-year dynamics of water use in the Lake Khanka basin during the period from 1960 to 2015.

then, in 2013–2015, a slight decline in water withdrawal followed, which was due to a decrease in rice crops during this period.

For the entire period of economic activity under consideration, water consumption also varied widely from 40 to 385 million m^3 per year. The highest values of irretrievable losses fall on the years of the greatest development of irrigated agriculture in the basin of Lake Khanka.

Comparison of the values of irrigated lands and water consumption for the period 1960–2015, necessary for assessing the water balance of the lake and the impact of irrigation on the water resources of Lake Khanka, is presented in **Table 1**.

According to [4], a distinctive feature of land use in the Chinese part of the Khanka basin is the high proportion of rice fields and arable land, which in total makes up 44.4% of the territory. At the same time, the area of rice fields is 2.2 times larger than that of the arable land occupied by other agricultural crops.

The data published by the Chinese side on water management infrastructure and water use in the Chinese territory of the basin are very incomplete and contradictory. An assessment of the water management situation in the Chinese territory of the lake basin was carried out using satellite data as part of a scientific project to study the rise in the water level in Lake Khanka and materials of the Joint Russian-Chinese Commission on the Rational Use and Protection of Transboundary Waters (June 2016) [2].

Since the late 1990s, a large-scale integrated water management system has been created in the Chinese part of the Khanka Lake basin, including HS, irrigation and reclamation canals, runoff diversion channels, and large irrigation systems, mainly rice ones. The main purpose of the water management system of the People's Republic of China in the basins of lakes Khanka, Small Khanka, and the Mulinghe River is to provide water for rice irrigation systems (RIS) and redistribute flood waters.

The Chinese part of the catchment area of Lake Khanka, including Lake Small Khanka, is relatively small, and economic activity on it cannot have a significant impact on the water regime and water balance of Lake Khanka. The main influence here belongs to the transfer of runoff from the Mulinghe River, which can then both enter Lake Khanka and, bypassing it, be fed to the RIS and discharged into the Sungach River.

The water management system in the Chinese territory of the Khanka Lake basin consists of the components shown in **Figure 3** (based on the Landsat 8 satellite image, May 2016) [2]:

1. Water divider on the Mulinghe River;

2. Dongdihe Canal (Musin);

3. Distribution reservoir;

4. Two discharge channels diverting water from the reservoir to the RIS and to the Sungach River;

5. Lake Small Khanka;

6. Three HS on the isthmus between the lakes Small Khanka and Khanka.

Lake Small Khanka serves as a storage reservoir for water supply to the RIS.

Year	Irrigated lands, ha	Water consumption, million m^3 per year	Year	Irrigated, ha	Water consumption million m^3 per year
1960	2700	19.6	1988	40,300	274.8
1961	3200	23.1	1989	36,300	239.2
1962	3700	26.6	1990	32,300	251.3
1963	4200	40.0	1991	28,700	210.1
1964	4700	44.1	1992	29,600	237.0
1965	5200	43.2	1993	28,700	267.5
1966	6400	57.6	1994	14,300	201.5
1967	6900	62.1	1995	10,400	124.4
1968	7900	63.2	1996	7500	85.1
1969	9800	88.2	1997	3000	74.4
1970	11,800	94.4	1998	4800	38.7
1971	13,700	108.8	1999	5000	76.2
1972	14,900	134.1	2000	6100	94.0
1973	16,300	146.7	2001	7500	111.2
1974	18,800	169.2	2002	6900	36.52
1975	20,900	188.1	2003	4600	51.62
1976	22,600	203.4	2004	1900	11.78
1977	25,800	206.4	2005	2800	16.8
1978	28,600	228.8	2006	3300	22.38
1979	31,100	248.8	2007	4800	22.4
1980	34,700	277.6	2008	7000	70.31
1981	35,100	280.8	2009	16,100	172.8
1982	36,200	289.6	2010	21,300	200.2
1983	39,300	314.4	2011	22,400	198.1
1984	41,100	328.8	2012	23,300	205.8
1985	52,200	384.6	2013	19,800	155.2
1986	48,200	284.7	2014	22,500	152.5
1987	44,200	324.0	2015	19,600	127.3

Table 1.
Irrigated areas and water consumption in the basin of Lake Khanka for the period 1960–2015.

At the same time, the presence of spillways with a total capacity of up to 200 m^3/s indicates that the Chinese side is designed to discharge the flood waters of the Mulinghe River, since Lake Small Khanka's own catchment area is very small and the flow from it cannot reach such values.

In conditions of medium water content, the runoff taken from the Mulinghe River in the amount of about 700 million m^3 per year is completely spent on the needs of rice cultivation. This value is estimated based on the irrigation norm of rice 700 m^3 per year per 1 mu (*author's note* 666.7 m^2). This volume may not affect the balance of

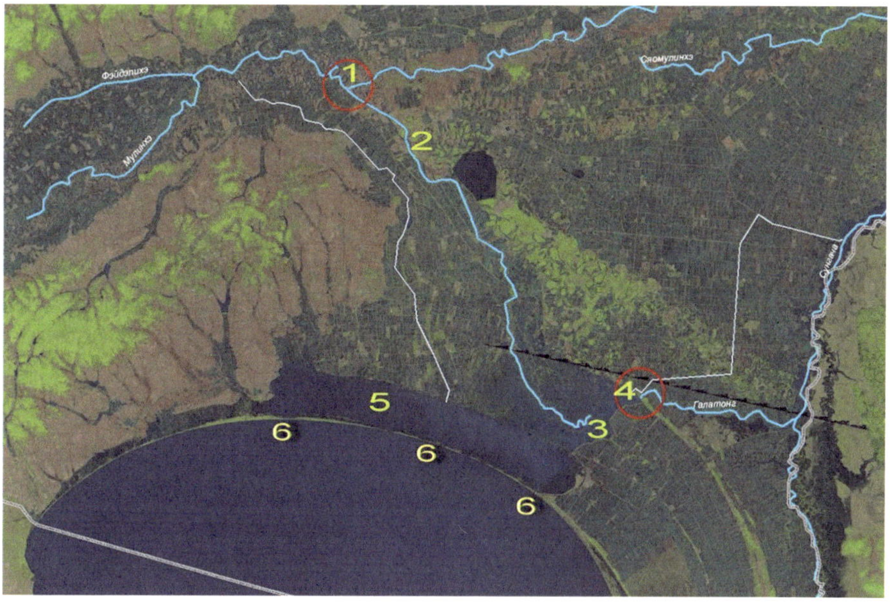

Figure 3.
The water management system in the Chinese side.

Lake Khanka, since discharges from irrigated areas are carried out mainly into the Sungach River, and additional injection from Lake Khanka is not carried out.

With a reduced water content of the Mulinghe River, it can be assumed that the river's own water resources will not be enough to cover the irrigation needs of the RIS, as a result of which water can be taken from Lake Khanka to Lake Small Khanka using pumping stations, the maximum capacity of which after reconstruction will be increased to 100 m^3/s.

In high-water years, an increased flow of water enters the Dongdihe (Musin) canal from the Mulinghe River, and the lock-regulator on the Mulinghe River limits the flow of the river below the canal, while the flow into the canal itself is not regulated. In order to avoid flooding of irrigated fields and residential areas, all excess water is probably discharged into Lake Small Khanka, and from there, through the waste hydraulic structures, into Lake Khanka. As noted above, the discharge value can reach 200 m^3/s. This is a design value that has a low probability. However, if we assume that during the high-water period 100 m^3/s will be discharged into Lake Khanka, then for 4 months of operation of spillways, an additional 1 km^3 of water will enter the lake, which may affect the level regime of the lake.

During the period under review, there is a gradual increase (with fluctuations) in total rice crops from 33,000 hectares in 1976 to 115,000 hectares in 2015. At the same time, the contribution to the development of irrigated agriculture in the study basin from the Russian Federation and China turned out to be different both in time and in magnitude. Until 1994, the area under rice cultivation in Russia exceeded that in China. In some years, Russia's contribution even reached 95%. However, since 1995, China has taken the lead in rice cultivation, and by 2004, the Russian Federation accounted for only 3% of the total area under rice in this basin. Only since the end of the first decade of this century, in connection with the revival of rice cultivation in Russia, has its contribution increased in recent years to 18% [3].

In addition to the noted differences in the areas under rice cultivation and trends in their changes, the sources of water for irrigation also differ. In the Russian part of the basin, the waters of the lake serve as a water source, and an increase in rice crops can lead to a decrease in the level due to water consumption. It should be noted that such a decrease, even during the period of maximum development of rice cultivation, did not exceed 10 cm. Irrigation water in the Chinese part of the basin is mainly supplied by the flow of the Mulinghe River. The use of the waters of Lake Khanka in China is possible only in dry years, which may lead to a decrease in its level. And, vice versa, in the years of increased water content in the Mulinghe River basin, through the network of canals, excess flood runoff of this river is discharged into Lake Small Khanka, and from it into a large lake, with a subsequent possible increase in the water level in it [3].

2.3 Scenarios for the development of agriculture and water use in the Russian territory of the Khanka Lake basin until 2030

An analysis of economic activity in the Khanka Lake basin showed that the main anthropogenic factor in the lake basin is irrigated agriculture, which requires large amounts of water. The volume of water withdrawn from the lake and, accordingly, water consumption in the lake basin directly depends on the prospective development of this industry. To develop scenarios for the development of such a water-intensive sector of the economy as irrigated agriculture, the existing national plans for the development of the agro-industrial complex of Russia [5, 6] and regional plans for the development of Primorsky Territory, a constituent entity of the Russian Federation, on the territory of which the Lake Khanka basin is located, were analyzed [7].

Based on all available information, three scenarios for the development of irrigated agriculture in the Khanka Lake basin up to 2030 are proposed.

The first scenario—the crisis one—is based on the premise that there will be no further increase in irrigation in the lake basin and that the areas under rice crops will remain unchanged until 2030.

In the second scenario—inertial—with the expansion of rice crops, it is planned to restore over 80% of the existing reclamation fund.

In the third scenario—the innovative one—the growth of irrigated lands is allowed due to the involvement in agricultural turnover of new plots located beyond the currently existing border of irrigation-developed territories. That is, it is supposed to fully develop all available irrigation-prepared irrigated lands in the basin under consideration.

To assess long-term plans for the state of irrigation and, consequently, irretrievable losses in the basin of Lake Khanka, the indicators given in [5–7] were taken into account and methodological approaches developed earlier at the State Hydrological Institute were applied in assessing irretrievable water consumption losses in irrigated agriculture [8].

For the initial level of rice cultivation, the average value for 2010–2015 was taken, the period of relatively stable irrigated land in the Khanka Lake basin, which is 21,500 hectares. During this period, the share of irrigated lands in the basin of Lake Khanka was estimated at 87% of all irrigated lands in Primorsky Territory. This ratio was adopted when calculating the prospective values of rice crops.

The following assumptions and norms were adopted to estimate the values of irretrievable losses from irrigated areas.

For the crisis scenario, according to which no further increase in irrigation in the lake basin and areas under rice until 2030 is expected, the calculation is made taking into account the current indicators of irretrievable water consumption, when water consumption from one hectare of irrigated land ranges from 7000 to 9000 m^3 per year (irrigation period). This value was obtained on the basis of available official information on water withdrawals and discharges in the lake basin in recent years.

For the inertial and innovative scenarios, the irrigation norms recommended for this region for rice, which range from 10,000 to 12,000 m^3/ha, are taken for calculation [9]. The value of 12,000 m^3/ha seems realistic for the inertial scenario in the case of expansion of rice crops and simultaneous reconstruction of rice irrigation systems through government programs and foreign and private investments in the development of rice cultivation in this region. An irrigation norm of 10,000 m^3/ha is included in the forecast for the period up to 2030 under the innovative scenario, taking into account the modernized irrigation system with lower losses during water transportation through irrigation canals.

The annual values of rice planting areas and water consumption in the Khanka Lake basin under three scenarios until 2030 are shown in **Table 2**.

Thus, in accordance with the three proposed scenarios for the development of irrigated agriculture in the Khanka Lake basin, the size of rice cultivation remains at the current level—about 22,000 hectares (crisis scenario) and increases to 46,000 hectares (inertial scenario) or up to 55,000 hectares (innovative scenario). At the same time, water consumption can be up to 200 million m^3 per year, more than 450 million m^3 per year and 550 million m^3 per year, respectively, according to the crisis, inertial, and innovative scenarios.

Year	Scenarios					
	Crisis		Inertial		Innovative	
	Irrigated land	Water consumption	Irrigated land	Water consumption	Irrigated land	Water consumption
2010–2015	21,5	173	21,5	173	21,5	173
2024	21,5	173	35,0	350	43,2	432
2025	21,5	173	36,8	368	46,1	461
2026	21,5	173	38,7	387	48,9	489
2027	21,5	173	40,5	405	50,0	500
2028	21,5	173	42,4	424	52,4	524
2029	21,5	173	44,2	442	54,2	542
2030	21,5	173	46,1	461	55,0	550

Table 2.
Annual values of promising rice crops (thousand hectares) and volumes of water consumption (million m^3 per year) in the Khanka Lake basin.

3. Conclusion

In the whole complex of the anthropogenic factors on the Russian part of the Lake Khanka basin, water use for irrigation is a principal factor. To quantify the impact of economic activity in the lake basin on the long-term dynamics of its level, the dynamics of water consumption was estimated for the entire period of development of rice cultivation in the lake basin—from the 1960s. It has been established that water consumption in different years here is up to 70% of the total water withdrawal.

The highest values of water consumption fall on the years of maximum development of irrigated agriculture in the basin of the Lake Khanka. During these years, water consumption reduced the water level by up to 10 cm for the period 1975–1994.

In modern conditions, when irrigation in the basin of the Lake Khanka has decreased by almost 3 times compared to the 1980s, the impact of the anthropogenic factor from the Russian part of the basin on the lake level regime can be neglected.

In the Chinese part of the basin, irrigation is also the dominant factor influencing the regime of the Small Khanka Lake and the Lake Khanka.

Assessment of the impact of the diversion scheme and the use of the Mulinghe river runoff to the basin of the Lake Khanka leads to the conclusion that Small Khanka Lake acts as a storage reservoir for the water use of the RIS. At the same time, the presence of spillways on the isthmus of Small Khanka Lake–Lake Khanka with a total capacity of up to 200 m^3/s indicates that the Chinese side is designed to discharge the flood waters of the Mulinghe River.

In accordance with the three proposed scenarios for the development of irrigated agriculture in the basin of Lake Khanka, the values of water consumption in the basin for the period up to 2030 were obtained depending on the size of rice cultivation. The lowest values of water consumption up to 200 million m^3 per year will be the case if the irrigated areas remain at the current level—about 22,000 hectares (crisis scenario). Water consumption of more than 450 million m^3 per year may occur with an increase in irrigated land to 46,000 hectares (inertial scenario). The maximum values of water consumption of about 550 million m^3 per year are possible in the implementation of the innovative scenario, in which all available irrigation-prepared lands will be developed—55,000 hectares in the Lake Khanka basin.

Author details

Jeanna A. Balonishnikova
State Hydrological Institute, St. Petersburg, Russia

*Address all correspondence to: jeannabalonishnikova@gmail.com

References

[1] Lake Khanka [Internet]. Geography Study Guide Wiki. Available from: https://geovillages.fandom.com/wiki/Lake_Khanka

[2] Georgievsky V, editor. Water Balance of Lake Khanka and its Changes under the Influence of Natural and Anthropogenic Factors. Spb. 272 p; 2022

[3] Balonishnikova JA, Tsytsenko KV, Kramareva LS. Water Resources Use in the Basin of the Lake Khanka. Water sector of Russia, Scientific/practical journal No. 3; 2019

[4] Egidarev EG, Bazarov KY, Mishina NV. Modern Land Use in the Basin of Lake Khanka. Vladivostok: Geographical Institute of the Far Eastern Branch of the Russian Academy of Science, Collection of scientific articles; 2019

[5] Forecast of the Long-Term Socio-Economic Development of the Russian Federation for the Period up to 2030. Moscow: Ministry of Economic Development of Russia; 2013

[6] Forecast of Scientific and Technological Development of the Agro-Industrial Complex of the Russian Federation for the Period up to 2030. Basic Provisions. Available from: https://prognoz2030.hse.ru/

[7] Socio-Economic Strategy for the Development of Primorsky Krai until 2025 (as amended on: 06/23/2015). Available from: http://docs.cntd.ru/document/494219477

[8] Shiklomanov I, editor. Water Resources of Russia and their Use. SPb; 2008. 600p

[9] Nosovsky VS, Nekras YV. Status and prospects for the development of rice cultivation in the Far East. Land Reclamation and Water Management. 2013;**4**:2-5

Chapter 8

The Impacts of Introduced Fish and Aquatic Macrophytes on the Ecology and Fishery Potential of Lake Victoria, Kenya

Job O. Omweno, Reuben Omondi, Fredrick M. Ondemo and Argwings Omondi

Abstract

Fish have been deliberately introduced into new ecosystems as a management tool, to argument overfished native stocks, to occupy vacant niches, and to create lucrative commercial fisheries. Lake Victoria has witnessed successful introductions of predatory Nile perch, *Lates niloticus* and four Tilapiine species (including Nile tilapia, *Oreochromis niloticus, Tilapia zilii*, and *Oreochromis leucostictus*). These introductions have negatively and positively impacted the fishery potential and ecology of native fisheries in the lake. The predation of native species by the voracious Nile perch has contributed to decimation and virtual disappearance of over 300 species of Haplochromines. In addition, competition for feeding and breeding areas and interspecific hybridization between exotic *O. niloticus* and the native Tilapiines have also yielded undesirable results such as disappearance of native *Oreochromis esculentus*. The most successful invasive plant introductions have been water hyacinth, *Eichhornia crassipes*, Nile cabbage, *Pistia stratiotes*, and dense water weed, *Egeria densa*. Proliferation of water hyacinth has led to increased shading and turbidity. The introduced species have manifested more pronounced deleterious effects on the native fisheries and their ecology in Lake Victoria. Therefore, future introductions of new species should be based on sound scientific research in order to minimize their unprecedented impacts in the new ecosystems.

Keywords: introduced species, aquatic ecosystem, macrophytes, ecology, fishery, Lake Victoria

1. Introduction

Lake Victoria, located in the eastern part of Africa, is one of the largest tropical freshwater lakes in the world. It supports a rich ecosystem and plays a vital role in the economies and livelihoods of the riparian countries, including Kenya [1]. Over the years, however, the introduction of non-native fish species and aquatic macrophytes has had significant impacts on the ecology and fishery potential of Lake Victoria [1, 2]. Species introductions have been used as a fishery management tool globally. This has occurred through intentional and accidental introductions across biogeographic and ecological boundaries [1]. Over 237 species have been introduced in more than 140 countries globally, out of which the African continent has recorded 147 introductions [1]. Kenya has recorded a total 14 fishery introductions. With a total of six fishery introductions, Lake Victoria is the second leading ecosystem in fishery introductions after Lake Naivasha. Currently, the commercial fishery of Lake Victoria (with an annual fishery output of approximately one million tons) is dominated by two non-native species, Nile perch *Lates niloticus*, Nile tilapia *Oreochromis niloticus*, and the native Silver cyprinid *Rastrineobola argentea* [2]. Before the decline and disappearance of native species, Lake Victoria had a diverse multi-species fishery dominated by native Tilapiines, *Oreochromis esculentus*, *Oreochromis leucosticus, Oreochromis variabilis*, the African lungfish, more than species of haplochromines [3]. Some species like the African carp, *Labeo victorianus,* however, declined long before the introduction of exotic tilapia and predatory Nile perch due to overfishing and habitat degradation [3]. Other factors contributing to fishery changes in Lake Victoria are nutrient pollution and massive wetland degradation that led to eutrophication. However, remarkable changes in fish biodiversity occurred after the introduction of Nile perch and exotic Tilapiines such as Nile tilapia (*O. niloticus*) whose ecological impact has been widely manifested [3, 4].

The introduction of non-native fish species, particularly the Nile Perch (*Lates niloticus*), in the 1950s, aimed to enhance fisheries and boost economic development in the region. While the introduction initially resulted in a boom in the fishery industry, it has led to numerous ecological consequences. The Nile Perch, a voracious predator, caused a decline in native fish species, leading to a loss of biodiversity and disruption of the natural food web [5]. Several endemic cichlid species, which were crucial for maintaining the ecological balance, have been pushed to the brink of extinction. In addition to the introduction of non-native fish, Lake Victoria has also been invaded by aquatic macrophytes, primarily the Water Hyacinth (*Eichhornia crassipes*). The rapid proliferation of these invasive plants has had severe ecological and socioeconomic implications [6]. Water Hyacinth forms dense mats on the lake's surface, reducing oxygen levels in the water, and blocking sunlight from reaching submerged vegetation. This, in turn, negatively impacts native aquatic plants and leads to the decline of fish populations, as they lose vital habitats and food sources. The impacts of introduced fish and aquatic macrophytes on the fishery potential of Lake Victoria are substantial [7, 8]. Furthermore, the presence of Water Hyacinth has posed significant challenges to fishing activities. The dense mats of this invasive plant clog fishing gear, impede navigation, and hinder access to fishing grounds. Fishermen have to spend more time and effort clearing the waterways, reducing their fishing productivity [9]. Consequently, the overall fishery potential of Lake Victoria has been compromised, leading to economic losses and food security concerns for the surrounding communities. The introduction of non-native fish species, such as the Nile Perch, and the invasion of aquatic macrophytes, particularly Water Hyacinth, have had profound impacts on the ecology and fishery potential of Lake Victoria in Kenya. Efforts to mitigate these impacts and restore the ecological balance of the lake are crucial for ensuring the sustainable use of Lake Victoria's resources and the well-being of the surrounding communities [8, 10]. Comprehensively, based on a systematic literature review, the main objective of this chapter was to explore and determine the impacts of introduced fish and aquatic macrophytes on the ecology and fishery potential of Lake Victoria.

2. Introduced fish species and their impacts on the ecology and fishery potential of Lake Victoria

Lake Victoria is known for its diverse fish species and its significance as a vital fishery resource. The lake is home to numerous native and introduced fish species, each having its own impacts on the lake's ecology and fishery potential [5]. One of the most notable native fish species in Lake Victoria is the Nile perch (*Lates niloticus*). Another important native fish species in the lake is the dagaa (*Rastrineobola argentea*), also known as the silver cyprinid [11]. Dagaa are small pelagic fish that form a critical part of the fishery in Lake Victoria. They are an important food source for humans and are also used as bait for larger fish species. The population dynamics of dagaa are closely linked to the ecological conditions of the lake, such as water quality and the availability of food resources [10, 11]. These introductions have been driven by the desire to diversify fisheries and provide alternative fishing opportunities. However, the presence of non-native fish species can have both positive and negative impacts on the ecosystem. While they may enhance fishery potential by providing new fishing opportunities, they can also compete with native species for resources and potentially disrupt the ecological balance of the lake [2, 5]. Overall, the diverse fish species in Lake Victoria, including both native and introduced species, have complex interactions with the lake's ecology and fishery potential. Understanding these interactions is crucial for sustainable management practices and maintaining the long-term health and productivity of the lake's fishery resources [12].

2.1 The Nile perch, Lates niloticus

The introduction of Nile Perch (*Lates niloticus*) to Lake Victoria has had significant impacts on the ecology and fishery potential of the lake. Originally introduced in the 1950s to enhance fisheries and promote economic development, the unintended consequences of this introduction have had far-reaching effects on the ecosystem [4, 5]. One of the most notable impacts of the Nile Perch introduction is the decline in native fish species. As a voracious predator, the Nile Perch feeds on a wide range of fish, including many endemic species that were once abundant in Lake Victoria. This predation pressure has led to a significant reduction in the populations of these native fish, some of which are now critically endangered or have even become extinct. The decline of these species not only disrupts the ecological balance of the lake but also poses a threat to the cultural and biodiversity values associated with them [4]. Furthermore, the introduction of Nile Perch has caused a disruption in the natural food web of Lake Victoria. Native fish species played essential roles in the ecosystem as both predators and prey. Their feeding habits helped control populations of smaller fish and maintained a balance in the lake's biodiversity. With the decline of these native species, the natural food web has been altered, leading to cascading effects on other organisms. This disruption can result in changes in nutrient cycling, phytoplankton dynamics, and overall ecosystem health [6]. The ecological impacts of Nile Perch introduction extend beyond the decline of native fish species. The feeding habits and behavior of the Nile Perch have also affected other aspects of the ecosystem. For instance, Nile Perch feed heavily on zooplankton, which are crucial for controlling algae populations. The reduction in zooplankton abundance due to predation by Nile Perch has resulted in increased algal blooms, leading to eutrophication and a decline in water quality. This has negative implications for other organisms, including fish and aquatic plants, which rely on clean and well-oxygenated water for survival [13]. In addition to the ecological consequences, the introduction of Nile Perch has had mixed effects on the fishery potential of Lake Victoria. Initially, the introduction resulted in a boom in the fishery industry. The Nile Perch grew rapidly and provided a valuable commercial catch, attracting fishing activities and generating economic benefits for local communities. The export of Nile Perch fillets became a significant source of foreign exchange for the riparian countries [14]. However, over time, the fishery potential of Lake Victoria has faced challenges. The high demand for Nile Perch has led to unsustainable fishing practices, including overfishing and the use of destructive fishing methods. This has put pressure on Nile Perch populations and led to a decline in their abundance. As a result, fish catches have become less predictable, and the fishing industry has become less reliable as a source of income and livelihood for local communities [7]. Furthermore, the dominance of the Nile Perch in the fishery has resulted in a loss of diversity in catch composition. Traditional fish species, which were once abundant and provided food security for local communities, have been overshadowed by the Nile Perch. This has raised concerns about the resilience and long-term sustainability of the fishery, as the overreliance on a single species increases vulnerability to disease outbreaks, environmental changes, and market fluctuations [7, 15, 16]. Other species severely impacted upon by the Nile perch invasion and predation are three catfishes (*Clarias gariepinus*, *Synodontis* spp., and *Schilbe altinialis*) and the African Lungfish (*Protopterus aethiopicus*) [6, 7, 17]. The introduction of Nile Perch to Lake Victoria has had profound impacts on the ecology and fishery potential of the lake. The decline of native fish species, disruption of the natural food web, and alteration of the ecosystem dynamics have altered the ecological balance and biodiversity of the lake. The fishery industry initially benefited from the introduction but is now facing challenges due to overfishing, loss of diversity, and market uncertainties. Balancing the conservation of native species, sustainable fishing practices, and the economic needs of local communities is essential [17, 18].

2.2 Introduced Tilapiines

The introduction of Tilapiines, particularly the Nile tilapia (*Oreochromis niloticus*) and the Nile tilapia hybrids, to Lake Victoria has had significant impacts on the ecology and fishery potential of the lake. In the 1920s, *Oreochromis variabilis* and *O. esculentus* were the main native Tilapiines contributing to the commercial fishery of Lake Victoria. Tilapiines were introduced to Lake Victoria in the 1950s as an additional fishery resource and to provide alternative fishing opportunities [15]. The Nile tilapia, in particular, was known for its adaptability, fast growth, and tolerance to a wide range of environmental conditions. However, the unintended consequences of this introduction have had profound effects on the lake's ecosystem. However, the native tilapia fisheries declined because the species could not stand high levels of overexploitation. As a result, Haplochromines and Silver cyprinid *Rastrineobola argentea* became the main target species. Research has also shown that the introduced stocks of *Tilapia zilli* have contributed to proliferation of free-floating macrophytes such as *Pistia stratiotes* [16]. According to Outa et al. [15], the total haplochromine catches of 650,000 tons accounted for 80% of the total lake catches in 1967. It also prompted the introduction of *L. niloticus* and *O. niloticus* which increased the lake's fish catches fivefold. However due to its advanced competitive strategies, Nile tilapia outcompeted the endemic Tilapiines *O. esculentus* and *O. variabilis* resulting to their displacement and total disappearance from the lake [8]. As a result of these disappearances, the Nile perch became the main target species in the lake fishery and fishing pressure using illegal gears was meted on the species. The decline of Nile perch populations in Lake Victoria has reportedly contributed to the recovery of some of the cichlids that initially disappeared as a result of its invasion [6]. Nevertheless, the *O. niloticus* contributed to increases in fish landings from the lake providing affordable and quality animal protein sources to the riparian communities [19].

Another notable impact of introduced Tilapiines is the alteration of the trophic structure of the lake. Tilapiines are generalist feeders that consume a variety of food sources, including plankton, algae, and detritus. Their feeding habits and high reproductive rates have resulted in increased competition for resources with native fish species. This has led to changes in the relative abundance and distribution of different fish populations, potentially displacing or reducing the numbers of native species [19, 20]. Moreover, the introduction of Tilapiines has led to changes in the lake's nutrient dynamics. Tilapiines are efficient grazers of algae, and their feeding activities can reduce algal biomass. While this may initially seem beneficial in controlling algal blooms, it can also have unintended consequences. Algae play a vital role in the lake's food web, serving as a food source for zooplankton, which, in turn, are consumed by other fish species. The reduction in algal biomass due to Tilapiine grazing can disrupt the balance of the food web and impact the availability of food for other organisms, potentially leading to cascading effects throughout the ecosystem [17]. Additionally, the introduction of Tilapiines has had mixed effects on the fishery potential of Lake Victoria. Initially, the introduction provided new opportunities for fishing and contributed to the expansion of the fishery industry. Tilapiines, particularly the Nile tilapia hybrids, are known for their fast growth and high reproductive rates, which made them an attractive target for commercial fishing. The increased availability of Tilapiines led to economic benefits for fishing communities and contributed to local livelihoods [6].

However, the proliferation of Tilapiines has also posed challenges to the fishery. The high reproductive capacity and aggressive behavior of Tilapiines have led to overpopulation and increased competition for resources. This has resulted in slower growth rates and smaller sizes of individual fish, reducing their market value. Moreover, the dominance of Tilapiines in the fishery has resulted in a loss of diversity in catch composition, as other native species are overshadowed and less targeted [16]. This raises concerns about the resilience and long-term sustainability of the fishery, as it becomes more vulnerable to environmental changes and disease outbreaks. The ecological and socioeconomic impacts of introduced Tilapiines highlight the need for careful management and conservation strategies in Lake Victoria [6, 7]. Efforts are being made to balance the conservation of native fish species, the control of invasive Tilapiines, and the sustainable use of the lake's resources. These include implementing fishing regulations, promoting sustainable fishing practices, and conducting research on the impacts of Tilapiine populations [21]. Therefore, the introduction of Tilapiines, particularly the Nile tilapia and its hybrids, to Lake Victoria has had significant impacts on the lake's ecology and fishery potential. The alteration of the trophic structure, changes in nutrient dynamics, and challenges to the fishery industry are among the consequences of this introduction. Balancing the conservation of native species, management of invasive Tilapiines, and sustainable fishing practices are crucial for ensuring the long-term health and productivity of Lake Victoria's ecosystem and the well-being of local communities [10].

3. Introduced aquatic macrophytes species and their impacts on the ecology and fishery potential of Lake Victoria

Macrophytes are aquatic plants that play a significant role in the ecology and fishery potential of Lake Victoria, the largest tropical lake in Africa. These plants, which include both native and invasive species, have diverse impacts on the lake's ecosystem. Native macrophytes in Lake Victoria, such as various species of submerged plants, floating plants, and emergent plants, provide important ecological functions. They serve as habitats and spawning grounds for fish, provide food sources, and contribute to the overall biodiversity of the lake [10]. Submerged plants, for instance, offer shelter for small fish and provide areas for the attachment of algae and other organisms. Floating plants, like water lilies, create sheltered areas for young fish, while emergent plants, such as papyrus, form dense stands along the lake's shoreline, offering nesting sites for birds and other wildlife [22]. However, the presence of invasive macrophytes, notably water hyacinth (*Eichhornia crassipes*), has had significant impacts on the ecology and fishery potential of Lake Victoria. Water hyacinth, originally introduced as an ornamental plant, has rapidly spread across the lake, forming dense mats that cover large areas of the water surface. These mats block sunlight, hampering the growth of submerged plants and leading to reduced oxygen levels in the water. The reduced oxygen levels can cause fish kills and negatively impact other aquatic organisms [23]. Additionally, the mats impede water flow, disrupt navigation, and reduce available habitat for fish and other wildlife, affecting fishery activities. Managing the presence of invasive macrophytes and promoting the growth and conservation of native macrophytes are essential for maintaining the ecological balance and fishery potential of Lake Victoria [24].

3.1 Water hyacinth (*Eichhornia crassipes*)

Water hyacinth (*Eichhornia crassipes*) is an invasive macrophyte species that has become a major ecological and economic concern in Lake Victoria, the largest tropical lake in Africa. Originally introduced as an ornamental plant, water hyacinth has rapidly spread across the lake, forming dense mats that cover large areas of the water surface. Water hyacinth (*Eichhornia crassipes*) is a free-floating macrophyte that established in the lake due to nutrient influxes [16]. Water hyacinth was reportedly introduced to Lake Victoria from the Ugandan sector through the mouth of River Kagera [25]. Water hyacinth's ability to reproduce rapidly and form dense mats has contributed to its status as an invasive species in Lake Victoria. The plant's floating nature allows it to easily spread across the water surface, aided by wind, water currents, and human activities. The absence of natural predators and competitors in the lake has further facilitated its uncontrolled growth. Water hyacinth thrives in eutrophic conditions, taking advantage of high nutrient levels resulting from agricultural runoff and untreated sewage discharge [20]. The proliferation of water hyacinth has had severe ecological consequences in Lake Victoria. The dense mats of water hyacinth block sunlight, reducing photosynthesis and oxygen levels in the water. This inhibits the growth of submerged aquatic plants and disrupts the balance of the lake's ecosystem. The reduced oxygen levels also lead to fish kills and negatively impact other aquatic organisms [10]. Additionally, water hyacinth alters the physical structure of the lake by impeding water flow, hindering navigation, and reducing the available habitat for fish and other wildlife. The mats create stagnant water pockets, which promote the breeding of disease-carrying mosquitoes, increasing the risk of malaria and other waterborne diseases among the human population living near the lake [22].

Lake Victoria is renowned for its vibrant fisheries, providing a source of livelihood for millions of people. However, water hyacinth has severely impacted the fishery potential of the lake. The dense mats of the plant reduce access to fishing grounds, making it difficult for fishers to cast their nets or use traditional fishing gear [23]. As a result, fish catches have significantly declined, leading to economic losses and food insecurity. The plant's dense growth also alters the food web dynamics in the lake. Water hyacinth outcompetes and displaces native aquatic plants, which serve as important habitats and food sources for fish. The reduction in the availability of suitable spawning grounds and food for young fish hampers their survival and growth [10, 26]. Consequently, fish populations, particularly those that rely on submerged vegetation, have declined. Water hyacinth also reduces fishing pressure in the intensively infested areas by blocking the movement of fishing crafts and deployment of the fishing gear. The weed has contributed to reduced primary productivity of the lake due to shading of phytoplankton by water hyacinth mats. In addition, the water hyacinth mats also restrict wind action on surface waters preventing the exchange of oxygen across the air-water interface [25]. They also deplete dissolved oxygen from the water column through the microbial breakdown of decomposing plant remains [16]. This creates anoxic conditions that cause deleterious effects to aquatic organisms and can lead to massive fish mortalities [17]. In addition, this range falls below the lethal limit of 2 mg l^{-1} that is detrimental to fish survival and can often cause massive fish mortalities [27]. Studies have shown that *Lates niloticus* and *Rastrineobola argentea* mainly occur in the oxygenated open waters because they are sensitive to low DO levels of less than 5 mg l^{-1} [8]. This excludes *L. niloticus* and *R. argentea* from areas covered by water hyacinth mats that have reportedly recorded low DO levels ranging from 1.32 to 3.68 mg l^{-1} [28]. In addition, the proliferation of water hyacinth has led to the recovery of hypoxia tolerant native fish species as catfishes, *haplochromines*, *Protopterus aethiopicus,* and *O. niloticus* that find refugia from predation beneath water hyacinth mats [7, 16]. According to Meerhoff et al. [12, 25], the abundance and diversity of fish species in Lake Victoria were higher in submerged vegetation, followed by water hyacinth and unvegetated littoral sites. The marked differences in fish diversities between open waters and water hyacinth infested areas can be attributed to exclusion of *L. niloticus* which preys upon most native species as well as food, shelter, and refugia provided to the fishes by water hyacinth mats [12]. Several methods (biological, chemical, and mechanical) have been used by the three riparian countries to control the weed with very little success. It has been suggested that reducing nutrient loads through influent rivers can curtail the proliferation of the water hyacinth within the lake [7, 12, 16]. Numerous strategies have been implemented to mitigate the impacts of water hyacinth in Lake Victoria. These include manual and mechanical removal, biological control using weevils (Neochetina spp.), and the development of eco-friendly technologies for harvesting and processing the plant. These efforts aim to control the spread of water hyacinth, restore ecological balance, and revive the fishery potential of the lake [12].

3.2 Dense waterweed (*Egeria densa* Planchon)

Like water hyacinth, Dense waterweed, *Egeria densa* is endemic to warm tropical and temperate lakes in South America [18]. The plant belongs to family Hydrocharitaceae of submerged monocotyledonous perennial aquatic plants

(**Figure 1**). Little has been documented about the ecology of these plant in Lake Victoria. Nevertheless, the plant has been observed to grow in the shallow inshore areas of the Kenyan sector, forming dense and thick mats with intertwining stems, and rooted 1–2 m below the water surface [21]. Furthermore, the plant persists as fragments that drift in the water column, which are propagated into thick and extensive marts that cover large and expansive areas of the water column [9, 18, 21]. This helps to absorb nutrients locked in the substrate, making them available to biota. The plant is successfully propagated in a submerged environment due to its physiological adaptations related to its metabolism [29].

These traits enable the plant to photosynthesize under low CO_2 concentration, non-optimal water temperatures, and different nutrient concentrations of water and sediments that affect plant metabolism and ultimately community structure and distribution of the plants [21, 30, 31]. For instance, *Egeria densa*, exhibits the C4 pathway and utilizes bicarbonates HCO_3^- in waters with low CO_2 levels, can tolerate high phosphorous levels, but is susceptible to iron deficiency. Despite being able to thrive in turbid environments [10, 30], the plant has been displaced, having low populations in areas covered by dense marts of Water Hyacinth.

Note: The presence of large populations of *E. densa* decreases water turbulence and resuspension of sediments, which increases the amount of light available in the water column. By sequestering nutrients from sediments, *E. densa* reduces phytoplankton biomass and increases zooplankton abundance and distribution by providing refugia against their predation (**Figure 2**). However, in the long-term plants may result in the increase in sediment height [9].

As a dominant species in the nutrient dynamics, *E. densa* frequently influences phytoplankton biomass by shading phytoplankton in the water column, and can provide refugia to zooplankton and fish escaping predation [22, 23, 32, 33]. Given its tendency to acquire nutrients from the water column, *E. densa* can reduce nutrient availability for phytoplankton. The highly invasive nature of Egeria results in the weed outcompeting and displacing native underwater vegetation such as floating pondweed and ribbon weed [24, 34, 35].

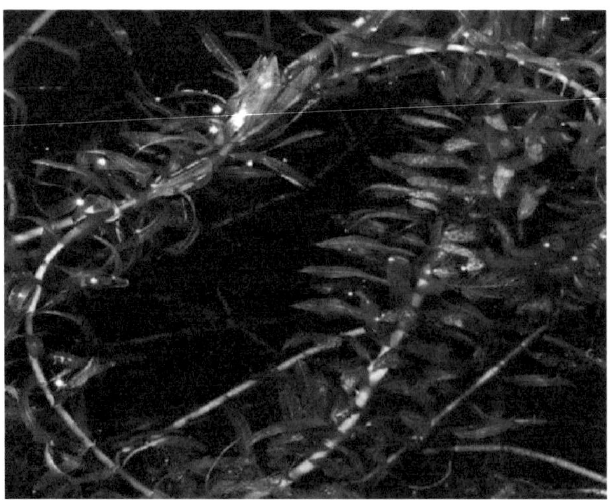

Figure 1.
Photograph of Egeria densa (Planchon) [9].

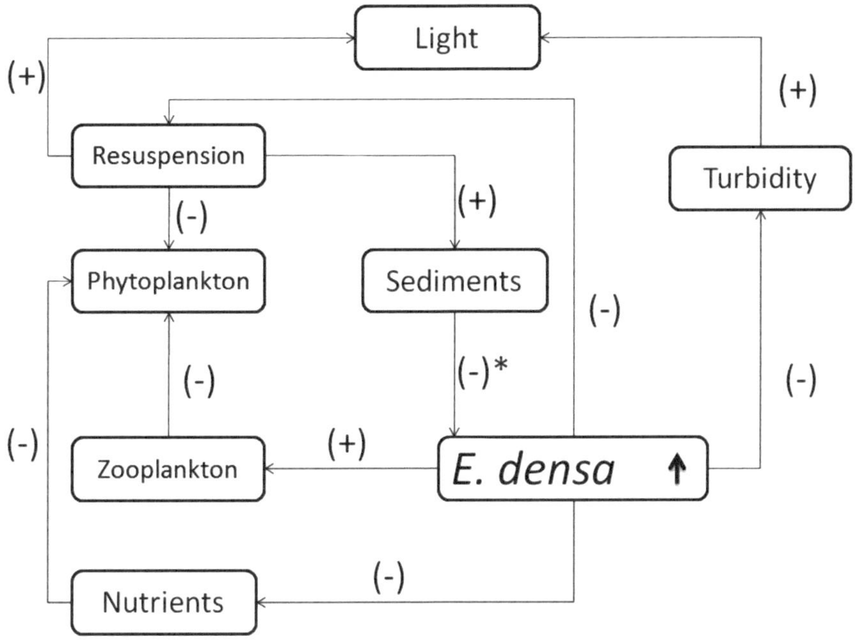

Figure 2.
Egeria densa as an ecosystem engineer.

4. The aftermath of species introductions

The decline and virtual disappearance of native fisheries in Lake Victoria can be attributed to overexploitation, destructive fishing, and introduction of non-native species [26]. The major cyprinids, *Labeobarbus altianialis*, and African carp *Labeo victorianus*, are currently redlisted among the critically endangered species by IUCN [36]. *Labeo victorianus* is a potamodromous fish that migrates from the lake to the major influent streams and tributaries such as Sondu, Kuja, and Mara during the rainy season to spawn in the floodplains [36]. According to Balirwa et al. [7], the contribution of *L. altanialis* to fish landings from Lake Victoria declined from 8173 tons in the 1980s to 152 tons in the late 1990s and early 2000. Before the establishment *L. niloticus* as a top predator, the catfishes *Bagrus docmak*, *Clarias gariepinus,* and *Schilbe intermedius* were the apex predators in the lake [15]. While *B. docmak* disappeared completely, *C. gariepinus* and *S. intermedius* were not greatly affected by this ecological change. Although S. *intermedius* can partly occupy open waters, it has been the main target species of the gill net fishery in the inshore areas, resulting to its decline over the years. On the other hand, the population of *C. gariepinus* is replenished by continuous recruitment of juveniles from major influent rivers. Populations of the African lungfish *Protopterus aethiopicus* in Lake Victoria have also reported a remarkable decline in the past, for instance, from 0.3 to 0.07 tons between 1986 and 1990 [37]. This could be attributed to loss of refugia caused by wetland conversion, and decreased recruitment, due to harvesting of the nest-guarding male lungfish [11]. Other possible reasons for this decline could be predation by the voracious Nile perch, declining food resources and habitat alteration [11, 37].

5. Conclusion and recommendations

In Lake Victoria, introduction of non-native cichlids such as Lates niloticus, *Oreochromis niloticus*, *Tilapia zilii*, and *Oreochromis leucosticus* had negative and positive impacts on the fishery potential and the ecology of the lake. These include interspecific competition among the introduced and native Tilapiines, hybridization of native tilapia by *O. niloticus*. The most remarkable change observed in Lake Victoria fisheries is the decline and total disappearance of endemic Tilapiines *Oreochromis esculentus* and *O. variabilis*, some catfishes (*Xenoclarias eupogon*), several species of Haplochromines and the cyprinids *Labeo victorianus* and *Labeobarbus altinialis*. This has significantly reduced the lake biodiversity, and prompted the introduction of non-native species such as *Lates niloticus* and four non-native Tilapiines including *O. niloticus* into the lake. There is a need to assess current fishery management strategies and formulate new ones based on sound scientific research, which can be implemented in order to prevent loss of biodiversity. On the other hand, some introduced species have led to increase in fish landings as well as utilization of unoccupied niches. Successful establishment and infestation of water hyacinth, *Eichhornia crassipes* outcompeted Nile cabbage, *Pistia stratiotes and* dense waterweed *Egelia densa* resulting to shading, increased turbidity, and reduced dissolved oxygen levels. This has been a major factor contributing to dwindling trends in *L. niloticus* and *R. argentea* fisheries in the lake. However, proliferation of the macrophyte has recently contributed to re-emergence of some native species such as catfishes, haplochromines, *O. niloticus,* and the African lungfish *P. aethiopicus*. Future introductions should be based on sound scientific research in order to minimize the effects of introduced species on native species in the new ecosystems.

Author details

Job O. Omweno[1,2*], Reuben Omondi[2], Fredrick M. Ondemo[1,2]
and Argwings Omondi[2,3]

1 Department of Agriculture, Livestock, Fisheries, Cooperative Development and Irrigation, Kisii County Government, Kisii, Kenya

2 Department of Environment, Natural Resources and Aquatic Sciences, Kisii University, Kisii, Kenya

3 Department of Agriculture and Environmental Science, Sigalagala National Polytechnic, Kakamega, Kenya

*Address all correspondence to: omwenojob@gmail.com

References

[1] Mboya DO, Manyala JO, Ngugi CC. Fish introductions and their impact on the biodiversity and the fisheries of Lake Victoria. Hydrobiologia. 2005;**3**:234-255

[2] Njiru M, Van der Knaap M, Taabu-Munyaho A, Nyamweya CS, Kayanda RJ, Marshall BE. Management of Lake Victoria fishery: Are we looking for easy solutions? Aquatic Ecosystem Health & Management. 2014;**17**(1):70-79

[3] Ogutu-Ohwayo R. The decline of the native fishes of lakes Victoria and Kyoga (East Africa) and the impact of introduced species, especially the Nile perch, Lates niloticus, and the Nile tilapia, *Oreochromis niloticus*. Journal of Environmental Biology of Fishes. 1990;**2**(1):118-131

[4] Goudswaard K, Witte F, Katunzi EFB. The invasion of an introduced predator, Nile perch (*Lates niloticus*, L.) in Lake Victoria (East Africa): Chronology and causes. Environmental Biology of Fishes. 2008;**81**(2):127-139

[5] Kitchell JF, Schindler DE, Ogutu-Ohwayo R, Reinthal PN. The Nile perch in Lake Victoria: Interactions between predation and fisheries. Ecological Applications. 1997;**7**(2):653-664

[6] Katunzi EFB, Zoutendijk J, Goldschmidt T, Wanink JH, Witte F. Lost zooplanktivorous cichlid from Lake Victoria reappears with a new trade. Ecology of Freshwater Fish. 2003;**12**(4):237-240. DOI: 10.1046/j.1600-0633.2003.00023.x

[7] Balirwa JS, Chapman CA, Chapman LJ, Cowx IG, Geheb K, Kaufman L, et al. Biodiversity and fishery sustainability in the Lake Victoria Basin: An unexpected marriage? Bioscience. 2003;**53**(8):703-715

[8] Njiru M, Ojuok JE, Okeyo-Owuor JB, Muchiri M, Ntiba MJ, Cowx IG. Some biological aspects and life history strategies of Nile tilapia, *Oreochromis niloticus* (L.) in Lake Victoria, Kenya. African Journal of Ecology. 2006;**44**:30-37

[9] Yarrow M, Marín V, Finlayson M, Tironi A, Delgado LE, Fischer F. The ecology of Egeria densa Planchon (Liliopsida: Alismatales): A wetland ecosystem engineer? Revista Chilena de Historia Natural. 2009;**82**:299-313

[10] Feijoo C, Garcia M, Momo F, Toja J. Nutrient absorption by the submerged macrophyte *Egeria densa* Planch.: Effect of ammonium and phosphorus availability in the water column on growth and nutrient uptake. Limnetica. 2002;**21**:93-104

[11] Outa N, Mungai D, Keyombe JLA. The impacts of introduced species on lake ecosystems: A case of lakes Victoria and Naivasha, Kenya. 2019;**5**(2):124-138. DOI: 10.31730/osf.io/b5ny

[12] Meerhoff M, Mazzeo N, Moss B, Rodriguez-Gallego L. The structuring role of free floating versus submerged plants in a subtropical shallow lake. Aquatic Ecology. 2003;**37**:377-391

[13] Witte F, Welten M, Heemskerk M, van der Stap I, Ham L, Rutjes HA. Major morphological changes in a Lake Victoria cihlid fish within two decades. Biological Journal of the Linnean Society. 2008;**94**:41-52

[14] Witte F, Van Oijen MJR. Taxonomy, ecology and fishery of Lake Victoria haplochromine trophic groups. 1990

[15] Outa NO, Yongo EO, Keyombe JLA, Ogello EO, Namwaya WD. A review on

the status of some major fish species in Lake Victoria and possible conservation strategies. Lakes & Reserv. 2020;**25**:105-111. DOI: 10.1111/lre.12299

[16] Njiru M, Othina A, Getabu A, Tweddle D, Cowx IG. The invasion of water hyacinth, *Eichhornia crassipes* Solms (Mart.), a blessing to Lake Victoria fisheries. In: Cowx IG, editor. Management and Ecology of Lake and Resevoirs Fisheries. Blackwell Science, Oxford, UK: Fishing News Books; 2002. pp. 255-263

[17] Ochumba PBO, Kibaara DI. Observation on blue -green algae blooms in the open waters of Lake Victoria, Kenya. African Journal of Ecology. 1989;**27**:23-34

[18] Consortium of California Herbaria. Consortium of California Herbaria. Berkeley, CA: Consortium of California Herbaria; 2014. Available from: http://www.gbif.org/dataset/4fa894f4-b6c6-4ec0-b816-9bb03b3ca106

[19] Ouma D, Manyala JO, Ngugi CC. Fish introductions and their impact on the biodiversity and the fisheries of Lake Victoria. Lake and Reservoir Management. 1993;**2**:118-127

[20] Ntiba MJ, Kudoja WM, Mukasa CT. Management issues in the Lake Victoria watershed. Lakes and Reservoirs: Research and Management. 2001;**6**:211-216

[21] Alfasane MDA, Khondker M, Islam MD, Bhuiyan MA. Egeria densa Planchon (Hydrocharitaceae): a new angiospermic record for Bangladesh. Bangladesh Journal of Plant Taxonomy. 2010;**17**(2):209-213

[22] Schriver P, Bogestrand J, Jeppesen E, Søndergaard M. Impact of submerged macrophytes on fish-zooplankton-phytoplankton interactions—large-scale enclosure experiments in a shallow eutrophic lake. Freshwater Biology. 1995;**33**:255-270

[23] Hamabata E, Kobayashi Y. Present status of submerged macrophyte growth in Lake Biwa: Recent recovery following a summer decline in the water level. Lakes & Reservoirs: Research and Management. 2002;**7**:331-338

[24] Takahashi K, Asaeda T. The effect of spring water on the growth of a submerged macrophyte Egeria densa. Landscape and Ecological Engineering. 2014;**10**:99-107

[25] Muli RJ. Environmental problems in Lake Victoria (East Africa). Lakes and Reservoirs: Research and Management. 1996;**2**:47-53

[26] Sitoki L, Gichuki J, Ezekiel C, Wanda F, Mkumbo OC, Marshal BE. The environment of Lake Victoria (East Africa): Current status and historical changes. International Review of Hydrobiology. 2010;**98**(3):209-223. DOI: 10.1002/iroh.201011226

[27] Chapman H. Water quality assessment: A guide to the use of biota, sediments and water in environmental modeling. Journal of Environmental Management. 1996;**3**:156-168

[28] Giraldo E, Garzon A. The potential for water hyacinth to improve the quality of Bogota River water in the Muña reservoir: Comparison with the performance of waste stabilization ponds. Water Science and Technology. 2002;**42**:103-110

[29] Dugdale TM, Clements D, Hunt TD, Butler KL. Survival of a submerged aquatic weed (*Egeria densa*) during lake drawdown within mounds of stranded vegetation. Lake and Reservoir Management. 2012;**28**:153-157

[30] Walsh GC, Dalto YM, Mattioli FM, Carruthers RI, Anderson LW. Biology and ecology of Brazilian elodea (*Egeria densa*) and its specific herbivore, Hydrellia sp., in Argentina. BioControl. 2013;**58**:1-147

[31] Randall JM, Marinelli J. Invasive Plants: Weeds of the Global Garden. Brooklyn, New York, United States of America: Brooklyn Botanic Garden; 1996

[32] Bini LM, Thomaz SM. Prediction of *Egeria najas* and *Egeria densa* occurrence in a large subtropical reservoir (Itaipu reservoir, Brazil-Paraguay). Aquatic Botany. 2005;**83**:227-238

[33] Center for Invasive Species and Ecosystem Health. EDDMapS: Early Detection and Distribution Mapping System. Tifton, GA: The University of Georgia; 2015. Availble from: http://www.eddmaps.org

[34] Plantlife. Large-flowered waterweed (*Egeria densa* (aka Elodia [sic] densa)). 2010. Available from: http://www.plantlife.org.uk/wild_plants/plant_species/large-flowered_waterweed [Accessed: March 14, 2023]

[35] Cavallim RM et al. Remote sensing water observation for supporting Lake Victoria weed management. Journal of Environmental Management. 2009;(3):136-148

[36] Ondemo FM, Getabu A, Gichana Z, Omweno JO. Length-weight relationships (LWRs) and condition factor of seven fish species in River Nyangweta Tributary, Kenya. Journal of Engineering Research and Sciences. 2022;**1**(3):193-199

[37] Goudswaard PC, Witte F, Katunzi E. The Tilapiine fish stock of Lake Victoria before and after the Nile perch upsurge. Journal of Fish Biology. 2002;**60**:838-856. DOI: 10.1006/jfbi.2002.1888

Chapter 9

Dynamic Limnology between Movements and Management

Maximilien Bernier

Abstract

The subject matter of this chapter revolves around the intersection of hydrodynamics in French lakes with the economic development of their respective territories. The aim is to understand the interactions between the dynamics of these lakes and territorial actors operating within these limnic areas. Each party is subjected to different management approaches, yet none of them take advantage of lake movements. These management strategies are organized based on the specific limnological mechanisms of each lake, including their temporal and spatial characteristics, as well as the risks they pose to human activities integrated into these lake ecosystems. This chapter will specifically focus on unused water and sediment movements as means to interact with local management practices. These phenomena serve as a foundation upon which all actors surrounding these lakes can rely to guide the protective measures implemented for their activities. This chapter aims to provide scientific insights to support decision-making for the actors operating in proximity to these lakes.

Keywords: hydrodynamics, lakes, limnology, management, movements

1. Introduction

During the inception of limnology, François-Alphonse Forel introduced the concept that human beings are not merely part of lake's ecosystem. Through their activities, humans have the capacity to interact significantly with a lake and impact its ability to provide ecosystem services [1]. From a legal perspective, French lakes, including those in Europe, are theoretically required to be managed in the same manner. However, the reality is different. Despite the fact that management approaches are heavily influenced by scientific knowledge, each manager operates according to their individual needs and challenges. Every entity present along the shores of these lakes contributes to human influence on their respective areas of interest.

From a management perspective, the administrative structuring of water resources in France is guided by a range of directives, including the Water Framework Directive (WFD-2000), Water Development and Management Master Plans, and municipal decrees. The overarching objective is to achieve a state of "good condition," which theoretically considers societal entities inhabiting the managed territory and their expectations. However, practical implementation deviates from this ideal. Despite aspirations for a systemic approach, the prevailing framework is characterized by

IntechOpen

rigidity, relying on naturalistic indicators while neglecting a diverse range of actors utilizing these aquatic environments, particularly for recreational purposes. Today, the number of stakeholders surrounding the lakes has diversified significantly and is still gradually increasing. Within the federation of municipalities of the Great Lakes, a multitude of sectors are represented, including military, oil, tourism, associations, regulatory authorities, and administrative entities. Consequently, the primacy of exclusively studying the dynamics of lakes as a reference for management indicators becomes increasingly questionable. Significant dynamic phenomena, such as flooding, water level fluctuations, navigation, erosion, or conservation of fauna and flora, come into play and shape management considerations. It is thus pertinent to explore dynamic behaviors of these lakes. Limnology facilitates a territorial approach that encompasses all relevant stakeholders integrating them into a comprehensive societal system. This approach considers the spatial and temporal dimensions, encompassing both natural and anthropogenic aspects. It involves establishing linkages between the environmental dynamics of biocenosis, biotope, and the socio-cultural practices imposed by human activities in relation to lakes.

In an approach to limnological geography [2], which places lakes at the centre of societal-environmental interactions across various spatial and temporal scales, the employ of dynamic limnology will use, in this chapter. This discipline seeks to understand the physical dynamics exerted by a lake within its own boundaries or towards its immediate surroundings. This dynamic perspective completes physical and biocenosis approach usually prefer to study water bodies to assist in management strategies in the face of global change. This study of lake dynamics (movements of water and sediments) was born with the creation of limnological sciences by F.-A. Forel, first studied the dynamics of the formation of Lake Geneva (Léman, in French), including geomorphological aspects and the appearance of "seiches" (variations in water level within the same lake) [3]. More recent researchers have furthered these investigations to enhance the quantification and mapping of lake physics at various scales [4]. We have opted for the dynamic approach due to its historical neglect in these limnic territories [5], especially by managers. The multitude of spatial approaches employed in contemporary limnology adds complexity to discerning the roles of different actors in lake management, without assigning undue importance to any perspective. To address this complexity and translate it into a diagnosis that prioritize specific parameters, our focus is on diagnosing water and sediment movements.

Our objective here is to produce a comprehensive diagnosis of the dynamic mechanisms occurring within these lakes to enhance their consideration by managers. By creating a diagnosis rooted in dynamic phenomena applied to specific territories, we aim to provide novel and readily applicable data on lake behaviours that can be easily understood by stakeholders and inform future management strategies. To achieve this, we will generate new cartographic data at the scale of the specific lake under study, irrespective of its characteristics, within the federation of municipalities of the Great Lakes located in the Landes, New-Aquitaine, France.

2. Methods

In order to develop a chapter that offers practical perspectives to managers, we will study the hydrodynamic processes of lakes using methodologies that can be easily replicated.

2.1 Presentation of studied lakes

The sites examined in this chapter concern two lakes situated in the federation of municipalities of the Great Lakes (**Figure 1**), within the Department of Landes, in

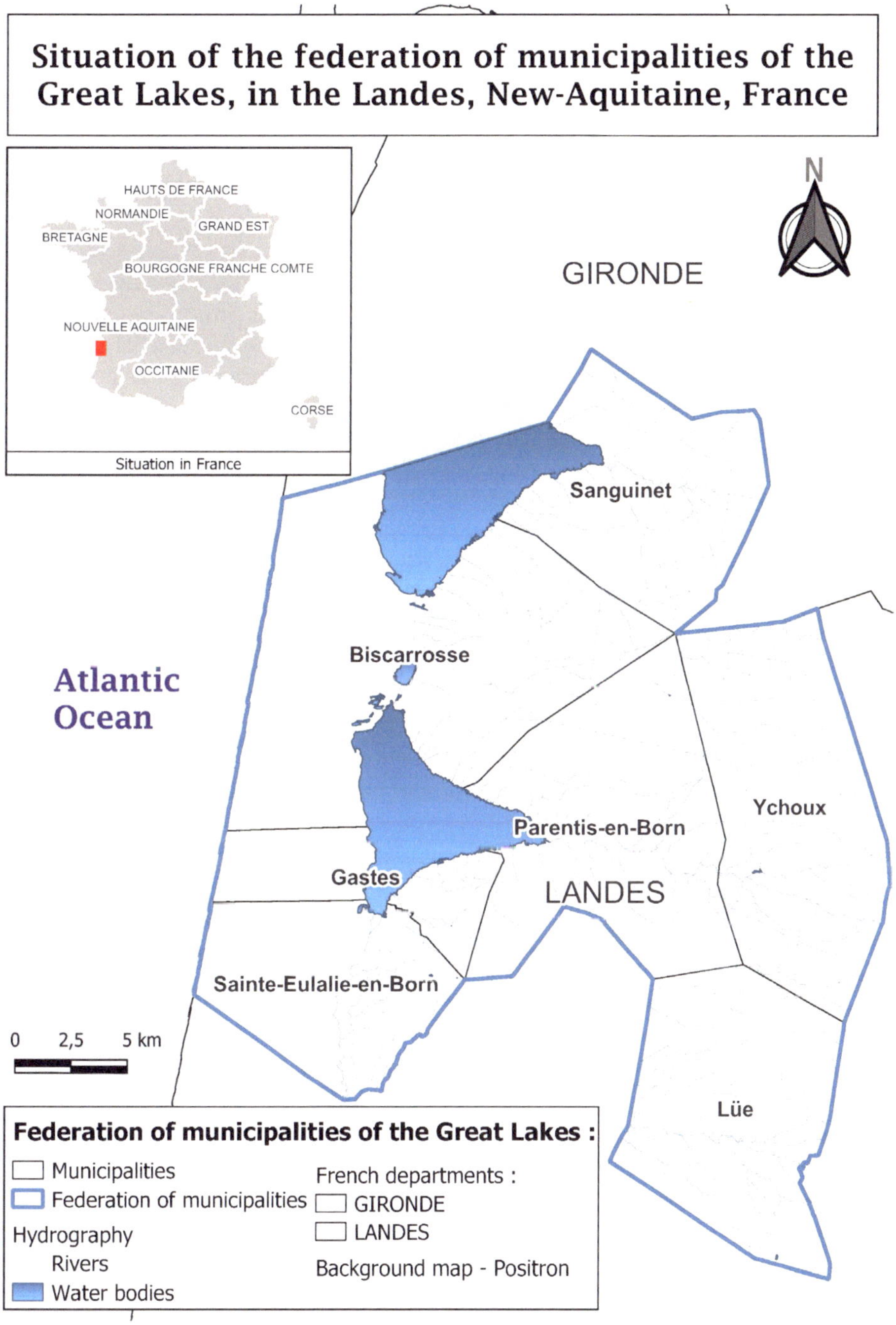

Figure 1.
Localization map.

the region of New Aquitaine (France). This study follows on from a French thesis that explores the following question: Can water movements and their effects on differentiated sediment deposits in lakes and ponds contribute to the maintenance of economic activities in their territory?

Most lakes along the Aquitaine coast are geographically separated from the Atlantic Ocean by coastal dune barriers. The formation and development of these water bodies are closely intertwined with the history of the dunes themselves. By examining the periods of mobility of the dune belts, it has been possible to determine the specific characteristics and evolution of these lakes. The available data indicate that these bodies of water originally existed as lagoons wide open to the ocean until around 1000 BC [6]. Subsequently, these lagoons progressively became obstructed due to the accumulation of sand transported by longshore drift and finally closed definitively in Gallo-Roman era.

These lakes within the federation of municipalities of the Great Lakes are commonly referred to as ponds of Parentis-Biscarrosse and Cazaux-Sanguinet in France. Historically, these study sites were neglected in favor of their proximity to the nearby ocean, leading to their derogatory classification as "ponds" at the time. However, according to Laurent Touchart's definition, these water bodies qualify as lakes. Locally, they are now recognized and referred to as lakes, which is technically accurate considering that their average depth exceeds six metres, as defined by the Ramsar Convention of 1971. Additionally, they cover an area exceeding 200 ha, which meets the criteria set forth by E. Jedicke for defining a lake. Therefore, in this chapter, we will refer to them as lakes. Studied lakes collectively encompass over 9340 ha, with a total volume of 750 million liters. More specifically, Lake Parentis-Biscarrosse holds an estimated volume of 250 million liters, while Lake Cazaux-Sanguinet is estimated to contain 500 million liters [7].

Parentis-Biscarrosse and Cazaux-Sanguinet lakes are located in the extreme northwest of the Landes department, close to the border with Gironde (a French department a little further north) and below Arcachon basin. These two natural lakes can be reached in around an hour's drive from Bordeaux. Lake Parentis-Biscarrosse is slightly smaller and situated a little further south than Lake Cazaux-Sanguinet. It covers an area of approximately 3540 ha at an altitude of 19 m above sea level, with a maximum depth of 20.5 m. Lake Cazaux-Sanguinet, on the other hand, covers an area of around 5600 ha, at an altitude of 20 m above sea level, with a maximum depth of 23 m [7].

These lakes also share the characteristic of being shared by different communes. Lake Parentis-Biscarrosse is located in the municipalities of Parentis-en-Born, Biscarrosse, Gastes and Sainte-Eulalie-en-Born. Its counterpart is spread across the communes of Biscarrosse, Sanguinet and Cazaux, which come under the jurisdiction of the Gironde town of La Teste-de-Buch.

To move beyond the confines of limnology solely for naturalistic or physicochemical purposes, we have incorporated dynamics (**Figure 2**) of these "lentic" features to address their economic and societal significance. This diagnostic approach considers the criteria mentioned above.

Methodologies employed in this study encompass a combination of surveys and field observations, complemented using Geographic Information Systems (GIS).

2.2 Bathymetries

Lakes exist within basins, and when examining their dynamics, bathymetry emerges as a fundamental approach for initial investigations. The Adour-Garonne

Figure 2.
Illustration of lake dynamics interest.

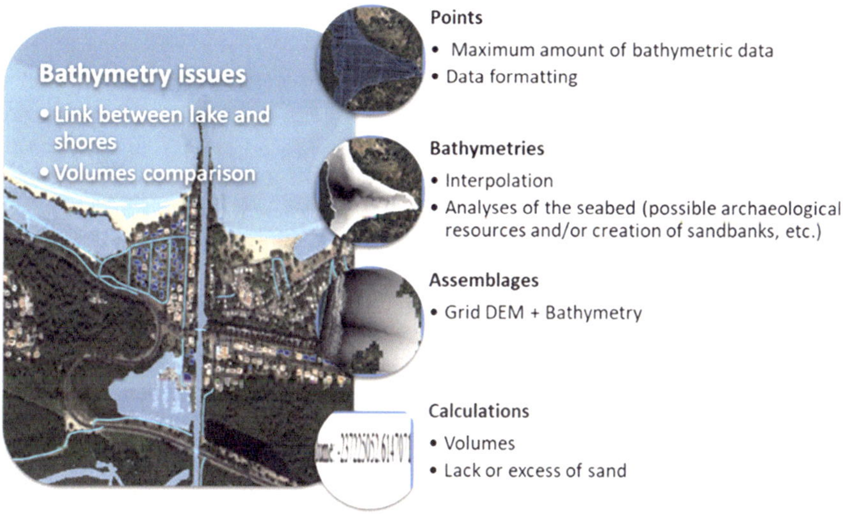

Figure 3.
Sediment movement's process.

water agency conducted bathymetric surveys of lakes under study in 2014, and we will leverage their existing data. Additionally, bathymetric information was supplemented with a Digital Terrain Model generated by the French National Geographical Institute (IGN). By utilizing pre-existing data available throughout French territory, this diagnostic approach can be replicated on other water bodies (**Figure 3**).

- Data formatting: a dataset was prepared involving the processing of approximately 37,000 points for each of these lakes.

- Bathymetry calculation: Bathymetry was determined either through echo sounder surveys conducted on water bodies under study or by using existing point data and employing TIN interpolation techniques.
- Spatial segmentation: Data were cut based on the area of the water body.
- Spatial projection: All calculations were performed in the appropriate spatial projection system, specifically EPSG: 2154 (French work).
- Integration of bathymetric data: Bathymetric data was merged with available digital terrain models.
- Volume calculations: Quantitative assessments of the volume were conducted.
- Integration with built elements and networks: These data were combined with information pertaining to constructed features and various networks, utilizing data from IGN's BD TOPO.
- Cartographic visualization: Cartographic representations were generated, depicting lake beds (including archeological features, sandbanks, etc.) and highlighting issues related to water levels.

This initial methodological approach enables us to understand dynamics of lake beds, as well as the water levels and volumetric characteristics of the water bodies studied.

2.3 Water movements on surface

Following the examination of lake depressions where certain movements occur, our attention will shift to movements generated by winds. The study of wind patterns serves to augment the previous analysis by elucidating the formation of waves on the lake surface. The investigation of these wind patterns (**Figure 4**) is conducted through the utilization of two distinct yet complementary methodologies.

- Potential wind races calculation [8]
- Effective fetch calculation [9]—determining the area where wind energy generates waves—for each studied water body using a regular grid.
- Calculation of wave height, period, and velocity.
- Determination of maximum fetch or wind races.
- Wind weighting based on an annual wind rose provided by MétéoFrance.
- Field verification of data accuracy, where a higher fetch or wind races correspond to increased coastal exposure to winds and erosion.

This secondary methodological approach allows for the acquisition of findings regarding the dynamics of the lake banks or beds and the effects of wave-induced water actions on lake surfaces.

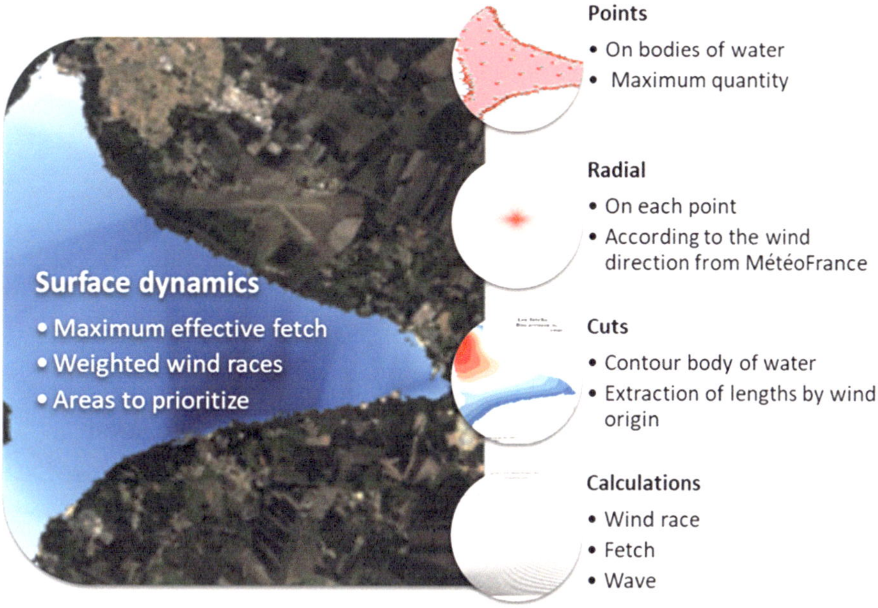

Figure 4.
Water movement process on lake's surface.

2.4 Water movements inside lakes

The movements taking place inside lakes are commonly called currents. Apparently, lakes are even less calm than we think. These currents are not visible to the naked eye but can be easily identified if we pay attention to the evolution of bathymetry, fish location or invasive plants.

The use of an acoustic Doppler current meter coupled with mapping software makes it possible to visualise currents by spatialising their dynamic cells within lakes. Their visualizations depend on whether the current metre is in a fixed station or if it moves on the water's surface. In a fixed station, the device is fixed on a support that is immersed at a predefined point upstream. Its aim is then to go from the lake bottom to the surface. On the other hand, the other device will be guided by our boat and will detect lake currents from the surface to the bottom. To detect the direction of the currents will be used by a compass linked to the device is calibrated directly in real-time by the device in a fixed station. In the mobile position, the compass must be calibrated before any measurement is taken.

- Calibrate the device according to water's salinity.
- Calibrate the compass.
- Prepare the field trip (weather, necessary equipment, and boats with the planned routes).
- Recovery of equipment if it operates in a fixed state.
- Export of data in a usable format under Geographic Information System.

- Assemble data in the projection system used.
- Display them with a map background.
- Interpolate all transects.
- Obtain a mesh representative of the lake (e.g., 10 m step).
- Join data (speed and direction) for each point in the mesh.
- Choose a spatial representation combining speeds and directions.
- Display data based on collected data and depth.
- Correlate data with other previously known movements and issues.

This third methodological approach enables the acquisition of results pertaining to the internal dynamics of lakes and directions of water flux.

These three methodological parts presented the dynamics that we seek to comprehend through a concise diagnosis of water and sediment movements. However, it is important to acknowledge that these dynamics do not exist in isolation from human interactions. Further societal investigations will soon complement this dynamic diagnosis of water bodies.

3. Results

To address societal concerns related to phenomena such as floods, water quality issues, erosion, and sedimentation, we will employ dynamic limnology and utilize the methodologies to assess the impacts in these two lakes with similar genesis.

3.1 Bathymetries

The bathymetric analysis of these lakes was conducted using GIS. Generated maps provide visual representations of submerged topography and features within the lakes. Through this data acquisition, updated volume calculations for lakes were obtained. Newly determined volumes exhibit a variation of approximately 5% compared to previous measurements, even for lakes with identical altimetry, as exemplified by Lake Parentis-Biscarrosse in **Figure 5**.

Nevertheless, these bathymetric analyses of these lakes revealed irregularities that correspond to archeological findings in certain instances, and potential archeological discoveries in others. The lake basins were subsequently incorporated into a digital model, facilitating adjustments to water levels to address the various concerns related to buildings and utility network infrastructures.

Water level projections (**Figure 6**) have been established based on historical records of significant flooding or low-water periods, with the aim of comparing them with water management practices regulated by the lake's altitude. The aim is to assess the need for better management of water levels.

This baseline data will then be used to analyze changes in sediment dynamics linked to the anthropization of the lake's banks.

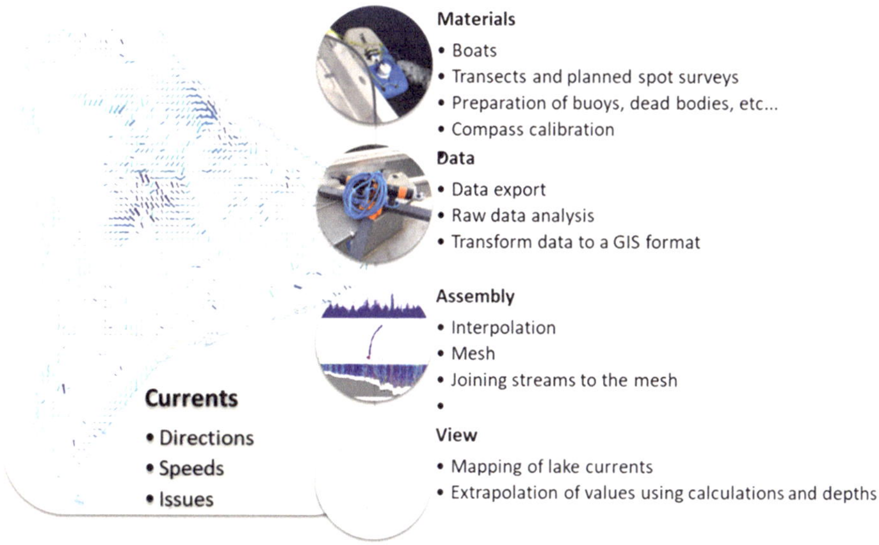

Figure 5.
Water movement process inside lakes.

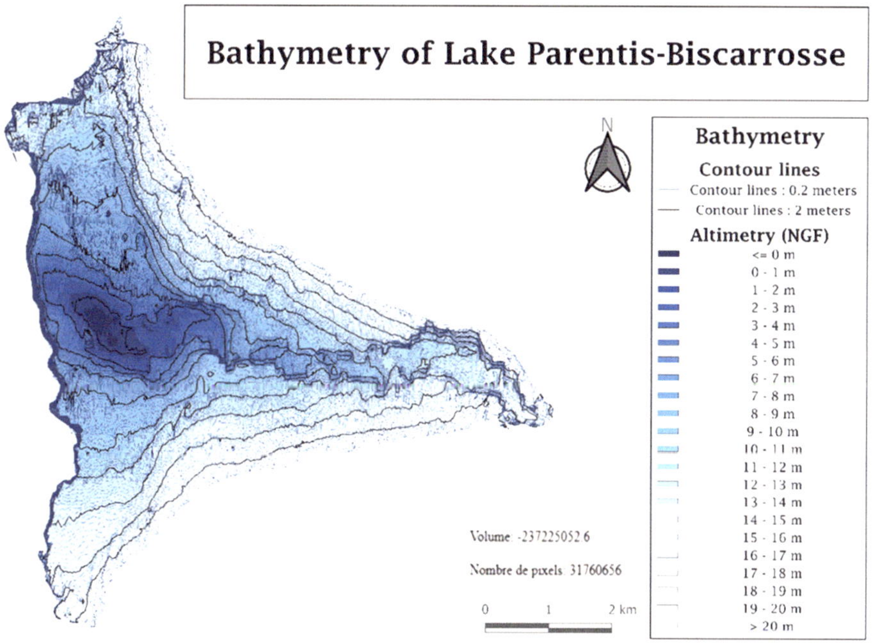

Figure 6.
Pixel-based bathymetry analysis and volume of lake parentis-Biscarrosse.

3.2 Water movements on surface

To study the relationship between wind and waves, we implemented the calculation method proposed by Håkanson and Jansson [9] to determine fetches. To carry out this approach, we selected points of interest in order to recreate fetch diagrams for each wind direction based on the wind rose data from MétéoFrance. Maximum fetches lengths were extracted, while the remaining values were weighted according to the frequency of the corresponding winds. The data obtained can be used to identify beaches subject to erosion due to wind-induced wave action.

The evaluation of fetches makes it possible to obtain theoretical values for the waves that reach the coast. This information helps managers to understand wave phenomena in their limnic territory. Moreover, these theoretical waves can be observed daily by anyone looking to enjoy the view from these lakes.

The analysis of lake dynamics related to wind-induced water movements has produced some interesting results, described in an article written jointly by Pascal Bartout, Laurent Touchart and myself [10]. The obtained data are presented in the form of graphs and maps (**Figures** 7 and **8**). Maps highlight the most exposed coasts, characterized by sectorial winds, which correspond to the eroded zones (illustrated in dark blue in **Figure 9**). Erosion can be understood through accompanying graphs that provide insights into the impacting wave characteristics. Conversely, areas that have experienced minimal wind exposition on these maps exhibit visible sediment accumulation (illustrated by light blue zones in **Figure 9**).

As revealed by the image captured prior to the 2022 summer season (**Figure 10**), the southern coast of Lake Cazaux-Sanguinet exhibits significant erosion, indicated by the dark blue color (**Figure 9**). As a result, this beach, previously used by tourists, is now closed to the public for safety reasons. These phenomena could have been anticipated and dealt with by implementing local policies aimed at reducing these risks. By identifying areas with a high potential for erosion, managers of the federation of

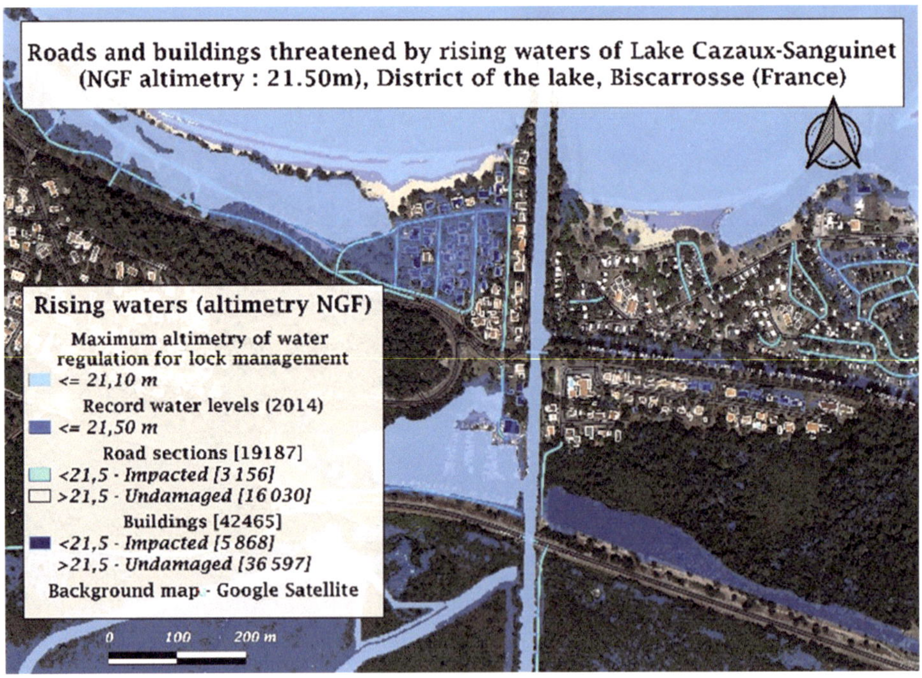

Figure 7.
Issues highlighted by GIS visualization.

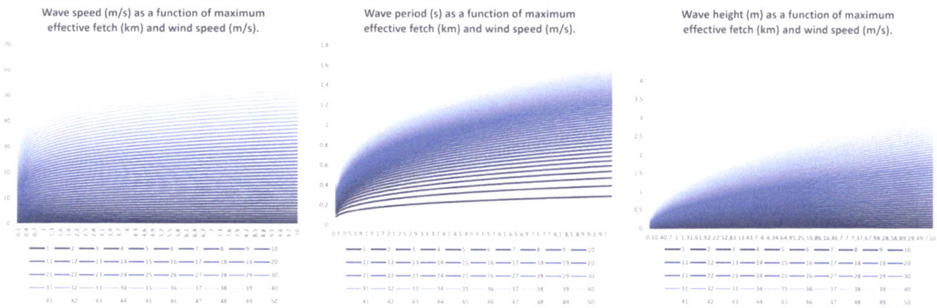

Figure 8.
Wave graphs according to fetches.

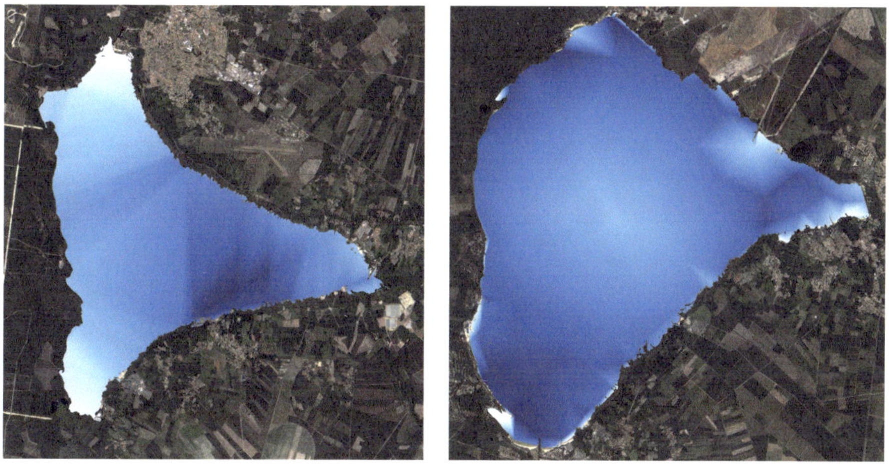

Figure 9.
Wind exposure maps by lake (weighted on the left for lake parentis-Biscarrosse and maximum on the right for lake Cazaux-Sanguinet).

municipalities of the Great Lakes can use these results to prevent risks and forecast the costs of environmental works to maintain uses. This study is, therefore, essential for lake managers to better manage and safeguard these lake environments.

Furthermore, it is noteworthy that invasive exotic plants, Lagarosiphon Major and Egeria Densa, present in these lakes exhibit a tendency to avoid areas characterized by intense winds and waves. These plants have a greater propensity to acclimatize themselves in protected areas from winds and waves. Additional factors and conditions play a role in the colonization of specific sectors by these plants, as it will be explored in the subsequent section.

3.3 Water movements inside lakes

Internal movements of lakes are commonly referred to as currents. Although not very visible, these currents manifest in the form of lake-specific cells moving in different directions and at speeds ranging from almost zero to several metres per second. This section highlights the results related to the correct interpretation of these currents in lake environments.

Figure 10.
Eroded coast.

3.3.1 Water movements directions inside lakes

Movements internal to the lakes are specific because of their directions. These currents must be detected thanks to currents metre in lakes that identify water moving in cells. Different directions and highs from few centimeters to some tens are remarkable. This section highlights the results related to the correct interpretation of these current directions in lake environments.

Cells can be identified by depth. Thanks to this representation by transects and georeferencing on GIS software, it is possible to map it.

With current data, we obtain significant values and representations with speeds associated (**Figure 11**). This makes it possible to have a global visualization of what happens in the lakes at a time T. This particularity allows us today to confirm that lakes are not simple basins filled with water and where water circulates from upstream to downstream, or tributaries to outfalls. In fact, we can follow tributaries from their entrance into lakes to the outfall. It appeared in **Figure 12** with purple water column next to the coasts. Entrance currents followed runs along the coast like an oceanic littoral drift [11], marking a new frontier that invasive plants do not overtake. Moreover, the rest is still moving as we can see a lot of different directions of water inside lakes (**Figure 13**).

Direction data is easily correlated thanks to date and time parameter with winds of the sector, easily identifiable thanks to a wind rose from MeteoFrance. Currents directions correspond to those of the winds and waves produced. Their directions seem to be affected by prevailing winds only on the first 20 cm (**Figure 14**).

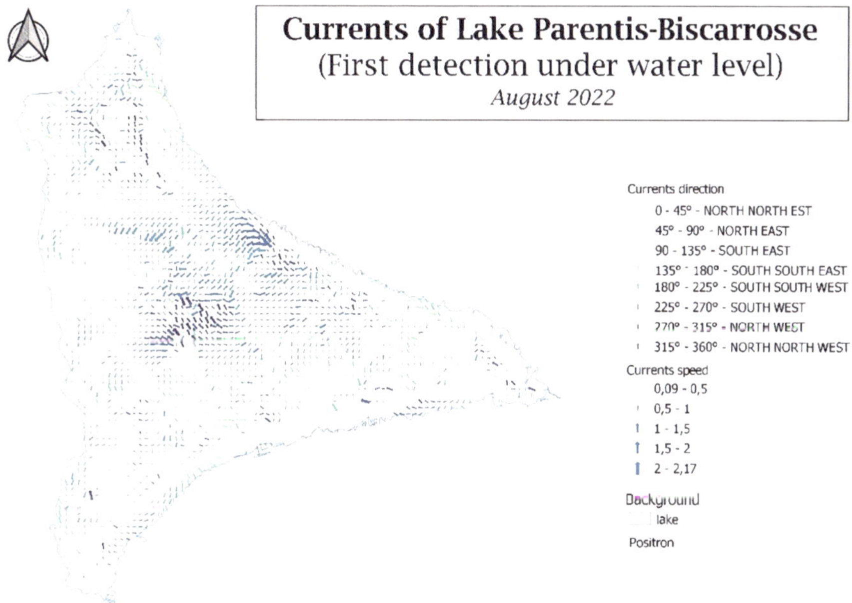

Figure 11.
One of currents map possible.

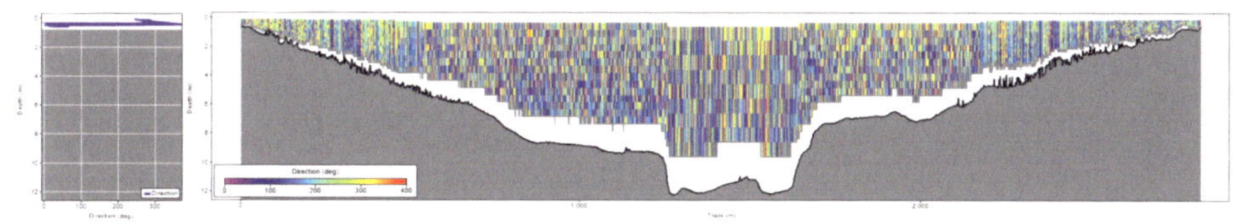

Figure 12.
Data directions untreated.

Figure 13.
Frontier between lake and invasive plants.

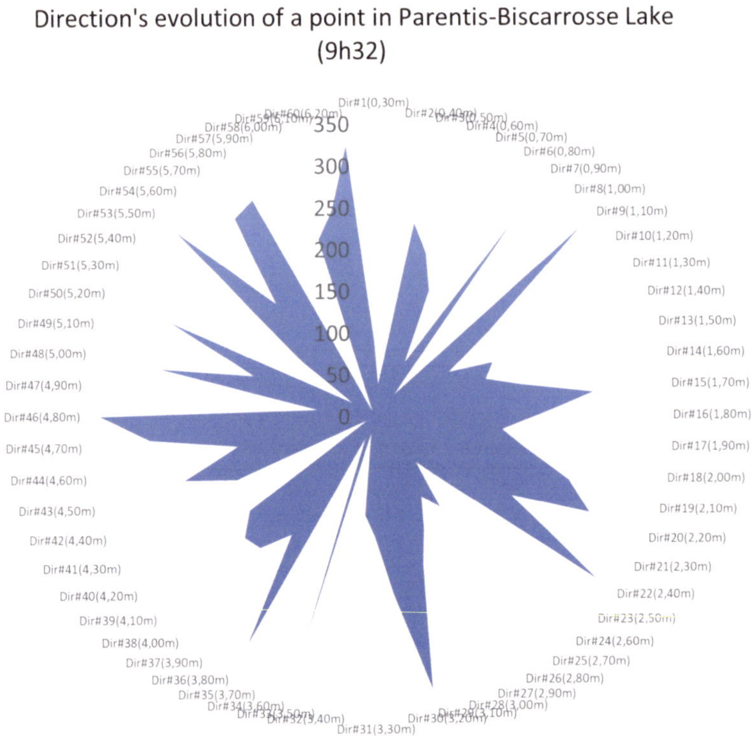

Figure 14.
Wind rose of the directions from currents.

Wind rose of currents in one point, taken from Parentis-Biscarrosse Lake gives evidence directions exiting inside lake and extracting from a current metre. Directions are oscillating inside water column, all day long but also any day (**Figure 15**).

Moreover, these directions on a step of 10 cm is significant to distinguish all movements inside the lakes despite the gales visible in **Figures 16** and **17**. A limit imposed by currents metre to constate that lakes are not calm and sleeping described in the literature.

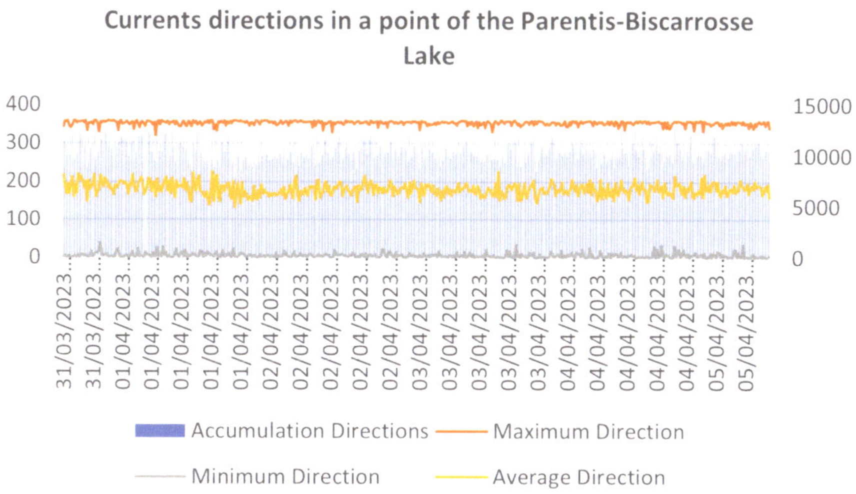

Figure 15.
Currents direction.

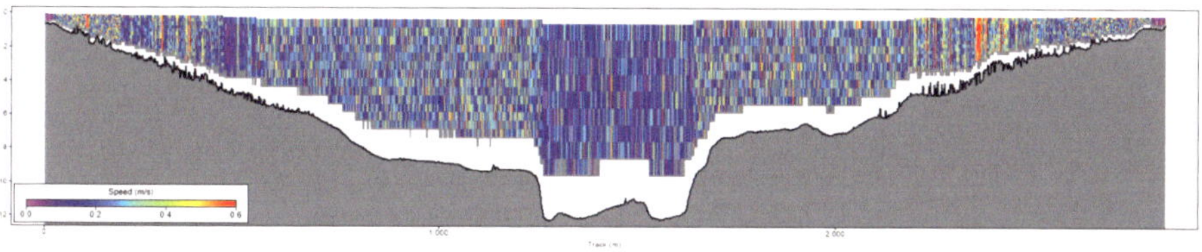

Figure 16.
Transect of speeds.

3.3.2 Water movements speeds inside lakes

Each current reveals a significant speed in cells detected by a current metre. Currents identified by the different measurement campaigns have speeds ranging from nearly 0.01 m per second to more than 4 m per second according to the wind (**Figure 17**).

As the previous figure shows speeds cells of currents, during a day without wind, in red, a tributary of Parentis-Biscarrosse that affects the entirety of the water column. The rest of the lake contains a lot of speeds in blocks that could be considered as a new typology of lake zonation.

Contrary to currents direction and homogeneity in changing, current speeds show irregularities because of gales.

With this graphic, the average speed is the same during all the period of measurement. Parentis-Biscarrosse Lake appears to be in a dynamic equilibrium situation at around a speed of 0.5 m per second, a speed corresponding to the value in metre per second of the largest tributary of the lake. Measurements over the month of June 2022 confirm this average balance at around 0.5 m per second average in this lake. Measurements on the lake, of similar genesis, Cazaux-Sanguinet, will they be able to confirm this hypothesis arises from the interpretation of the results of the lake of Parentis-Biscarrosse?

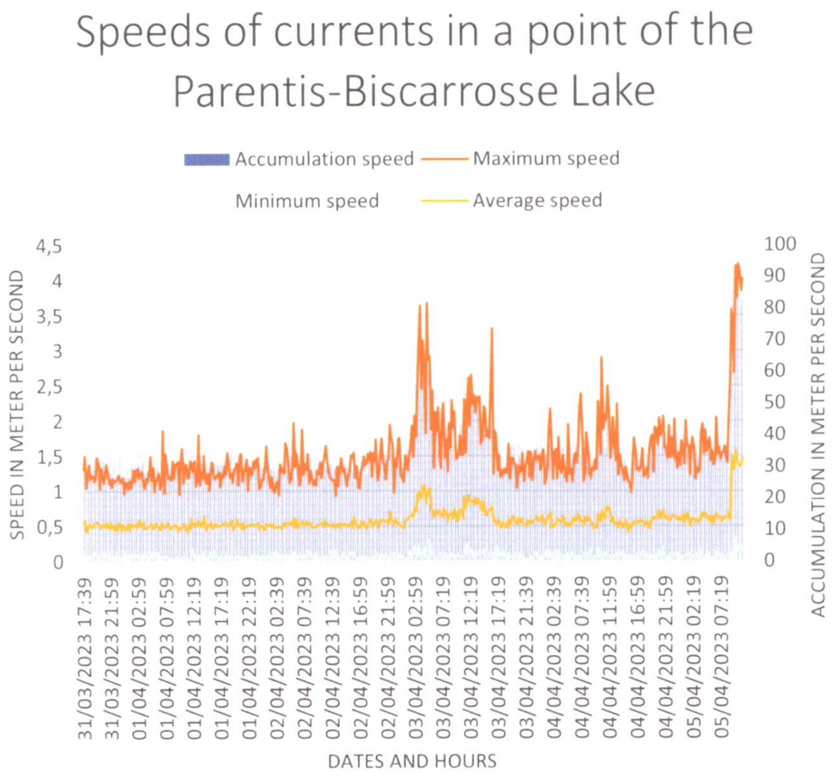

Figure 17.
Currents speeds graphic.

Nevertheless, balance is here of several increases in the average speed in connection with gales exceeding 60 km per hour, over the few days of measurements.

Therefore, a concise assessment of water body dynamics can be achieved utilizing open-source GIS software, requiring no specialized equipment other than for current measurements. The outcomes generated through these distinct methodologies serve as a bridge between local stakeholders and the lakes, facilitating the guidance and planning of actions on these limnic territories.

4. Discussions

These lakes seem to be full of vitality. Are they alive? And just as blood circulates in the human body, water circulates in the lakes, and in an almost random way but under the main effect of winds or rivers.

With these results, the findings obtained enable actors and managers of these lakes to comprehend most phenomena that impact their limnic territories. They can perceive these lakes or ponds as living systems, dependent on various mechanisms that require monitoring to ensure their proper functioning. Armed with this knowledge, each stakeholder can establish guidelines for the preservation of their activities. By analyzing dynamic movements of water and sediments within these water bodies, action plans can be devised to combat invasive plants and safeguard banks and infrastructure. An improved understanding of these dynamics facilitates the anticipation of water body evolution. Based on this information, what measures will the stakeholders in these limnic territories undertake next?

That section on currents brings here, a lot of future evolution. Movement inside lakes are manifold but useful to manager of this lake. These currents can guide jobs for lake as new zonation to advantage some part rather than other?

Currents are permanent, but their direction is not dependable. Could these currents help human being to energetic transition?

This concise chapter awaits further completion with available information pertaining to the subject of these water bodies. It will soon be augmented with comprehensive lake data.

5. Conclusion

This chapter presents a significant advancement in the integration of dynamic limnology by providing a diagnostic approach based on lake movements. It highlights the previously underemphasized dynamic aspects of water bodies, which have not received sufficient attention from managers due to the predominant focus on biocenosis. The inclusion of spatial criteria for dynamics now allows for a comprehensive consideration of the entire system, including human interactions, within the limnic territory. Through these studies, various lake users who previously overlooked the significance of movements occurring within these water bodies can now better comprehend the intricacies of the environment in which they operate. The understanding of erosion, sediment deposition, floods, invasive plant development, and other issues can be improved by incorporating dynamic criteria related to water and sediment movements. This dynamic perspective facilitates the creation of decision-support maps for stakeholders in their respective limnic areas. At the scale of water bodies, focusing on dynamics raises questions about the future evolution of these aquatic environments. Indeed, dynamic characteristics of a water body can vary from one region to another, as observed in these two lakes of similar origin and shape in the northern region of Landes, France. Therefore, this diagnostic approach warrants consideration for every water body due to its ease of implementation. It offers valuable insights and enhances management practices by facilitating a deeper understanding of these complex environments through the lens of lake dynamics.

Acknowledgements

The thanks for this chapter go directly to the federation of municipalities of the Great Lakes (in the Landes, France) and all users or managers of these lakes. Of course, I thank university of Orleans, my thesis directors, and university Mont Blanc – Savoy to that opportunity to study misunderstood lakes movements with materials and knowledges.

Author details

Maximilien Bernier
EA 1210 CEDETE – Laboratory of Geography, University of Orleans, France

*Address all correspondence to: maximilien.bernier@etu.univ-orleans.fr

References

[1] Baulaz Y, Dorioz JM. Exploration Fonctionnelle Du Territoire Des Grands Lacs Alpins Par Le Prisme Des Services Écosystémiques. 2021;**25**

[2] Bartout P. L'incompréhension de la place prise par les plans d'eau dans l'Union européenne et ses conséquences réglementaires. Norois. 2015;**235**:17-36

[3] Cvijic J. La limnologie générale de Mr Forel. geo. 1901;**10**(49):70-72

[4] Choffel Q, Touchart L, Bartout P, Al DM. Le centre thermique, nouvel outil de compréhension du bilan thermique et de l'évolution spatio-temporelle de la température d'un étang? norois. 2018;**246**:57-73

[5] Bartout P, Touchart L. Le territoire limnique, une alternative à la gouvernance des plans d'eau par masses d'eau? vertigo. 2017;**17**:3. Disponible sur: http://journals.openedition.org/vertigo/18692

[6] Grousset F. Chapitre 2 : Du climat du passé au climat du futur. In: Le Treut H. Les impacts du changement climatique en Aquitaine. Presses Universitaires de Bordeaux; 2013. pp. 41-60. Disponible sur: http://books.openedition.org/pub/621

[7] Moreira S, Laplace-Treyture C, Eon M, Jan G. Rapport d'activité du suivi scientifique des plans d'eau de Carcans-Hourtin. Lacanau: Cazaux-Sanguinet et Parentis- Biscarrosse; 2015. p. 106

[8] Papon P, Maleval V, Nedjaï R. Le bilan sédimentaire en lac : l'influence de la course du vent sur l'érosion (Lake sedimentary balance : the fetch influence on the erosion). bagf. 2005;**82**(2):213-223

[9] Håkanson L, Jansson M. Principles of Lake Sedimentology; 2002. p. 316

[10] Bernier M, Bartout P, Touchart L. Dynamic limnology between actors and managements. 2022; 64-74. Disponible sur: http://aerapa.conference.ubbcluj.ro/2022/pdf/7%20BERNIER%2064-74.pdf

[11] Dehouck A, Lafon V, Sénéchal N, Froidefond JM, Almar R, Castelle B, et al. Evolution morphodynamique interannuelle du littoral sud de la Gironde. RFPT. 2014;**197**:31-42

Section 3

The Limnological Characteristics of Lakes

Chapter 10

Temperature Regime, Dynamics and Water Balance of Two Crater Lakes in the Nevado de Toluca Volcano, Mexico

Anatoliy Filonov, Iryna Tereshchenko, María del Refugio Barba-López, David Avalos-Cueva and Cesar Monzon

Abstract

Volcanic lakes are ecosystems in which thermodynamic processes have a complex relationship with atmospheric variables. This study presents the results of an analysis of the thermal regime and dynamics of two high-altitude tropical lakes located in the crater of the Nevado de Toluca volcano in Mexico, at an altitude of more than 2200 m above sea level. Joint meteorological and hydrological measurements taken in two adjacent lakes revealed strong diurnal fluctuations in water temperature, which are caused by wind-induced internal gravity waves and free seiches oscillations. During the daytime, heating occurs in the near-surface layer of the lakes, which creates a thermocline at a depth of 2–3 m, but it is washed out at night. The heat penetration into the lakes is significantly different due to differences in water transparency and algae density, despite the small distance of only 200 m between the lakes separated by a 100-m high lava dome. Temperature and level fluctuations were analyzed using spectral analysis. The numerical model used in Lake El Sol allowed for the first-ever evaluation of the circulation and the impact of wind circulation regimes on lake-level fluctuations. Analyzing such physical processes is crucial in assessing the chemical and biological processes occurring in this reservoir. Field measurements uncovered unexpected temperature changes near the lake bottom, along with heat exchange between the bottom water layer and bottom sediments (during winter, sediments emit heat to the water column). The estimated heat fluxes through the lake bottom were less than 0.3 W/m^2 during winter and less than 0.1 W/m^2 for the rest of the year.

Keywords: Mexico, Nevado de Toluca volcano, lakes El sol and La Luna, thermal structure, hydrodynamic modeling, lakes' level fluctuation, water balance

1. Introduction

Lake ecosystems exhibit a strong interrelationship between physical, chemical, and biological processes. A comprehensive understanding of these processes would enable the establishment of better management policies. Due to the dependence of the thermal and dynamic behavior of lakes on variations in atmospheric parameters, and in order to understand their correlation, wind speed, wind direction, solar radiation, and ambient temperature on the outer slopes and interiors of the volcano crater were analyzed.

High mountain lakes are located above the tree line, where the climatic conditions are extreme [1]. Mountainous regions around the world, including the Andes, the Alps, the Himalayas, and the Rockies, commonly feature bodies of water nestled within their peaks [1, 2].

The lakes, due to their location in high mountain areas, have specific characteristics in terms of their environment and climate. Their basins have steep slopes, and water inputs can come from glaciers, snowmelt, or strong rainfall [3–5]. Furthermore, the water temperature in high mountains is often low because of their high altitude and exposure to solar and ultraviolet radiation [6–8].

High mountain lakes can also be sensitive to environmental perturbations due to their geographic location, which limits water flow and interaction with other water bodies [9, 10]. Hence, they are important indicators of climate change and water quality in high mountain regions [11, 12].

It is also possible to establish a direct relationship between lake water level variations and the balance between precipitation and evaporation in the lake basin [13, 14]. The daily variations are the main factors determining the heat balance in tropical aquatic ecosystems.

Mountain lakes located in the craters of volcanoes at higher altitudes experience particularly complex interactions between the atmosphere and the lakes. As a result, turbulent flows can arise and affect the transport of matter, nutrients, pollutants, and phytoplankton dynamics [10]. However, these interactions have not been extensively studied in tropical high mountain systems. The tropical high mountain lakes in Mexico provide a natural laboratory to investigate the effects of atmospheric forcing on lake hydrodynamics and productivity, and to explore potential climatic impacts.

The volcano Nevado de Toluca is the fourth highest volcano in the mountains of Mexico and is classified as a strombolian volcano, with alternating explosions and lava emissions [15]. Several authors have reported that the volcano is situated at the convergence of three fault systems, and the most recent eruption occurred approximately 10,445 $\pm$ 95 years ago [16–18]. The region encompassing Nevado de Toluca has undergone an extensive and intriguing geological history spanning over 1.5 million years [18, 19]. This area is situated within the Trans-Mexican Volcanic Belt and is characterized by a diverse array of volcanic rocks in its composition [17, 18, 20]. These rocks comprise andesite, basaltic andesite, dacite, and trachyte/trachydacite. Additionally, it is noteworthy to mention the presence of the Tilzapotla Formation, which consists of rhyolites, rhyodacites, and deposits of pyroclastic flows. These volcanic rocks originated from subvolcanic magmatic chambers, and their diversity reflects the intricate dynamics of magma emplacement processes [18].

In the Nevado de Toluca area, highly significant volcanic deposits have been identified, which can be attributed to a catastrophic eruption that resulted in lahars or mudflows of considerable magnitude [18]. During the Late Pleistocene, a remarkable Plinian eruption from the Nevado de Toluca volcano gave rise to a complex sequence of pyroclastic deposits known as the Upper Toluca Pumice [17]. There are two lakes in the crater, El Sol and La Luna, located at 4200 m above sea level. They are separated by a lava dome and are considered to be evidence of a moderate eruption that took place 3140 $\pm$ 195 years ago [19]. Both lakes are separated by 200 m within the interior of the volcano and have been dated to have the same age, origin, and hydrological regime. However, several studies have described significant differences in the physicochemical and biological characteristics of lakes [21, 22].

For this research, we utilized *in situ* measurements of meteorological and water column variables over long-term and high-frequency periods. These measurements enabled us to determine the nonlinear dynamics of the crater lakes of the Toluca Volcano. For this purpose, we assume that the coupling properties between the atmosphere and the water column in El Sol and the Moon are different due to the characteristics caused by the crater walls and the dome. The analysis of the data revealed the existence of complex, coupled forcings at small spatiotemporal scales that are rarely considered in the literature. Forcings can cause substantial variations in lake productivity, which corroborate the meteorological and thermal findings described by Refs. [21, 22]. Interdisciplinary analysis of lake dynamics is essential in improving our understanding of how these ecosystems modulate productivity and subsequently impact climate.

This research aims to make a significant contribution to the understanding of the lakes located in the Nevado de Toluca volcano region. Through a comprehensive and in-depth approach, including precise measurements of meteorological variables, numerical modeling of circulation and currents, and analysis of heat transfer patterns in sediments, we endeavor to gain a profound understanding of the hydrodynamic processes and water balances in these high mountain ecosystems. What sets this research apart is its multidisciplinary approach and meticulous examination of crucial factors, such as solar radiation, thermal stratification, and wind influence. The outcomes of this study will enhance our comprehension of the dynamics of high mountain lakes, thereby establishing a solid scientific foundation for the effective management and conservation of these invaluable aquatic ecosystems.

2. Study area

Limnology has been focused for several decades on the study of high mountain lakes, which are considered to be excellent sentinels of current global change because of their particular limnological characteristics. These lakes tend to occur in clusters, with several to many lakes forming lake districts, as mentioned by previous studies [1, 23]. In this study, we will focus on our interest in El Sol and La Luna lakes, which are two high mountain lakes located on the Nevado de Toluca volcano at coordinates 19°10′N 99°45′W and an altitude of 4200 m above sea level. During the warm rainy season, temperatures can reach up to 11°C in the water column. However, during the winter season, temperatures are closer to 4°C. The waters of the lake are transparent and well-oxygenated, with concentrations close to saturation (~7 mm/L) throughout the water column [24]. Lakes have a homogeneous mixture and a very distinctive polymictic pattern [24].

Figure 1 shows the location and configuration of Nevado de Toluca volcano. It has a horseshoe shape that surrounds two central lakes, with a circular dome reaching a height of approximately 100 m in between. Due to the difference in height between the western and eastern walls of the volcano, wind currents enter the crater, with the former being 345 m higher than the latter [24]. It has been reported that the lakes exhibit a limited buffering capacity, leading to an acidic pH range of 4.9–5.6 [25]. Additionally, the conductivity levels were recorded between 18 and 24 μS c/m, with low levels of dissolved and suspended organic matter.

The maximum depth recorded in El Sol is 15 m, while its average depth is 6 m. Additionally, its area covers 237,321 m^2, with a length of 795 m and a width of 482 m

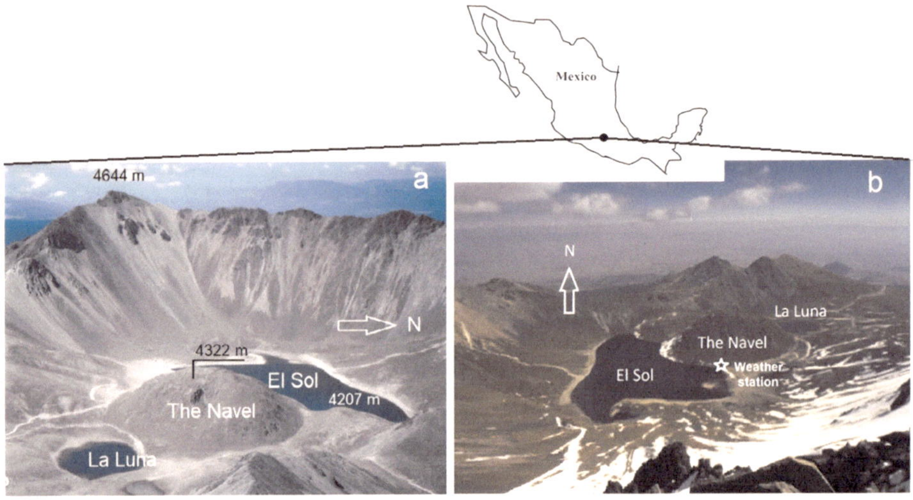

Figure 1.
Location and panoramic photos of Toluca Volcano (a) and (b).

[21, 24]. Similarly, La Luna has a maximum depth of 10 m and an average depth of 5 m. Its surface area is 31,083 m^2, with a length of 227 m and a width of 209 m [24, 26]. The climate in Nevado de Toluca varies from cold to semi-cold and humid, with average monthly temperatures ranging between 2 and 12°C and an annual average of approximately 3.8°C [27]. The average annual precipitation is between 1200 and 2000 mm, concentrated between the months of May and September. In winter, there is precipitation in the form of snow, while after March, it occurs in the form of rain [24].

3. Data and methods

3.1 Field measurements and instruments

Field measurements in the lakes were conducted during two separate experiments. The first experiment took place from October to September 2010, while the second one occurred from May 2017 to September 2019. It is worth noting that the latter experiment was executed with greater precision and utilized additional equipment. During the first experiment, four buoys were deployed as shown in **Figure 2a**. On buoys numbered 1–3, five HOBO Water TempProV2 sensors from Onset Computer Corp. were installed with an accuracy of $\pm$0.2°C at depths of 1 to 5 m. On buoy number four, 10 sensors were installed at each meter up to a depth of 10 m, and an upward-facing current meter (ADP SONTEK 1000 kHz) was anchored on the bottom (**Figure 2b**). The sampling interval for all sensors was 1 minute. Also, in the two lakes, water level meters (tide and wave recorder SBE-26, Sea-Bird Electronics Inc.) were installed at the bottom of the coastal zone to record high-frequency water variations (seiches) with a sampling frequency of 1 second and a level resolution of 0.1 mm for 7 hours (**Figure 2b**).

The first theoretical period of the waves can be estimated by modeling the theoretical period of seiche in a rectangular basin using Merian's formula, where the first mode is given $= \frac{2L}{\sqrt{gH}}$, where H is the mean depth and L is the length of the lake.

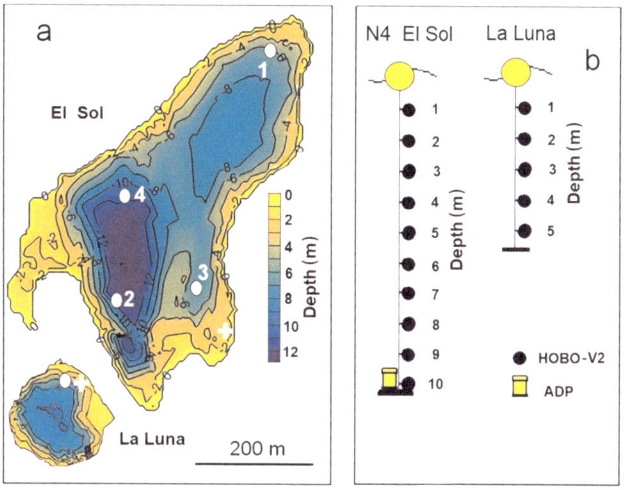

Figure 2.
(a) Bathymetric maps from September 28, 2010, and may 9, 2017. The white circles indicate the positions of the moorings with HOBO thermographs and SBE-26 (white cross) in the two lakes. (b) Shows the location of the instruments in the water column and the position of the ADP (modified from ref. [26]).

3.2 Measurements in lakes in 2018: 2019

To measure the temperature of two lakes, a HOBO© Water TempProV2 thermistor chain was deployed in the deepest part of each lake. The measurement period lasted for over 2 years, from May 10th, 2017 to September 8th, 2019. In Lake El Sol, the thermistor chain consisted of 13 sensors, which were connected to a floating weather station. Meanwhile, Lake El Sol and Lake La Luna had thermistor chains deployed in their deepest parts to measure their temperature. A HOBO© Water TempProV2 thermistor chain was used for this purpose. The measurement period lasted for over 2 years, from May 10th, 2017 to September 8th, 2019. Lake El Sol's thermistor chain had 13 sensors and was connected to a floating weather station. On the other hand, Lake La Luna's chain had 12 sensors. To monitor the temperature of Lake La Luna, several thermographs were placed vertically from the surface to a depth of 3 m, with seven devices every 0.5 m, and then at 1 m intervals to the bottom. The temperature was recorded every 15 minutes. An innovative approach was also employed in this study by installing a floating weather station in the center of Lake El Sol.

The study utilized data from two weather stations. One of the stations was installed directly in the crater of the El Sol Lake volcano, while the other station, registered under the National Meteorological Service of Mexico (NMS 00015062, 19° 7.171′ N; 99° 44.851′O), was situated on the slope near the volcano surveillance post. The straight-line distance between the two weather stations is 1.7 km, and the crater station sits approximately 47 m higher than the slope station. Both stations recorded wind speed and direction, relative humidity, barometric pressure, solar radiation, and rainfall.

Figure 3 shows the structural diagram of the anchoring system for the floating weather station. The system consists of a metallic structure on which the weather station is mounted, with the anemometer positioned 2 m above the water level at the upper end of the structure. Whereas the other end of the structure was submerged at a depth of 4 m, supported by a 30 kg weight, and three buoys were attached to the

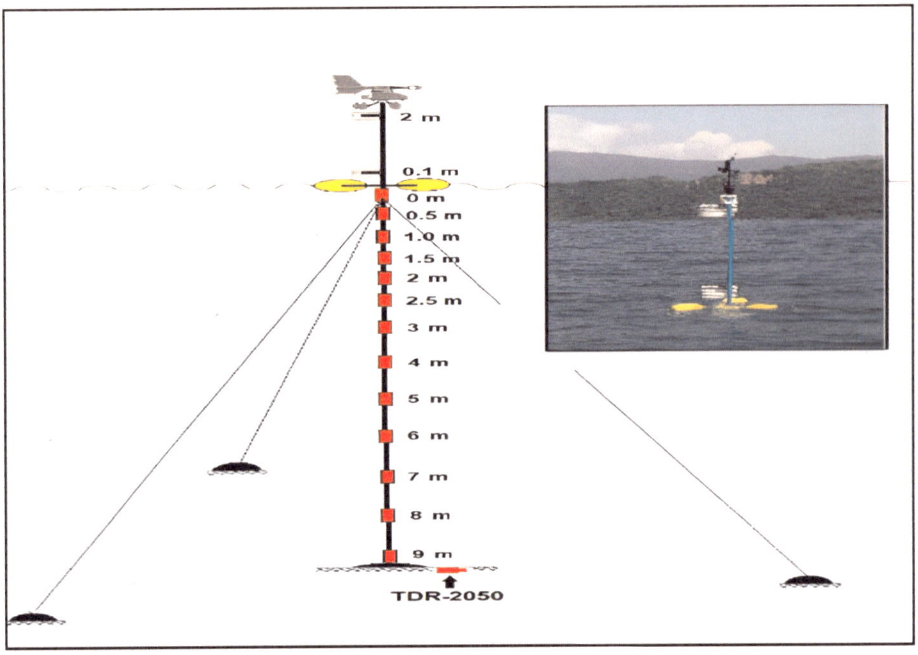

Figure 3.
Structural scheme of a floating weather station. The photograph in the right corner shows the part of the station above the water level.

structure to maintain the stability of the weather station and avoid recording errors caused by wind. Fifteen HOBO thermographs were installed in the same mooring. Two of them were placed on the water's surface, at a height of 0.1 and 2 m, and were protected inside two white, heat-reflecting plastic cylinders that were nested one inside the other. The remaining thermographs were placed below the lake level at depths of 0, 0.5, 1, 1.5, 2, 2.5, 3, 4, 5, 6, 7, 8, and 9 m.

Measurements of wind speed and air temperature were taken at a frequency of 15 minutes from June 19 to October 25, 2019. However, wind speed records were not available for the last 2 months of the period. It has been asserted that in shallow lakes, heat flux reaches the bottom [28, 29]. Hence, in this study, we will analyze the impact of this phenomenon on the bottom stratification of two lakes. We used an RBR TDR-2050 temperature and depth recorder to measure the heat fluxes between the water and the lake bottom sediment. This equipment is calibrated to an accuracy of ±0.002° C (ITS-90) over a range of −5 to +35°C and was buried at a depth of 15 cm in the lake bottom sediments. In addition, we used an instrument that had a depth measurement range of 0–50 m, ensuring an accuracy of 2.5 mm in our level measurements. Moreover, spectral analysis was employed to study the spatiotemporal variation of temperature within the lakes.

3.3 Spectral analysis

Spectral analysis was conducted on the measured water temperature series to assess the spatiotemporal variability of temperature fluctuations in Lake Sol y la Luna. The automatic spectral functions were calculated using the fast Fourier transform, and subsequently, the periodograms of the frequencies were smoothed. The functions were calculated to determine the spatial and temporal relationship of temperature fluctuations in both lakes [30–33]. The estimate of the spectrum of the smoothed periodogram was obtained as follows:

$$\hat{S}_{xx}(\omega) = \int_{-\infty}^{\infty} S_{xx}(\omega')Z(\omega - \omega')d\omega' \quad (1)$$

where $Z(\omega)$ is a smoothing function, and $S_{xx}(\omega)$ is defined as the auto-periodogram:

$$S_{xx}(\omega) = \frac{1}{T}C_x(\omega)C_x^*(\omega) \quad (2)$$

where $C_x(\omega)$ is the amplitude spectrum:

$$C_x(\omega) = \int_0^T x(t)e^{-2\pi i\omega t}dt \quad (3)$$

of the time series $x(t)$; T is the total length of the series; ($*$) denotes the complex conjugate of the amplitude spectrum. The standard algorithms described in the literature [30–33] were used to calculate the confidence intervals for all spectral estimates. The number of degrees of freedom υ, was found as $\upsilon = 2(2F + 1)$; where F is the half-width of the filter that is used to smooth the periodograms. The mean square amplitudes of the harmonics of the dominant peaks in the spectra were found from the values of the spectral density.

3.4 Hydrodynamic model

Delft3D is an open-source numerical model developed by WL/Delft Hydraulics and the Delft University of Technology [34]. It includes implementations of several mathematical models for different physical phenomena (currents, transport, wave propagation, morphological developments, etc.). The Delft3D model solves the Navier-Stokes equations for an incompressible fluid, under the shallow water and the Boussinesq assumptions using a finite difference scheme.

This model is a tool made up of different modules (Delft3D-FLOW, Delft3D-WAVE, Delft3D-MOR and Delft Dash Board) open source and with a free version. It is made up of several program modules that focus mainly on coastal and river systems and that has been well evaluated in work on lentic systems [35–37]. In the present study, the model is used to predict the circulation in the lake system. The model includes the depth-averaged horizontal momentum equations:t

$$\begin{aligned}
&\frac{\partial u}{\partial t} + \boldsymbol{u} \bullet \nabla u + g\frac{\partial \eta}{\partial x} - fv + \frac{u\|\boldsymbol{u}\|}{C^2(d+\eta)} - \frac{F_x}{\rho(d+\eta)} - \nu\nabla^2 u = 0 \\
&\frac{\partial v}{\partial t} + \boldsymbol{u} \bullet \nabla v + g\frac{\partial \eta}{\partial y} + fu + \frac{v\|\boldsymbol{u}\|}{C^2(d+\eta)} - \frac{F_y}{\rho(d+\eta)} - \nu\nabla^2 v = 0
\end{aligned} \quad (4)$$

the depth-averaged continuity equation:

$$\frac{\partial \eta}{\partial t} + \frac{\partial(d+\eta)u}{\partial x} + \frac{\partial(d+\eta)v}{\partial y} = Q(d+\eta) \quad (5)$$

and the vertical momentum equation, which reduces to the hydrostatic pressure relationship *via* the Boussinesq approximation: $\frac{\partial p}{\partial z} = -\rho g$, where u, v are the depth-averaged velocity in the x and y directions, C is the Chézy coefficient, d is the water depth, η is the free surface elevation above the reference plane (at $z = 0$), $\boldsymbol{u}$ is two-dimensional current vector, with $\| \bullet \|$ the Euclidean norm, Q are sinks or sources of water, f is the Coriolis force, $F_{x,y}$ is the Reynolds stress, g is the gravity, ν is the horizontal eddy viscosity, p is the pressure, and ρ is the water density.

Modeling was done only for Lake El Sol since no meteorological measurements were made for the other lake. The RFGRID module of Delft3D was used to generate the grid file. A mesh grid of 61 x 60 cells and 1 layer were used in the horizontal and vertical direction, respectively. Bathymetry measurements obtained during the field campaigns were used to interpolate the depth file by using the QUICKIN module of Delft3D [26]. The time step allowed was $\Delta t = 0.6$ seg. The model was run for 34 days, with the first 4 days used as warm-up in the model. The other physical parameters had typical values: gravity, 9.81 m/s^2; water density 1000 kg/m^3; air density, 1 kg/m^3; uniform Chézy roughness, 65 $m^{1/2}/s$; background horizontal and vertical eddy viscosity and diffusivity of 1 and 10 m^2/s, respectively. Delft3D is a model that has been tested and used globally for the study of lake systems [35, 37, 38]. In addition, it has been used specifically in the hydrodynamic analysis of Mexican crater lakes [36, 39].

4. Results

4.1 Meteorological variables

Detailed information about the meteorological conditions at the Nevado de Toluca volcano can be found in various recent publications [14, 15, 18]. For our study, we utilized a three-year dataset of solar radiation, air temperature, and precipitation from 2017 to 2019, which was collected at the NMS meteorological station. This time frame corresponds to the period of our fieldwork within the volcano crater (**Figure 4a–c**). However, wind data could not be included due to the poor performance of the speed sensor.

In a recent study [24], researchers analyzed an eight-year dataset of meteorological parameters, including wind, collected at the NMS meteorological station located on the southern outer slope of the Nevado de Toluca volcano. The results showed that wind on the slope of the volcano is relatively weak and predominantly blows in two directions: 80° (southeast) and 200° (southwest) (**Figure 5a** and **b**). The wind speed throughout the year did not exceed 4–5 m/s, but during some winters, an increase in the average speed up to 10 m/s was observed (**Figure 5a** and **b**).

The total solar radiation entering the crater was lower than at the NMS station due to the obstruction of the sun's rays by the high walls of the crater in the morning and evening hours. Consequently, solar radiation input is closely related to temperature changes in the lakes, although the impact of wind currents cannot be overlooked, especially in Lake La Luna where mixing is more intense than in Lake El Sol. The latter lake is protected by the lava dome [24].

4.2 Water temperature, rainfall, evaporation, current, and level fluctuations

The meteorological regime around the Nevado de Toluca volcano has been extensively described in several recent publications [22, 24, 26]. According to these

DOI: http://dx.doi.org/10.5772/intechopen.112196

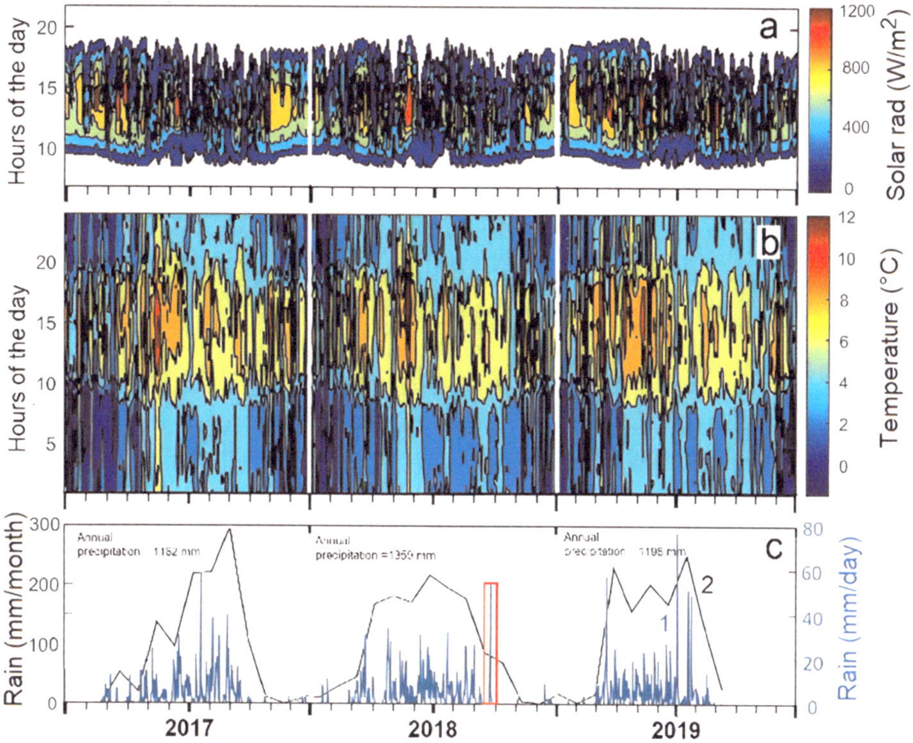

Figure 4.
Hourly variability of solar radiation (a), air temperature (b), and precipitation (c) measured at the NMS station during the period of 2017–2019.

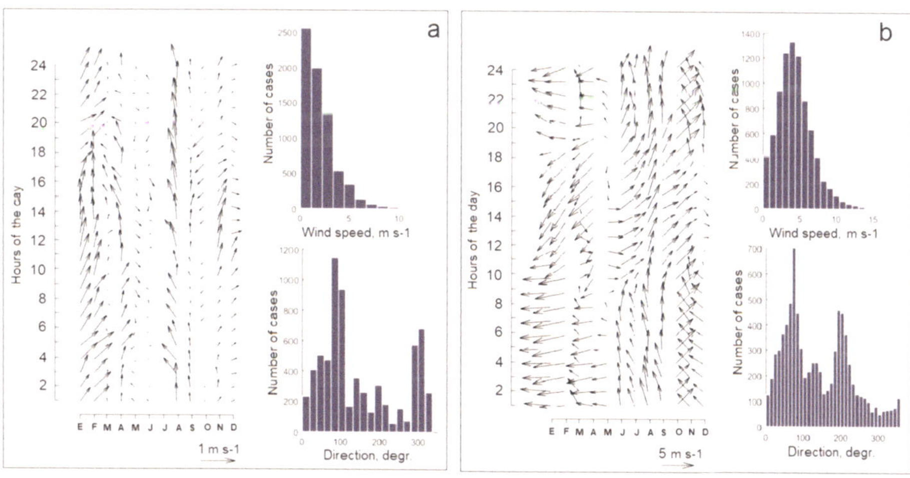

Figure 5.
Monthly dynamics of wind speed and direction vectors (a) inside and outside (b) the volcano crater for the 24-hour cycle of January 1 to December 31, 2007. The figures also show wind speed and direction histograms for the entire observation period on the right-hand side (modified from ref. [24]).

publications, the slope of the volcano receives a maximum of 1000 W/m2 of solar radiation around 3 pm, and there is no radiation from 6 pm to 9 am the following day. Solar energy penetrates 20% less into the crater of a volcano due to its high walls, as reported in a study [24]. The supply of solar energy to the crater lakes is at its peak during March and April, and at its lowest between June and August, as well as during the rainy season when cloud cover reduces energy penetration. Fluctuations in air temperature are highest in April and May, particularly in the early afternoon, while the lowest temperatures occur during the rainy and winter seasons.

Another recent study [22] suggested that the difference in lake stratification (**Figure 6a** and **b**) during summer may be due to Lake El Sol's lower transparency, which absorbs more heat at the surface compared to the more transparent Lake La Luna. It also suggested that vortex diffusion, not wind mixing, causes heat exchange between the near-surface layer and deeper waters.

Although seiches fluctuations in lakes are a common occurrence, they are weak in the crater lakes studied due to their small size and depth, as well as protection from wind impact by the crater walls. Spectral analysis revealed two seiche oscillations in Lake El Sol with periods of 167 and 81 seconds and corresponding amplitudes of about 2 and 1.5 mm. The first oscillation was along the main axis of the lake, while the second was perpendicular to this axis. Lake La Luna had only one free oscillation with a period of 53 seconds and an amplitude of about 3 mm.

To examine the relationship between precipitation, evaporation, runoff area, and the level of Lake El Sol, the authors used data from the NMS weather station and time series of water level measured by the TDR-2050 instrument. The results of these experiments are presented in **Figure 7a–e**. Precipitation was the main source of water entering the lakes, which had small runoff areas due to their specific orography. The catchment area of Lake El Sol was only 2.17 km^2, while Lake La Luna's catchment area was 2.0 km^2. Average annual precipitation was 1.2771 m^3, and evaporation was 0.9708 m^3, with the largest average monthly evaporation in February (112 mm/month) and the lowest in June (63 mm/month).

Currents measured in the upper 2-m layer at the floating weather station in September–October 2010 showed that wind currents only reached this depth and did not

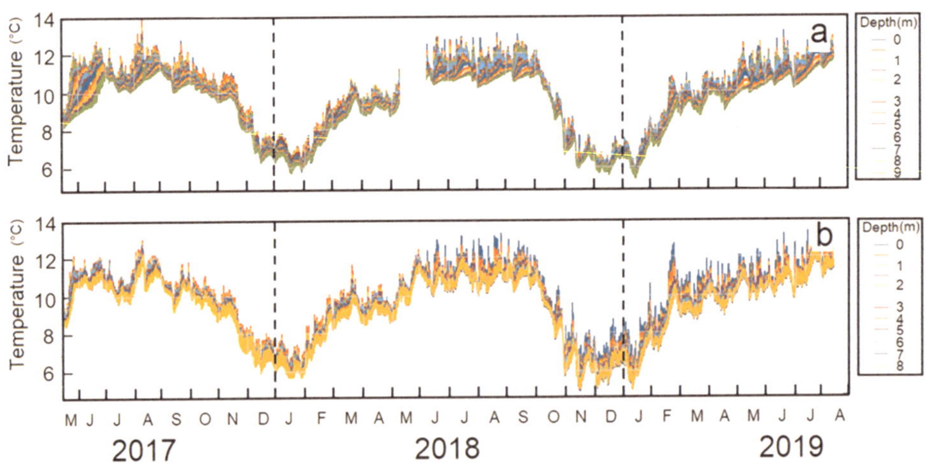

Figure 6.
Hourly variability of temperature at 13 depths of El sol Lake (a) and at 12 depths of La Luna Lake (b) for the same time period. The instrumental depths are clearly indicated in the legend.

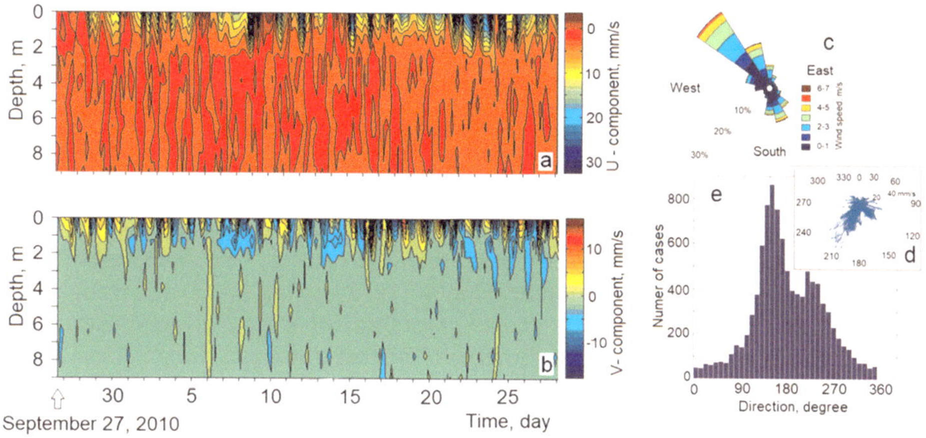

Figure 7.
Zonal (a) and meridional (b) components of the currents were measured by the acoustic doppler profiler (ADP) in mooring 4, located in El sol Lake. The wind rose from the NMS station data is shown in (c), while (e) displays a histogram of the directions of the currents in the upper 1-m water layer. Additionally, a vector histogram of the same data is presented in (d).

penetrate deeper (**Figure 7a** and **b**). The currents were weak, not exceeding 3–5 cm/s, and did not display strict daily periodicity as temperature fluctuations in the surface layers of the lake (**Figure 7a** and **b**). **Figure 7c** and **d** also displays the wind direction and speed during the experiment.

The level of Lake El Sol's annual variation was relatively smooth, sometimes disturbed by small jumps due to precipitation. The annual minimum temperature occurred in early January 2018 at the water-bottom sediment boundary in the deep-water part of the lake, reaching 4.8°C and increasing slightly to 5.7°C at the bottom before dropping again to 4.5°C in early February (**Figure 8**). In addition, as shown in **Figure 8**, it can be observed that during the summer season, the bottom temperature was approximately 10°C.

4.3 Lake level fluctuations and processes occurring at its bottom

A precise time series spanning over 13 months was obtained by using the TDR-2050 RBR temperature and level meter, which was submerged in the sludge at the bottom of the lake (**Figure 8**). This was made possible due to the highly accurate temperature and level sensor, with an accuracy of ±0.002°C and 0.05% full scale for

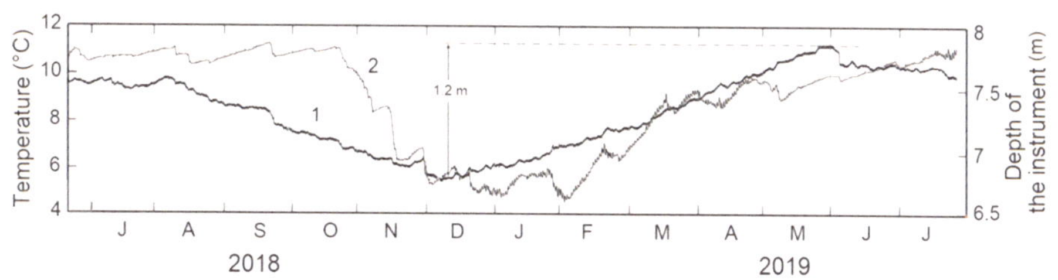

Figure 8.
Annual series of temperature fluctuations of bottom sediments (1) and the water level of Lake El sol (2) in the area surrounding the floating meteorological station have been recorded.

pressure, making it two orders of magnitude better than the temperature measured using the Hobo V2.

In the winter months, the water temperature near the bottom of the lake drops to a minimum of 4.7°C, which is close to the maximum density temperature of water. The diurnal variation is prominently observed at the lake's bottom, with a larger range from November to May (0.2–0.3°C) than the remaining months of the year (**Figure 8**).

Using temperature measurements taken from both the bottom of a body of water and the sediment at the bottom, heat fluxes were determined near the interface of the water and sediment using the gradient method [28, 29]. This was done through the use of the formula $Q = -\lambda\, \partial T/(\partial z)$, where Q represents the heat flow in W/m^2 near the boundary between the water and sediment, λ represents the coefficient of molecular thermal conductivity of water in the 0–10 m layer (which is 0.5813 W/(m°C)), and $\partial T/(\partial z)$ represents the temperature gradient.

According to the calculations, the bottom of the lake had a steady heat flow during the entire observation period. The heat transfer from the lower layers of water to the sediment beneath was measured to be below 0.1 W/m^2 in the spring, summer, and autumn months, which is much lower compared to the solar radiation that enters the lake's surface [24]. This solar radiation can reach up to 1000 W/m^2, but gets absorbed and dissipated within the water column. However, during winter, the heat flow changed direction with heat moving from the sediment to the water column, and the heat flux measured during this period was almost 0.3 W/m^2.

4.4 The water balance of the lake depending on precipitation and evaporation

The water equilibrium of a crater lake relies on several factors, including the inflow of water from precipitation, as well as the loss of water due to processes such as evaporation, runoff, and gravity filtration. On Nevado de Toluca, the steep slopes act as collectors of atmospheric precipitation, which helps to maintain a hydrological balance through the processes of evaporation and filtration. The altitude of the volcano also facilitates groundwater retention, which is essential in determining the water balance of closed water bodies. To achieve this, physiographic and meteorological characteristics, such as the watershed area, precipitation rates, evaporation, and water seepage volume, must be considered. While precipitation has been measured in the Nevado de Toluca basin and the surface area of the basin is known, it is important to study the volumes of water lost due to seepage from the bottom.

The slopes on the southern and western sides of the volcano's crater are steep, which results in shading over the lake. As per a study [24], this shading reduces the penetration of solar energy into the lake by roughly 20% during the morning and afternoon hours compared to daytime hours. We have taken this factor into account during our calculations, and we have determined that the annual evaporation rate is 940.1 mm.

Table 1 presents information obtained from our field measurements and estimates of the water balance of Lake El Sol during the dry season. According to the data, the area of the lake during this season is 200,330 m^2, while the runoff area is significantly larger, at 2,170,000 m^2. These results suggest that Lake El Sol's water level largely depends on runoff to maintain itself during the dry season, as well as on annual precipitation, which does not exceed 50.8 mm. However, the evaporation rate reaches 3.33 mm/day, indicating that the lake loses a significant amount of water due to evaporation.

Annual measurements of mean values (using CNA measurements) reveal that the average area of the lake during both the rainy and dry seasons is 201,165 m^2.

Dry season area of lake El Sol	200,330 m^2
Runoff area	2,170,000 m^2
Evaporation rate (November 2018)	3.33 mm/day
Rain case	50.8 mm
Real level rise	0.11 m
Accumulated in runoff area	110,236 m^3
Expected level rise with evaporation	0.550 m
Expected level rise without evaporation	0.586 m

Table 1.
Annual average values as extracted from ref. [26].

Furthermore, the runoff area amounts to 2,170,000 m^2, indicating that a considerable amount of water flows into the lake from its catchment area.

Moreover, the rainfall rate reported by the NMS is 1.359 m, while the calculations show a rate of 1.2771 m. Similarly, the NMS records an evaporation rate of 0.941 m, whereas the calculations indicate a rate of 0.9708 m. These disparities could be attributed to differences in measurement techniques.

The accumulated volume in the lake is 907,060 m^3, which represents the total amount of water in the lake at a given time. The expected elevation due to evaporation is 4.509 m, but the actual elevation only increases by 1.20 m. This difference could be due to other factors that affect the water balance of the lake, such as outflow.

4.5 Hydrodynamics of lake El sol

Thermal stratification is the primary hydrodynamic force in high mountain lakes. However, other factors, such as wind shear, precipitation, and seismic activity, also contribute to the system. Furthermore, the interaction of these forcings can generate complex hydrodynamic processes that significantly affect circulation, mixing, and water quality in high-mountain aquatic ecosystems. Importantly, thermal stratification occurs due to the difference in water density in different layers of the lake, which is caused by solar radiation. Wind shear can break the thermal stratification, allowing vertical mixing of water and generating currents on the lake's surface. Precipitation can also affect lake hydrodynamics, especially in regions where rainfall is intense, and lakes have a direct connection to rivers and streams. Seismic activity can generate waves and disturbances in the water that affect the circulation and mixing of water in the lake.

Figure 9 shows the results of numerical modeling performed using the Delft3D model. The objective of this simulation was to complement the findings described in [26], which focused on the current field in Lake El Sol. In the previous study, we employed a current meter (ADP Sontek) and determined that wind-induced currents are restricted to the uppermost layer of the lake, extending up to a depth of 2 m, and do not exceed speeds of 15 mm/s. Consequently, we have restricted the presentation of current simulation outcomes to the upper layer of the lake.

The model inputs were created using accurate temperature, wind direction, and wind speed data recorded at the surface of the lake. This was made possible through the installation of a floating weather station, which is marked with a red circle in the upper-left panel and located in the deepest part of the lake.

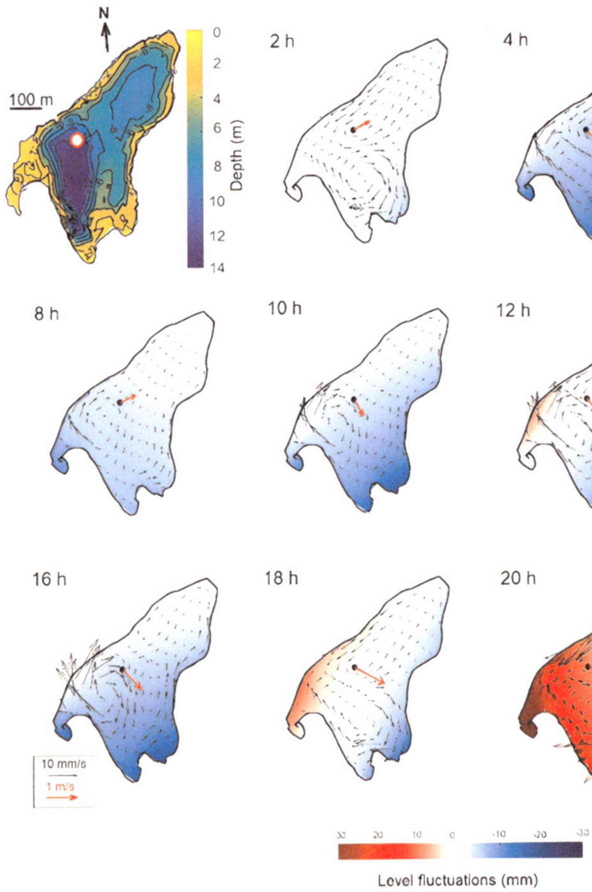

Figure 9.
Modeling results of current and water level fluctuations in Lake El sol based on observations taken on September 5, 2018. The rectangles illustrate the simulated currents and level fluctuations for every 2 hours. The black and red patterned arrows in the lower left corner represent the scales for current velocities and measured wind, respectively.

The modeling results indicate the existence of two distinct closed circulation patterns, each with an opposite direction of circulation. The weaker circulation, with speeds ranging from 2 to 4 mm/s, appears in the central-northeast regions of the lake and rotates counterclockwise (**Figure 9**). Variations in this circulation are directly linked to the wind patterns recorded at the floating meteorological station. On the other hand, the stronger circulation, with clockwise rotation and speeds ranging from 8 to 10 mm/s, is observed in the deep southern part of the lake (**Figure 9**).

Furthermore, it has been observed that the level fluctuations of Lake El Sol vary up to a maximum height of ±25 mm. However, no numerical analysis was conducted for Lake La Luna because it is impossible to install a floating meteorological station in this body of water.

5. Discussion and conclusions

During the period from 2017 to 2019, various meteorological variables were examined in the Nevado de Toluca volcano region, including solar radiation and air temperature outside the crater. The results obtained were consistent with the values reported in previous studies (e.g., [24, 26]), which covered the period from 2000 to 2007. With regards to the annual precipitation in the area adjacent to the volcano, values of 11,622 mm, 1359 mm, and 1198 mm were recorded during the studied period, which falls within the normal range of precipitation reported in Ref. [26] for the same area.

On the other hand, the magnitude and direction of the wind directly over the surface of Lake El Sol turned out to be much lower than outside the crater, due to the presence of the volcano walls and their shielding effect on the lake. This process has been well described in other works [40–44].

Although it is known that there is a decrease in temperature with altitude (for dry air, 9.9°C/km), the air is rarely dry and the actual relationship between temperature and altitude varies both temporally and spatially depending on climatic conditions (humidity, wind, and radiation) and topography [45]. In our study, the air temperature showed a diurnal oscillation, typical of tropical zones [46, 47]. In addition to a daily variation, a seasonal variation was also observed in air temperature, with greater amplitude from September to May (2 m above the lake surface), and amplitudes below 0.1 m above the lake surface for these same months. For both time series of air temperature, maximum and minimum values (−9°C, 22°C) fall within the range of temperatures observed in other mountainous areas [46, 48].

The behavior of meteorological variables, specifically air temperature, affects the resistance of the water column to mixing processes [49]. Air temperature is a key factor in temperature changes in lakes [48, 50]. Spectral analysis has been used to study these relationships in alpine lakes [51], Mexican lakes [26, 36, 52, 53], and particularly in lakes El Sol and La Luna [24]. The spectrum was dominated by diurnal and semidiurnal oscillations, with high coherence values (0.920) between air temperature and lake temperature. Frequency spectra analysis was also used to describe fluctuations in the lake level, where diurnal and semidiurnal oscillations were observed (24- and 12-hour), as well as 4- and 8-day periods.

During our measurement period, the temperature of Lake El Sol fluctuated between 6 and 14°C, while Lake La Luna recorded temperatures between 6 and 13°C. These values are similar to those reported by [24]. Additionally, they are consistent with the maximum values previously reported by Alcocer, Roberson, Oseguera, and Lewis [22] for these same lakes.

Through *in situ* measurements, we observed unexpected changes in temperature near the bottom of the lake. We detected a heat exchange between the bottom water layer and the sediments that is distinct from the exchange pattern described in Ref. [54] for Lake Biwa (Japan). During winter, the sediments release heat to the water column, resulting in this unique pattern. These heat transfer patterns can often be complex and arise from a wide range of characteristics. The steady-state temperatures of volcanic lakes are primarily determined by the magnitude of the volcanic heat influx relative to the lake's surface area [55].

The level of the lake is a parameter that depends on rainfall and evaporation. In our lake, it was observed that the level increased less than what theoretical calculations would suggest [56]. Numerous studies have demonstrated links between water levels and climate variability or change [57, 58]. However, some studies suggest that although water levels and climate cycles may be correlated, it is challenging to isolate the effect of individual forcing or attribute water-level trends to that forcing [59, 60].

In our study, we analyzed the lake level data over a two-year time series and observed a discrepancy, for the first time at this site, between the measured levels and theoretical calculations. Our findings emphasize the challenges of planning and maintaining equipment in high mountain areas, as well as the complex relationships among meteorological drivers, thermal regimes, and water balances in these unique mountain lake ecosystems.

For the first time, we were able to observe the free surface level of Lake El Sol and the circulation patterns in its surface layer resulting from wind stress using a numerical model. Since the hydrodynamic behavior of this body of water largely dominates and regulates the chemical and biological processes within it, a structured analysis of these processes is crucial to evaluate and understand the lake system as a whole.

Acknowledgements

The authors would like to express their gratitude to each participant of the expedition who made it possible to complete this work. Specifically, we would like to extend our thanks to Dr. Luis Alberto Oseguera, Dr. Omar Mireles-Loera, Arturo Orozco-Estrada, and Carlos Villarreal-Olavarrieta. Additionally, we would like to thank the Comisión Estatal de Parques Naturales y de la Fauna (CEPANAF) within the Secretaría de Ecología of the Government of the State of Mexico for providing the permit for scientific research at the Área de Protección de Flora y Fauna Nevado de Toluca. We are also grateful to the Mexican National Science Foundation (CONACyT) for providing financial support for projects No. 466674-F and 262979.

Author details

Anatoliy Filonov, Iryna Tereshchenko, María del Refugio Barba-López, David Avalos-Cueva* and Cesar Monzon
Department of Physics, University of Guadalajara, Guadalajara, Jalisco, Mexico

*Address all correspondence to: david.avalos@academicos.udg.mx

References

[1] Catalan J, Donato Rondón JC. Perspectives for an integrated understanding of tropical and temperate high-mountain lakes. Journal of Limnology. 2016;**75**:215-234. DOI: 10.4081/jlimnol.2016.1372

[2] Catalan J, Camarero L, Felip M, Pla S, Ventura M, Buchaca T, et al. High mountain lakes: Extreme habitats and witnesses of environmental changes. Limnetica. 2006;**25**:551-584

[3] Hinder B, Gabathuler M, Steiner B, Hanselmann K, Preisig HR. Seasonal dynamics and phytoplankton diversity in high mountain lakes (Jöri lakes, Swiss Alps). Journal of Limnology. 1999;**58**: 152-161

[4] Hobbs WO, Telford RJ, Birks HJB, Saros JE, Hazewinkel RR, Perren BB, et al. Quantifying recent ecological changes in remote lakes of North America and Greenland using sediment diatom assemblages. PLoS One. 2010;**5**: e10026

[5] Oleksy I, Sadro S, Kopáček J. Mountain lakes: Eyes on global environmental change. Global and Planetary Change. 2019;**178**:77-95

[6] Pastorino P, Prearo M. High-Mountain lakes, indicators of global change: Ecological characterization and environmental pressures. Diversity. 2020;**12**:260

[7] Zhang Y, Shi K, Zhou Q, Zhou Y, Zhang Y, Qin B, et al. Decreasing underwater ultraviolet radiation exposure strongly driven by increasing ultraviolet attenuation in lakes in eastern and Southwest China. Science of the Total Environment. 2020;**720**:137694. DOI: 10.1016/j.scitotenv.2020.137694

[8] Durán-Romero C, Medina-Sánchez JM, Carrillo P. Uncoupled phytoplankton-bacterioplankton relationship by multiple drivers interacting at different temporal scales in a high-mountain Mediterranean lake. Scientific Reports. 2020;**10**:350. DOI: 10.1038/s41598-019-57269-y

[9] Group MRIEW, Pepin N, Bradley RS, Diaz HF, Baraer M, Caceres EB, et al. Elevation-dependent warming in mountain regions of the world. Nature Climate Change. 2015;**5**:424-430. DOI: 10.1038/nclimate2563

[10] Jeppesen E, Jensen JP, Søndergaard M, Lauridsen T. Trophic dynamics in turbid and Clearwater lakes with special emphasis on the role of zooplankton for water clarity. In: Walz N, Nixdorf B, editors. Shallow Lakes'98: Trophic Interactions in Shallow Freshwater and Brackish Waterbodies. Dordrecht: Springer Netherlands; 1999. pp. 217-231

[11] Elser JJ, Wu C, González AL, Shain DH, Smith HJ, Sommaruga R, et al. Key rules of life and the fading cryosphere: Impacts in alpine lakes and streams. Global Change Biology. 2020; **26**:6644-6656. DOI: 10.1111/gcb.15362

[12] Catalan J, Pla-Rabés S, Wolfe AP, Smol JP, Rühland KM, Anderson NJ, et al. Global change revealed by palaeolimnological records from remote lakes: A review. Journal of Paleolimnology. 2013;**49**:513-535. DOI: 10.1007/s10933-013-9681-2

[13] Battarbee RW, Grytnes J-A, Thompson R, Appleby PG, Catalan J, Korhola A, et al. Comparing palaeolimnological and instrumental evidence of climate change for remote mountain lakes over the last 200 years.

Journal of Paleolimnology. 2002;**28**: 161-179. DOI: 10.1023/A: 1020384204940

[14] Vigliani S. Las aguas celestiales. Nevado de Toluca. Cuicuilco. 2009;**16**: 265-273

[15] Quesada JFA, Martin Del Pozzo AL, Blanco JL. Volcanic hazards zonation of the Nevado de Toluca Volcano, Central Mexico. Natural Hazards. 2007;**41**: 159-180. DOI: 10.1007/s11069-006-9029-6

[16] Cas R, Wright J. Volcanic Successions Modern and Ancient: A Geological Approach to Processes, Products and Successions. Netherlands: Springer; 2012

[17] Arce JL, Macías JL, Vázquez-Selem L. The 10.5 ka Plinian eruption of Nevado de Toluca volcano, Mexico: Stratigraphy and hazard implications. Geological Society of America Bulletin. 2003;**115**:230-248. DOI: 10.1130/0016-7606(2003)115<0230: TKPEON>2.0.CO;2

[18] García-Palomo A, Macías JL, Arce JL, Capra L, Garduño VH, Espíndola JM. Geology of Nevado de Toluca Volcano and surrounding areas, central Mexico. 2002; **MCH089**:26

[19] Macías JL, García P, Arce J, Siebe C, Espíndola J, Komorowski J, et al. Late Pleistocene-Holocene cataclysmic eruptions at Nevado de Toluca and Jocotitlan volcanoes, central Mexico. Brigham Young University Geology Studies. 1997;**42**:493-528

[20] Bloomfield K, Valastro S. Late Pleistocene Eruptive History of Nevado de Toluca Volcano, Central Mexico. Geological Society of America Bulletin. 1974;**85**:901

[21] Alcocer J, Oseguera LA, Escobar E, Peralta L, Lugo A. Phytoplankton biomass and water chemistry in two High-Mountain Tropical Lakes in Central Mexico. Arctic, Antarctic, and Alpine Research. 2004;**36**:342-346. DOI: 10.1657/1523-0430(2004)036[0342:PBAWCI]2.0.CO;2

[22] Alcocer J, Roberson J, Oseguera LA, Lewis WM. Rhythmic episodes of heating and cooling control thermal stratification of two tropical high mountain lakes. Aquatic Sciences. 2020; **82**:58. DOI: 10.1007/s00027-020-00731-1

[23] Catalan J, Curtis CJ, Kernan M. Remote European mountain lake ecosystems: Regionalisation and ecological status. Freshwater Biology. 2009;**54**:2419-2432. DOI: 10.1111/j.1365-2427.2009.02326.x

[24] Barba-López MdR, Filonov A, Alcocer J, Ladah L, Tereshchenko I. Complex interactions between meteorological parameters and the thermal regime of two tropical, high altitude crater lakes (Nevado de Toluca, Mexico). Science of the Total Environment. 2021;**771**:145310. DOI: 10.1016/j.scitotenv.2021.145310

[25] Caballero-Miranda M. The diatom FLORA of two acid lakes IN Central MÉXICO. Diatom Research. 1996;**11**: 227-240. DOI: 10.1080/0269249X.1996.9705381

[26] Filonov A, Barba-López MdR, Ladah L, Tereshchenko I, Palacios-Hernández E, Alcocer J, et al. The dynamics and temperature regime of the crater lakes in the Nevado de Toluca volcano, Mexico. Limnologica. 2022;**92**: 125938. DOI: 10.1016/j.limno.2021.125938

[27] García, Amaro E. Modificaciones al Sistema de Clasificación Climática de

Köppen. Universidad Nacional Autónoma de México. 2004:92

[28] Gavrilenko G, Zdorovennova G, Zdoroven-nov R, Palshin N, Mitrohov A, Terzhevik AY. Teplopotok na granice voda-donnye otlozheniya v nebolshom ozere [Heat flux at the water-sediment interface in a shallow lake]. Trudy KarNC RAN [Transactions of Kar-RC of RAS]. 2015;**1**:3-9

[29] Golosov S, Kirillin G. A parameterized model of heat storage by lake sediments. Environmental Modelling & Software. 2010;**25**:793-801. DOI: 10.1016/j.envsoft.2010.01.002

[30] Bendat JS, Piersol AG. Random Data: Analysis and Measurement Procedures. 4th ed. Hoboken, New Jersey: John Wiley & Sons; 2012

[31] Freiberger WF, Jenkins GM, Watts DG. Spectral analysis and its applications. Journal of the American Statistical Association. 1969;**64**. DOI: 10.2307/2286100

[32] Koniāev, K.V.e. Spectral Analysis of Physical Oceanographic Data; Division of Ocean Sciences. Vol. 80. Washington, DC: National Science Foundation; 1990

[33] Thomson RE, Emery WJ. Data analysis methods in physical oceanography. Elsevier Science. 2014; **1**(3):728

[34] Delft3D-FLOW, D. Simulation of Multi-Dimensional Hydrodynamic Flows and Transport Phenomena, Including Sediments. Vol. 690. Deltares: The Netherlands; 2020. p. 725

[35] Baracchini T, Hummel S, Verlaan M, Cimatoribus A, Wüest A, Bouffard D. An automated calibration framework and open source tools for 3D lake hydrodynamic models. Environmental Modelling & Software. 2020;**134**:104787. DOI: 10.1016/j.envsoft.2020.104787

[36] Gasca-Ortiz T, Pantoja DA, Filonov A, Domínguez-Mota F, Alcocer J. Numerical and observational analysis of the hydro-dynamical variability in a small Lake: The case of Lake Zirahuén, Mexica. Water. 2020;**12**: 1658

[37] Kranenburg W, Tiessen M, Veenstra J, de Graaff R, Uittenbogaard R, Bouffard D, et al. 3D-modelling of Lake Kivu: Horizontal and vertical flow and temperature structure under spatially variable atmospheric forcing. Journal of Great Lakes Research. 2020;**46**:947-960. DOI: 10.1016/j.jglr.2020.05.012

[38] Razmi AM, Barry DA, Bakhtyar R, Le Dantec N, Dastgheib A, Lemmin U, et al. Current variability in a wide and open lacustrine embayment in Lake Geneva (Switzerland). Journal of Great Lakes Research. 2013;**39**:455-465. DOI: 10.1016/j.jglr.2013.06.011

[39] Cardoso-Mohedano J-G, Sanchez-Cabeza J-A, Ruiz-Fernández A-C, Pérez-Bernal L-H, Lima-Rego J, Giralt S. Fast deep water warming of a subtropical crater lake. Science of the Total Environment. 2019;**691**:1353-1361. DOI: 10.1016/j.scitotenv.2019.07.183

[40] Baron JILLS, Caine NEL. Temporal coherence of two alpine lake basins of the Colorado front range, U.S.A. Freshwater Biology. 2000;**43**:463-476. DOI: 10.1046/j.1365-2427.2000.00517.x

[41] Bitan A. Lake Kinneret (sea of Galilee) and its exceptional wind system. Boundary-Layer Meteorology. 1981;**21**: 477-487. DOI: 10.1007/BF02033595

[42] Kossmann M, Sturman AP, Zawar-Reza P, McGowan HA,

Oliphant AJ, Owens IF, et al. Analysis of the wind field and heat budget in an alpine lake basin during summertime fair weather conditions. Meteorology and Atmospheric Physics. 2002; **81**:27-52. DOI: 10.1007/s0070 30200029

[43] McGowan HA, Owens IF, Sturman AP. Thermal and dynamic characteristics of alpine Lake breezes, Lake Tekapo, New Zealand. Boundary-Layer Meteorology. 1995;**76**:3-24. DOI: 10.1007/BF00710888

[44] Santo MA, Toffolon M, Zanier G, Giovannini L, Armenio V. Large eddy simulation (LES) of wind-driven circulation in a peri-alpine lake: Detection of turbulent structures and implications of a complex surrounding orography. Journal of Geophysical Research: Oceans. 2017;**122**: 4704-4722. DOI: 10.1002/ 2016JC012284

[45] Barry RG. Mountain Weather and Climate. London: Taylor & Francis; 2013

[46] Labaj A, Michelutti N, Smol J. Annual stratification patterns in tropical mountain lakes reflect altered thermal regimes in response to climate change. Fundamental and Applied Limnology. 2018;**191**:267-275

[47] Michelutti N, Labaj AL, Grooms C, Smol JP. Equatorial mountain lakes show extended periods of thermal stratification with recent climate change. Journal of Limnology. 2016;**75**

[48] Livingstone DM, Lotter AF, Walkery IR. The decrease in summer surface water temperature with altitude in Swiss Alpine Lakes: A comparison with air temperature lapse rates. Arctic, Antarctic, and Alpine Research. 1999;**31**: 341-352. DOI: 10.1080/15230430.1999. 12003319

[49] Idso SB. On the concept of lake stability 1. Limnology and Oceanography. 1973;**18**:681-683. DOI: 10.4319/lo.1973.18.4.0681

[50] Livingstone DM, Lotter AF, Kettle H. Altitude-dependent differences in the primary physical response of mountain lakes to climatic forcing. Limnology and Oceanography. 2005;**50**: 1313-1325. DOI: 10.4319/ lo.2005.50.4.1313

[51] van Haren H, Dijkstra HA. Convection under internal waves in an alpine lake. Environmental Fluid Mechanics. 2021;**21**:305-316. DOI: 10.1007/s10652-020-09774-2

[52] Serrano D, Filonov A, Tereshchenko I. Dynamic response to valley breeze circulation in Santa María del Oro, a volcanic lake in Mexico. Geophysical Research Letters. 2002;**29**: 27-21-27-24. DOI: 10.1029/ 2001GL014142

[53] Soares LMV, Silva TFdG, Vinçon-Leite B, Eleutério JC, Lima LCd, Nascimento NdO. Modelling drought impacts on the hydrodynamics of a tropical water supply reservoir. Inland Waters. 2019;**9**:422-437. DOI: 10.1080/ 20442041.2019.1596015

[54] O'Neill K, Ashton GD. Bottom Heat Transfer to Water Bodies in Winter. U.S. Army Engineer Waterways Experiment Station: Hanover, New Hampshire; 1981

[55] Pasternack GB, Varekamp JC. Volcanic lake systematics I. Physical constraints. Bulletin of Volcanology. 1997;**58**:528-538. DOI: 10.1007/ s004450050160

[56] Williams ZC, Pelletier JD. Self-affinity and surface-area-dependent fluctuations of lake-level time series.

Water Resources Research. 2015;**51**: 7258-7269. DOI: 10.1002/2015WR017254

[57] Mercier F, Cazenave A, Maheu C. Interannual lake level fluctuations (1993–1999) in Africa from Topex/ Poseidon: Connections with ocean–atmosphere interactions over the Indian Ocean. Global and Planetary Change. 2002;**32**:141-163. DOI: 10.1016/ S0921-8181(01)00139-4

[58] Wilcox DA. Lake-Level Variability and Water Availability in the Great Lakes. U.S. Geological Survey: Reston, VA; 2007

[59] Pasquini AI, Lecomte KL, Depetris PJ. Climate change and recent water level variability in Patagonian proglacial lakes, Argentina. Global and Planetary Change. 2008;**63**:290-298. DOI: 10.1016/j.gloplacha.2008.07.001

[60] Sellinger CE, Stow CA, Lamon EC, Qian SS. Recent water level declines in the Lake Michigan–Huron system. Environmental Science & Technology. 2008;**42**:367-373. DOI: 10.1021/ es070664+

Chapter 11

Characterization of Lake Kivu Water Chemistry and Its Environmental Impacts

Francois Hategekimana, Theophile Mugerwa, Digne Edmond Rwabuhungu Rwatangabo and Young-Seog Kim

Abstract

Among the world's lakes, Lake Kivu, a rift lake in the western branch of the Eastern African Rift System, has significant reserves of dissolved chemicals. However, no research has been done on their vertical variation in lake and how they affect the environment. This proposed chapter will review earlier research to better understand the origin of Lake Kivu's chemical composition and its effects on the aquatic environment. Water samples were collected using Niskin bottles at various depths, as well as in various locations away from Nyamyumba hot spring sources. Hach kits and procedures were used to conduct chemical analyses on water samples. This study found that the majority of chemical concentrations rise with depth, primarily as a result of the deposition of organic matter. The sewage water from residential buildings, hospitals, runoff from agricultural activities, and rock-water interaction through dissolution process are the possible sources of chemicals discovered in Lake Kivu water. The levels of chemicals in the water of Lake Kivu at this time are less polluting and damaging to the aquatic environment. Therefore, it is important to implement a continuous monitoring strategy to stop eutrophication and other diseases linked to water pollution in humans.

Keywords: Lake Kivu, chemicals, water stratification, environment, eutrophication

1. Introduction

Over the last few decades, global awareness of lake eutrophication and aquatic ecological degradation has grown. The process of eutrophication refers to the change in the chemical properties of water caused by the accumulation of excess nutrients such as nitrogen and phosphorus. In this process, phytoplankton and other microorganisms are rapidly produced leading to the deterioration of water quality which is detrimental to aquatic ecology [1–3].

Water body chemistry results mainly from the interplay between diverse hydrological, geochemical, and biological processes controlled by either natural or anthropogenic factors such as: lithology, climate, vegetation, relief, agricultural activities, and domestic or industrial discharge [4–7].

In this research, Lake Kivu which is a rift lake located in the western branch of the East African Rift System was considered (**Figure 1**). Like other lakes, Lake Kivu contains marine organisms (e.g. fishes) that should be environmentally protected through continuous monitoring of its water chemistry.

It is well known that the chemistry of water bodies varies with depth [8]. Numerous studies (e.g., [9–11]) were conducted to comprehend the chemistry of Lake Kivu while only focusing on surface chemistry; however, none of these studies ever made the connection between the vertical variation in chemistry and environmental pollution.

The water of East African rift lakes contains large amounts of dissolved gases such as carbon dioxide and methane, especially in Kivu Lake which is located between Rwanda and the Democratic Republic of Congo (DRC) (**Figure 1**). Lake Kivu is one of the largest carbon dioxide and methane gas reservoirs on earth [12, 13]. It was observed that the amount of CO_2 and CH_4 dissolved phase in Lake Kivu is 300 km^3 and 55 km^3 at standard temperature and pressure (gas volume at 0°C and 1 atm [14]. According to ref. [15], Methane is formed in three ways: (1) mantle-derived, (2) thermal maturation of organic matter, and (3) bacterial degradation of organic matter at the shallow depths. In the case of methane in Lake Kivu water column, the dominant methane-forming process is bacteria-mediated methanogenesis of CO_2 which is originated from volcanic activities.

Other than those gases, Lake Kivu contains several other chemicals including silica, phosphate, iron, sulfide, and ammonia which vary with depth [8, 9, 16]. Lake Kivu water is stratified where the divide between mixolimnion and monomolimnion occurs at a depth of around ~65 m [17] and ~ 45 m [8], respectively.

Lake Kivu waters are drained by different fluid sources and/or biochemical processes controlling water chemistry (i.e., water-rock interactions, bacterial activity) that is not homogeneously distributed all over the entire lake [8, 18, 19]. Furthermore, Hategekimana et al. [16] studied the chemistry of Nyamyumba hot springs located

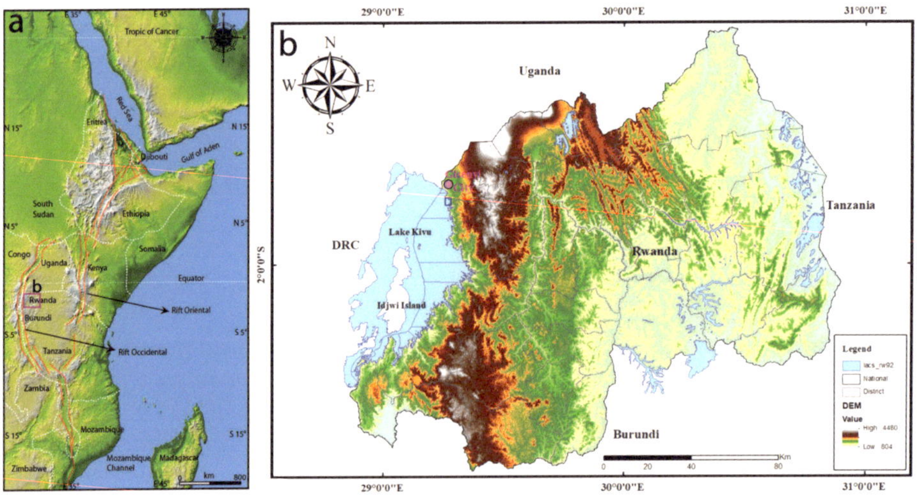

Figure 1.
Location map of Rwanda: (a) a map of Africa for Rwanda localization; (b) an elevation map of Rwanda in meters. The study area is shown by a blue rectangle. DRC stands for the Democratic Republic of Congo.

along the shores of Lake Kivu and interpreted that there is a contribution of these hot springs to the chemistry of Lake Kivu water.

In this chapter, we reviewed earlier research findings from Lake Kivu to identify the causes of its water chemistry while considering vertical chemistry variations and their corresponding environmental issues, particularly for aquatic and human life.

2. Geological, geochemical, and hydrogeological background of Lake Kivu

The lake's surface area is 2370 km^2, and its drainage basin is 4940 km^2 excluding the lake [20]. The majority of the drainage basin is made up of a river-active region (4255 km^2) that is dominated by humic ferralsols in the southwest, humic acrisols in the east, and haplic acrisols in the northwest and southeast [21].

Lake Kivu is one of the rift lakes located in the western branch of the East African Rift System. Different geodynamic processes, such as faulting and magmatism, influenced and contributed to the formation of the lake and various structures along Lake Kivu's margin (**Figure 1**).

The East African Rift System resulted from two mantle plumes beneath the Afar and Kenyan Plateau [22–24]. The tectonic uplift and an extension led to the creation of the East African Rift (EAR) [25], the best example of an active rift system. The plateaus are dynamically supported by the convective activity under the asthenosphere [26], providing heat transfer for partial melting of the lithospheric mantle. All of these parameters make the East Africa Rift System (EARS) a very good potential area for geothermal resources [27]. The western branch of the East African Rift System has a limited and localized volcanic product with a more diverse chemistry than the eastern branch.

The Kivu rift valley is composed of deep lacustrine basins and structural heights which are overlain by volcanic rocks [28] indicating the presence of a mantle plume beneath the lithosphere. The northern basin of Lake Kivu contains about 0.5 km of sediments which overlie a basement believed to be of crystalline rocks of Precambrian age [9].

Lake Kivu's deep waters are known to have high concentrations of carbon dioxide and methane. A trace amount of nitrogen is present at all depths, and the amount of dissolved oxygen decreases with depth. The pH of oxygenated waters is around 9, while that of anoxic waters is below 7.

Furthermore, the distribution of principal cations shows that salt content increases with water depth, and they are at a relatively uniform concentration level at a given depth in an independent geographical location [9].

3. Material and methods

Seven water samples were collected from the surface to a depth of 390 meters, with at depths of (0 m, 40 m, 90 m, 240 m, 290 m, 340 m, 390 m) using Niskin bottles suspended on a calibrated cord in Lake Kivu near Gisenyi city (**Figure 1**), with the assistance of a Kilindi boat. Other 14 samples were collected at different locations from Hot spring sources to Lake Kivu for the comparison of the chemistry of Hot springs and Lake Kivu waters.

Conductivity, temperature, and depth (CTD) sonde data were also gathered. Utilizing Hatch kits, water samples were examined in the lab of the Lake Kivu Monitoring Program.

With increasing salinity, conductivity—a metric of water's capacity to conduct electricity—increases [18]. Using a CTD Sonde, the conductivity of water samples was assessed in seven different locations. Comparing the dissolved ions and factors facilitating the dissolution at various locations is made easier by measuring the conductivity at each location.

Using Hach test kits and procedures, the concentrations of sulfate, iron, ammonia, silica, and phosphate as well as the alkalinity of the water were determined in water samples.

The silico-molybdate method (silica, high range [0–75.0 mg/L]) was used to determine the silica content. The samples were warmed to room temperature before being analyzed to determine the silica concentration. The silica standard solution contained 50 mg/L of SiO_2. Under acidic conditions, the sample's silica and phosphate reacted with the molybdate ion to form yellow silico-molybdic acid complexes and phosphormolybdic acid complexes. In order to dissolve the phosphate complexes, citric acid was added. The remaining yellow complex was measured to obtain the silica content.

Furthermore, a digital titrator was used to measure the alkalinity, or the water's ability to resist acidification. The sample bottles for phosphate analysis, in contrast to other chemicals, were first cleaned with a 1:1 hydrochloric acid solution and rinsed with deionized water. The mixing bottle was filled with the sample. One pillow of phenolphthalein indicator powder was added, then blended. Drops of a standard solution of sulfuric acid, 0.035 N, were added. After each drop, the solution was thoroughly blended to achieve colorlessness from pink. The alkalinity of phenolphthalein was calculated as $CaCO_3$ by multiplying the number of drops until the color changed by 20. One pillow of bromocresol green-methyl red powder was added, and the mixture was stirred. Drops of a standard solution of sulfuric acid, 0.035 N, were added. After every drop, the solution was thoroughly mixed once more, and drops continued to be added until the color changed from green to pink. The total number of drops for the whole procedure was calculated and multiplied by 20 to obtain the total alkalinity (methyl orange) as $CaCO_3$.

Using the method described above, a standard solution containing 500 mg/L of $CaCO_3$ was used. Water samples were collected and stored in plastic containers that had been acid-cleaned before the iron concentration was measured. No acid was added because the water samples were analyzed right away. The analysis made use of the Ferro Ver method for iron (0–3.00 mg/L). By adding 100 mg/L of Fe to 100 mL of deionized water and diluting 1.00 mL of iron standard solution, 1.0 mg/L of iron standard solution was prepared. The test was then conducted using the AccuVac Ampuls method.

Phosphorus concentration was determined by adding 25 mL of sample to a 25 mL sample cell. A 1 mL calibrated dropper was used to add the molybdate reagent. The addition of 1 mL of the amino acid reagent solution came next. The reaction took 10 minutes to complete and the sample was thoroughly mixed. The sample was added to the sample cell in a volume of 25 mL. The timer beeped, and mg/L PO_4 was shown. The cell holder was filled with the blank. The instrument cap was placed over the sample cell, and after pressing "zero," the cursor moved to the right and the reading of 0.0 mg/L PO_4 appeared. The prepared sample was put into the cell holder, then the instrument cap was put on top of it. After selecting "read," the cursor shifted to the right and the sample's final concentration was shown. To determine the concentration of each sample, this was finished. Reactive (0–30.0 mg/L PO_4 3) method was used for phosphorus analysis. A 50 mg/L as PO_4 3 phosphate standard solution was pipetted into a 50 mL volumetric flask to create a 10.0 mg/L phosphate standard. Deionized water was used to dilute the sample to the desired volume. The procedure was carried out as described above, and a concentration of 10 mg/L was obtained.

By adding 5 mL of the sample to two tubes, the concentration of ammonia was also calculated. The left opening of the color comparator box received one tube, and the second tube received the ammonia salicylate reagent powder pillow. The powder was shaken out of the tube completely. An ammonia cyanurate reagent powder pillow was added after a short while, shaken, and after 15 minutes, a green color appeared. In the color comparator box, which was held up in front of the light source, the second tube was placed. By rotating the color disc, color matching was discovered, and the outcome was displayed in the scale window. Ammonium ions (NH_4^+) and unionized ammonia (NH_3) are the two different forms of ammonia that are found in water.

So, this method measures both NH_4^+ and NH_3 as ammonia nitrogen (NH_3–N). One mg/L NH_3–N of nitrogen ammonia standard solution was used. The mg/L NH_3 in the sample was calculated as follows: • mg/L NH_3 = ((mg/L NH_3–N × percent NH_3 of water sample at a given temperature and pH) ÷ 100) × 1.2 [29].

To calculate the salinity of the water samples, a CDC401 conductivity probe was employed. The probe was then dried with a lint-free cloth after being rinsed with deionized water. The shroud was put in place. The sensor was fully inserted into the sample when the probe was inserted. Shaking the probe removed air bubbles. It was shaken, then stirred while the salinity was measured.

Furthermore, the Sulfa-Ver 4 method for sulfate (0–70 mg/L) was used to complete the sulfate analysis. The method was initially calibrated using a 50 mg/L sulfate standard solution. Pipetting 1 mL of a Pour Rite ampule standard for sulfate (2500 mg/L) into a 50 mL volumetric flask produced the standard solution. Deionized water was used to dilute the sample. In this procedure, barium sulfate precipitate is created when sulfate ions react with the metal in Sulfa-Ver 4, a sulfate reagent. The stabilizing agent in Sulfa-Ver 4 holds the suspended precipitates, and the turbidity formed is proportional to the sulfate concentration.

4. Results and discussion

4.1 Lake Kivu water chemistry variation with depth

Hategekimana et al. [8] concluded that Lake Kivu water chemistry varies significantly along the depth profile. Lake Kivu exhibits a unique vertical density stratification that is driven by dissolved gasses and the influx of saline groundwater. Except for anaerobic microbial processes, biologic activity only occurs in the mixoliminion (upper 60–65 m of the lake) [8, 17].

According to Hategekimana et al. [8], Lake Kivu water chemistry varies significantly along the depth profile.

Phosphorus concentration range from 0 to 3 mg/L in Lake Kivu. It shows an abrupt change at around the depth of 45 m (**Figure 2a**). This concentration also changes at 240 m deep to the constant value of 3 mg/L. According to ref. [30], phosphate concentration between 40 and 120 mg/L can lead to environmental pollution. The concentrations recorded in this study are below that range indicating that Lake Kivu water is less prone to pollution.

Like other chemicals, silica concentration also shows a constant value up to 45 m deep. In the deeper part, silica content increased downward up to the depth of 390 m

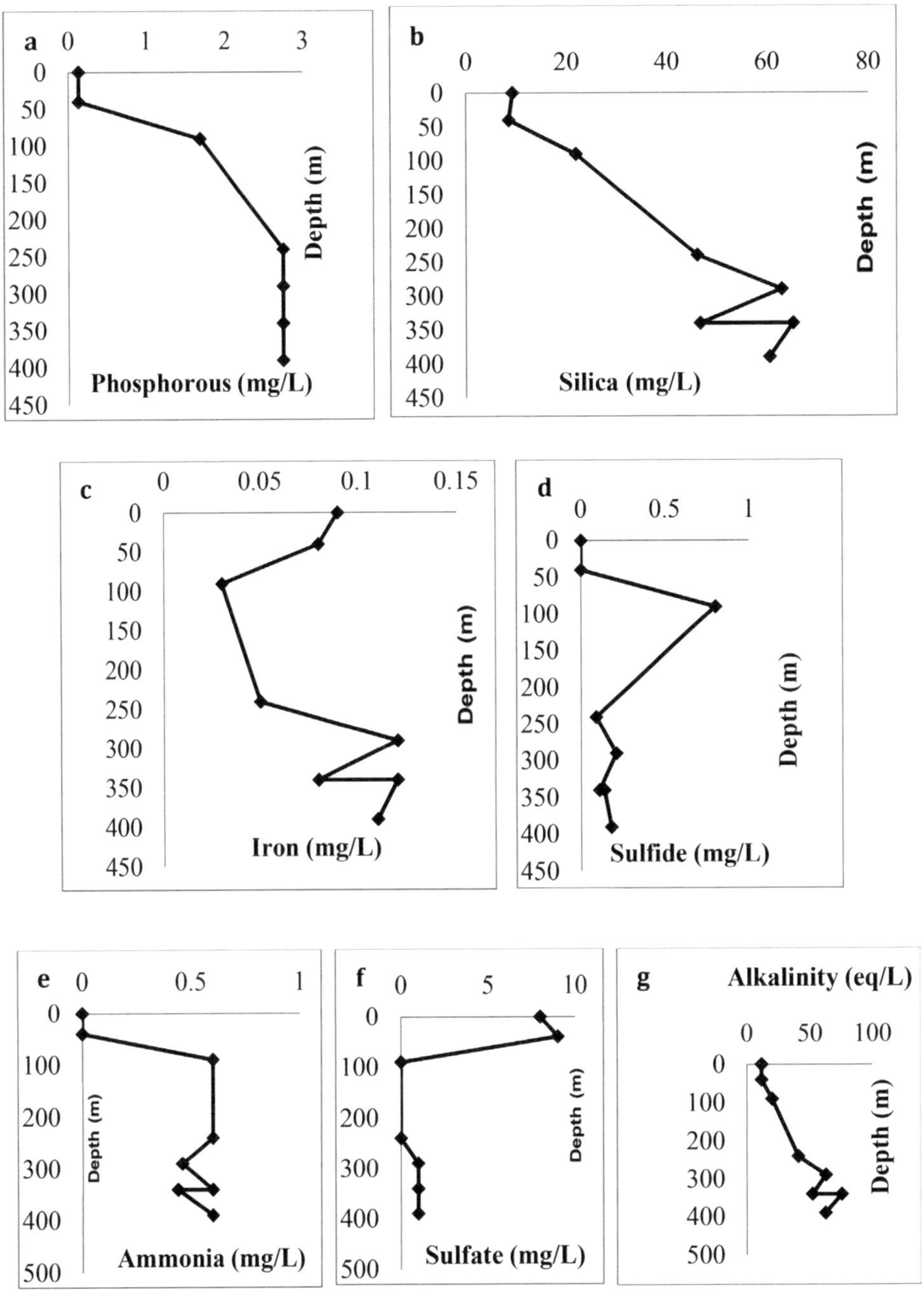

Figure 2.
The variation of Lake Kivu water chemistry with depth (modified from [8]).

(**Figure 2b**). The increase in silica concentration could be probably related to the sinking of dead diatoms [31].

Iron concentration in Lake Kivu water decreases with depth up to 90 m deep as shown in **Figure 2**. The concentration then increased downward up to the depth of 390 m (**Figure 2c**).

Sulfide and sulfate concentrations showed a decrease and increase respectively in the range from 0 to 45 m deep. This increase in sulfate concentration can be explained as the result of sulfide oxidation in an oxic environment. In contrast, the concentration of sulfate decreases in the region below 100 m. This reduction of sulfate increases the concentration of sulfide in Lake water (**Figure 2d** and **f**). This change in concentrations can be used to determine the boundary between oxic and anoxic zones in Lake Kivu which is estimated at 45 m deep also consistent with the boundary set by Roland et al. [32].

Up to 45 m from the water surface, the ammonia concentration is 0 mg/L. The concentration suddenly increased with depth up to 100 m. Ammonia concentration is constant up to 390 m deep (**Figure 2e**). The increase in ammonia could be probably related to the deposition of organic matter in the deeper part.

The alkalinity increases with depth (**Figure 2g**) indicating the ability of Lake Kivu water to withstand the acidity. The alkalinity in Lake Kivu is in the range recommended by WHO [16, 33].

The concentration of dissolved oxygen reduces with depth but is still above the critical level of 3 mg/L for fish (**Figure 3a**); [34]. The decrease in dissolved oxygen is the result of bacterial oxygen consumption during the decomposition of organic substances including from public sewage and agricultural farms [10, 34].

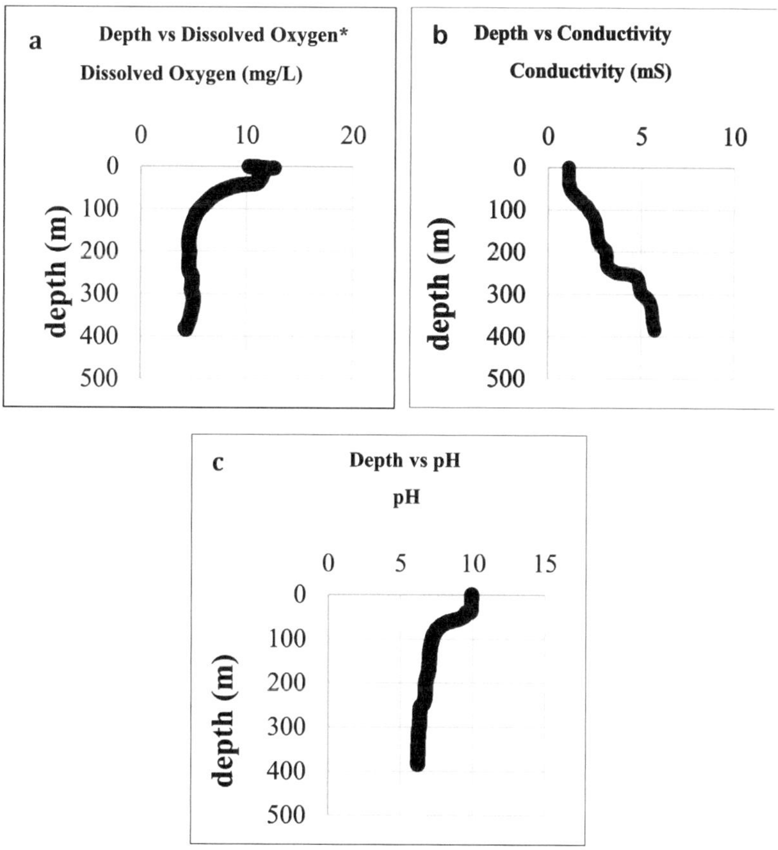

Figure 3.
The variation of Lake Kivu water chemistry with depth (modified from [8]).

Conductivity also increases with depth (**Figure 3b**). The increase in conductivity in Lake Kivu is probably related to the increased salinity [16]. The pH in **Figure 3c** indicated a decrease from 10 at the surface and tends to be neutral which is better for swimming.

4.2 The source of Lake Kivu water chemistry

This research disclosed that Lake Kivu water chemistry is mostly derived from sewage water from resident houses, hospitals, runoff from farming activities, and water-rock interaction through the dissolution process [8, 16].

In fact, Lake Kivu is located in the vicinity of two densely populated cities; Gisenyi (Rwanda) and Goma (DRC). Therefore, the chemistry of Lake Kivu water could be associated with the urbanization in those two cities.

A significant amount of wastewater is disposed of without proper treatment as a result of urbanization. To increase agricultural yields, farming activities also need more fertilizers. In addition, rivers carry domestic sewage and rainwater from agricultural fields to Lake Kivu. As a result, chemicals like nitrogen and phosphorus are present in higher concentrations in water. According to Ref. [35], higher concentrations worsen the aquatic environment and impair lakes' functionality, causing eutrophication.

Additionally, it was discovered that because the water from Nyamyumba Hot Springs is directly discharged into Lake Kivu, it affects the chemistry of the Lake (**Figure 4**) [16].

The higher concentration of silica (9.2 mg/L) found closer to the shoreline is probably the result of an influx of sediment from weathered bedrock (**Figure 2b**), especially close to stream and river outlets. Due to the dissolution of diatoms below the photic zone and the precipitation of calcite after it has been dissolved in water, deeper waters have higher concentrations of silica than surface waters [16].

Dissolved oxygen levels show a decrease near the water's surface that may be caused by photosynthetic processes (**Figure 3a**).

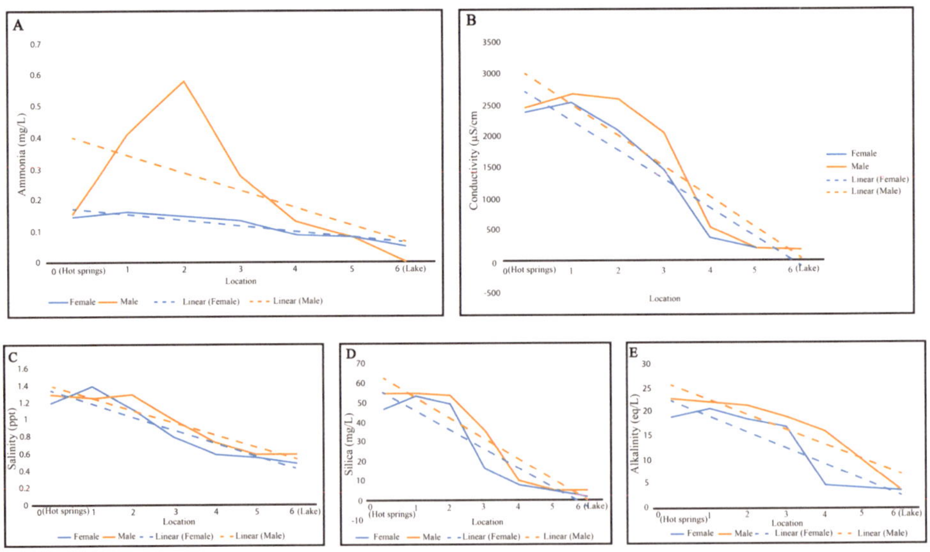

Figure 4.
The comparison of chemical concentrations in water from Lake Kivu and hot springs was adapted from [16].

With the inflow of saline groundwater, the conductivity, which reflects the salinity of the water, is increasing downward (**Figure 3b**). Saline groundwater is most likely alkaline, and alkalinity rises in the monimolimnion most likely because of calcium carbonate precipitation in the upper levels of the water column and mixolimnion dissolution (**Figure 3b**). Due to phytoplankton deposition in deep waters, the concentration of phosphorus typically rises with depth as depicted in **Figure 1**.

Figure 3 shows that as the age of deep waters increases, iron concentrations at a depth of about 100 m decrease. Due to the remineralization of iron, the concentration rises in the intermediate zone. After that, iron is taken out of the water and added to the sediments, where it cannot mix back in and cause replenishment. The main factor contributing to the increase in ammonia in the lake water may be the use of fertilizers, which cause runoff into waterways from farmlands (**Figure 2e**). Alkalinity was used to calculate the partial pressure of CO_2, but it is currently difficult to evaluate how accurate this calculation was.

The CO_2 flux calculation indicates that CO_2 moves upward in the water column, with the exception of the deepest measurements where it is more likely to move downward from 340 to 390 m.

The data collected by Schmidt in February 2004 and June 2018 differ from one another. At greater depths, the change is typically significant.

The pH level of lake water rises as a result of higher algal and plant growth that is influenced by rising temperatures or excess nutrients from farmland runoff and wastewater streams. The pH drops downward as the temperature rises, and the presence of carbon dioxide increases acidity, which in turn causes the pH to drop.

The sulfide concentration suggests that phytoplankton numbers are declining, temperatures are rising [36], there is an increase in organic matter in the sediment [37], the sediments are iron-poor [38], and the water is deeper and has less oxygen. Due to the implementation of control measures, the concentration of sulfide decreased between 2004 and 2018 and was a result of the slower organic matter sedimentation rate.

Over time, the phosphate concentration is lowering. This is because of the managed runoff, measures taken to keep livestock out of water sources, and a manure management plan.

When a significant amount of wastewater from homes and hospitals is dumped directly into a lake, it adds chemicals to the lake's water body, like phosphate and nitrates. Additionally, the use of fertilizers causes agricultural runoff, which raises the phosphate concentration in water. According to ref. [39], phosphorus increases the productivity of plankton and aquatic plants, which in turn feed larger organisms like zooplankton, fish, humans, and other mammals. The aquatic life that consumes phytoplankton and zooplankton will be greatly impacted by the gradual decrease in phosphate concentration. Some organisms will vanish at a later time.

4.3 Environmental impacts

According to this study, the concentration of the majority of chemicals in Lake Kivu water rises with depth. Aquatic life is impacted by the increased chemicals [8]. The increased pollutants in the water cause an overabundance of phytoplankton, which causes eutrophication. According to Ref. [8], when nitrogen and phosphorus levels rise in water, algae and other microorganisms grow erratically, which reduces the amount of oxygen in the water. This buildup of nutrients is essential for the eutrophication of Lake water. In order to prevent eutrophication in Lake Kivu, consistent monitoring should be considered.

The chemical concentration levels in Lake Kivu are lower than those in hot springs [16]. Because the chemical concentrations were below the WHO-recommended ranges, they concluded that swimming in hot springs water was safe. As a result, swimming is also safe in Lake Kivu.

5. Conclusions

In Lake Kivu, where water chemistry varies greatly with depth, the mixolimnion and monolimnion boundary was found at a depth of about 40 meters, which is 20 meters closer compared to the previous researches. Hospitals and residential wastewater both add nitrates and sulfates to Lake Kivu's water supply. Additionally, farmlands contribute to an increase in the concentration of phosphates in water through water runoff. Increased chemical concentrations encourage the growth of plankton and other marine organisms, which can cause eutrophication, the depletion of available oxygen, and adverse effects on aquatic life. In order to safeguard Lake Kivu's aquatic ecosystem, we advise the relevant organizations to continue monitoring the lake's chemical state.

Acknowledgements

We wish to express our gratitude to the Geological Structures and Geo-Hazard Research Lab (GSGR), Department of Earth and Environmental Sciences, Pukyong National University, for supporting this research and We thank all members of the laboratory for their assistance with data analysis and interpretation.

Conflict of interest

The authors declare no conflict of interest.

Author details

Francois Hategekimana[1,2]*, Theophile Mugerwa[1],
Digne Edmond Rwabuhungu Rwatangabo[1] and Young-Seog Kim[2]

1 Departments of Geology, College of Science and Technology, University of Rwanda, Kigali, Rwanda

2 Geological Structures and Geo-Hazard Research Lab, Department of Earth and Environmental Sciences, Pukyong National University, Busan, Korea

*Address all correspondence to: francoishate@pukyong.ac.kr

References

[1] Organization for Economic Co-Operation and Development (OECD). Eutrophication of waters: Monitoring, assessment and control. In: Final Report. OECD Cooperative Program on Monitoring of Inland Waters (Eutrophication Control, Environment Directorate). Paris: OECD; 1982

[2] Nixon SW. Eutrophication and the macroscope. In: Eutrophication in Coastal Ecosystems: Towards better understanding and management strategies Selected Papers from the Second International Symposium on Research and Management of Eutrophication in Coastal Ecosystems, 20-23 June 2006, Nyborg, Denmark. Netherlands: Springer; 2009. pp. 5-19

[3] van Beusekom JE. Eutrophication. Handbook on Marine Environment Protection: Science, Impacts and Sustainable Management. 2018:429-445

[4] Karikari AY, Bernasko JK, Bosque EKA. Hamilton an assessment of water quality of Angaw River in southern – Eastern coastal plains of Ghana. West African Journal of Applied Ecology. 2007;**2**:77-87

[5] Li S, Gu S, Tan X, Zhang Q. Water quality in the upper Han River basin, China: The impacts of land use/ land cover in riparian buffer zone. Journal of Hazardous Materials. 2009;**165**:317-324. DOI: 10.1016/j.jhazmat.2008.09.123

[6] Viers J, Dupré B, Gaillardet J. Chemical composition of suspended sediments in World Rivers: New insights from a new database. Science Total Environment. 2009;**407**:853-868. DOI: 10.1016/j.scitotenv.2008.09.053

[7] Rezende-Filho AT, Valles V, Furian S, Oliveira CM, Ouardi J, Barbiero L. Impacts of lithological and anthropogenic factors affecting water chemistry in the Upper Paraguay River Basin. Journal of Environmental Quality. 2015;**44**(6):1832-1842

[8] Hategekimana F, Ndikuryayo JD, Habimana E, Mugerwa T, Christian K, Digne R. Lake Kivu water chemistry variation with depth over time, Northwestern Rwanda. Rwanda Journal of Engineering, Science, Technology and Environment. 2020;**3**(1):1-20

[9] Degens ET, von Herzen RP, Wong HK, Deuser WG, Jannasch HW. Lake Kivu: Structure, chemistry and biology of an East African rift lake. Geologische Rundschau. 1973;**62**:245-277

[10] Olapade OJ, Omitoyin BO. Anthropogenic pollution impact on physico-chemical characteristics of Lake Kivu, Rwanda. African Journal of Food, Agriculture, Nutrition and Development. 2012;**12**(5):6517-6536

[11] Ross KA, Gashugi E, Gafasi A, Wüest A, Schmid M. Characterisation of the subaquatic groundwater discharge that maintains the permanent stratification within Lake Kivu, East Africa. PloS One. 2015;**10**(3):e0121217

[12] Tietze K. 1978: Geophysikalische Untersuchung des Kivusees und seiner außergewöhnlichen Methangaslagerstätten.Schichtung, Dynamik und Gasgehalt des Seewassers. Kiel: Universität Kiel

[13] Halbwachs M, Tietze K, Lorke A, Mudaheranwa C. Investigations in Lake Kivu (East Central Africa) after the Nyiragongo eruption of January 2002,

specific study of the impact of the sub-water lava inflow on the lake stability. European Community Humanitarian Office. 2002:1-56

[14] Schmid M, Halbwachs M, Wehrli B, Wüest A. Weak mixing in Lake Kivu: New insights indicate increasing risk of uncontrolled gas eruption. Geochemistry, Geophysics, Geosystems. 2005;**6**(7):1-11

[15] Selley RC. Elements of Petroleum Geology. Oxford: Gulf Professional Publishing; 1998

[16] Hategekimana F, Mugerwa T, Nsengiyumva C, Byiringiro FV, Rwatangabo DER. Geochemical characterization of Nyamyumba Hot Springs, Northwest Rwanda. Applied Chemistry. 2022;**2**:247-258. DOI: 10.3390/appliedchem2040017

[17] Schmid M, Wüest A. Stratification, mixing and transport processes in Lake Kivu. Lake Kivu: Limnology and Biogeochemistry of a Tropical Great Lake; 2012. pp. 13-29

[18] Tassi F, Vaselli O, Montegrossi G, Huertas AD. Water and gas chemistry at Lake Kivu (DRC): Geochemical evidence of vertical and horizontal heterogeneities in a multi basin structure. Geochemistry, Geophysics, Geosystems. 2009;**2009**:10

[19] Gomaa MM. Salinity and water effect on electrical properties of fragile clayey sandstone. Applied Water Science. 2020;**10**:116

[20] Ballatore T. Lake Basins: Topography and Surface Water. Lake Basin Action Network. 2012. Available from: http://www.lakebasin.org

[21] Muvundja FA, Wüest A, Isumbisho M, Kaningini MB, Pasche N, Rinta P, et al. Modelling Lake Kivu water level variations over the last seven decades. Limnologica. 2014;**47**:21-33

[22] Ebinger C. The active solid earth. In: EGU General Assembly Conference Abstracts. 2016. EPSC2016-5428

[23] Furman T, Bryce J, Rooney T, Hanan B, Yirgu G, Ayalew D. Heads and Tails: 30 Million years of the Afar plume. Vol. 259 No. 1. London: Geological Society Special Publications; 2006. pp. 95-119

[24] Rogers NW, Macdonald R, Fitton JG, George RWW, Smith M, Barreiro BA. Two mantle plumes beneath the East African rift system; Sr, Nd and Pb isotope evidence from Kenya Rift basalts. Earth and Planetary Science Letters. 2000;**176**:387-400

[25] Wood DA. Structure, Paleolimnology and Basin History of the East Kivu Graben, Lake Kivu, Rwanda. Syracuse, NY, USA: Syracuse University; 2014

[26] Ebinger C, Bechtel T, Forsyth D, Bowin C. Effective elastic plate thickness beneath the East African and Afar Plateaux, and dynamic compensation of the uplifts. Journal of Geophysical Research. 1989;**94**:2883-2901

[27] Bahati G, Pang Z, Ármannsson H, Isabirye EM, Kato V. Hydrology and reservoir characteristics of three geothermal systems in western Uganda. Geothermics. 2005;**34**:568-591

[28] Pouclet A, Bellon H, Bram K. The Cenozoic volcanism in the Kivu rift: Assessment of the tectonic setting, geochemistry, and geochronology of the volcanic activity in the South-Kivu and Virunga regions. Journal of African Earth Sciences. 2016;**121**:219-246

[29] Hach. Ammonia Nitrogen Test Kit. 2015. Available online: https://www.

hach.com/p-nitrogen-ammonia-test-kit-model-ni-sa/2428700.DOC326.98.00007 [Accessed: February 28, 2015]

[30] Ajani P, Lee R, Pritchard T, Krogh M. Phytoplankton dynamics at a long-term coastal station off Sydney, Australia. Journal of Coastal Research. 2001;**2001**:60-73

[31] Lewin JC. The dissolution of silica from diatom walls. Geochimica et Cosmochimica Acta. 1961;**21**(3-4):182-198

[32] Roland FA, Darchambeau F, Morana C, Crowe SA, Thamdrup B, Borges AV. Anaerobic methane oxidation in an East African great Lake (Lake Kivu). Biogeosciences Discussions. 2016;**2016**:1-27

[33] World Health Organisation. Guidelines for Drinking Water: World Health Organization Health Criteria and Other Supporting Information. Geneva: WHO; 1998. pp. 62-315

[34] Zdenka S, Lloyd R, Machova J, Vykusova E. Water Quality and Fish Health. Rome: FAO; 1993. p. 59

[35] Le C, Zha Y, Li Y, Sun D, Lu H, Yin B. Eutrophication of lake waters in China: Cost, causes, and control. Environmental Management. 2010;**45**:662-668

[36] Koch MS, Erskine JM. Sulfide as a phytotoxin to the tropical seagrass Thalassia testudinum: Interactions with light, salinity and temperature. Journal of Experimental Marine Biology and Ecology. 2001;**266**(1):81-95

[37] Govers LL, de Brouwer JH, Suykerbuyk W, Bouma TJ, Lamers LP, Smolders AJ, et al. Toxic effects of increased sediment nutrient and organic matter loading on the seagrass Zostera noltii. Aquatic Toxicology. 2014;**155**:253-260

[38] Marbà N, Duarte CM, Holmer M, Calleja ML, Álvarez E, Díaz-Almela E, et al. Sedimentary iron inputs stimulate seagrass (*Posidonia oceanica*) population growth in carbonate sediments. Estuarine, Coastal and Shelf Science. 2008;**76**(3):710-713

[39] Ricklefs RE, Schluter D. Species diversity: Regional and historical influences. In: Ricklefs RE, Schluter D, editors. Species Diversity in Ecological Communities. Chicago: University of Chicago Press; 1993. pp. 350-363

Chapter 12

Studies on *Xerophilic*, *Acidiphilic*, and *Alkaliphilic Fungi* in Wadi El-Natrun

Hassan Abdel Motagly Abdel Mougod Gouda, Abdel-Aal Hassan Moubasher, Mady Ahmed Ismail and Nammat Abd el Gowad Hussein

Abstract

The present study is an unprecedented extensive survey of mycobiota of Wadi El-Natrun depression, Western Desert of Egypt, which is a hypersaline extreme environment. The study was confined to the eight main lakes of the Wadi during six seasons in the years 2007/2009. In general, 159 species, in addition to four species varieties assigned to 50 genera, were recovered during the current investigation. The widest spectra of species were recorded in the genera *Aspergillus* (22 species +2 varieties), *Penicillium* (19), *Fusarium* (17), and *Acremonium* (8). The widest spectrum of species was recorded in El Zugm Lake (82 species) while the lowest was in Fasida (51). Also, the control medium contributed the widest spectrum of species (95 species) while 10% NaCl medium had the lowest (46 species), with the wider spectrum also being recorded in winter and spring seasons and the narrowest during summer. Total of 40 isolates of the most commonly encountered species from different sources, lakes, and isolation media were tested for their capabilities of producing cellulase, protease, lipase, phosphatase, xylanase, and pectinase enzymes. Most isolates had the capabilities of producing cellulase (96%), protease (86.8%), lipase (92.3%), and phosphatase (100%) but with different degrees; however, only 3 out of 20 isolates tested were xylanolytic (15%) and only one out of 38 was pectinolytic.

Keywords: Wadi El-Natrun, hypersaline, mycobiota, extreme environment, enzymes

1. Introduction

The Wadi El-Natrun, one of the most important alkaline environments, is situated on the western side of the Nile Delta of Egypt and includes some water bodies characterized by high salinity. Wadi El-Natrun climate is dry, has low and very variable rainfall, dry summer, high evaporation, and low humidity. It contains eight principal lakes for a distance of about 30 km; from south to north: Fasida, Umm Risha, Rosetta, Abu Gubara, Hamra, El Zugm, Al Beida, Khadra, and Al Gaar, noting that Abu Gubara and Hamra form one lake in the summer [1]. Extreme environments are populated by groups of fungi that are specifically adapted to these particular conditions and these are usually referred to as extremophilic fungi. Extremophiles fungi can be grouped according to the conditions in which they thrive into thermophiles and psychrophiles (which grow at the extremes of temperature ranges), acidophiles and alkaliphiles (extremes of pH), halophiles (high salt), barophiles or piezophiles (high pressures), and xerotolerant, which tolerate very low water activity [2, 3]. The active and stable nature of the microbial enzymes lead to their wide-spread use in various industries and applications [2].

1.1 Aim of the work

The study was confined to the eight main lakes of the Wadi during six seasons in the years 2007/2009. The study comprised the following aspects:

- Chemical analysis of different substrates studied (mud, salt crusts, and water).
- Isolation and identification of mycobiota of soil, mud, salt crusts, water, and air of the investigated lakes on sucrose agar medium as Osmophilic/osmotolerant; on salt agar medium as halophilic/halotolerant; on media adjusted at different acidic pHs as acidiphilic/aciditolerant; or alkaline pHs as alkaliphilic/alkalitolerant.
- Capabilities of the most common fungi of Wadi El Natrun of producing a wide range of enzymes, including cellulase, protease, lipase, phosphatase, pectinase, and xylanase under acidic, alkaline or saline conditions.

2. Materials and methods

2.1 A: Collection of samples

Samples (reclaimed soil, salt crusts, mud, water, and air) were collected during January 2007 – May 2009, from eight lakes (Fasida, Umm-Risha, Rosetta, Hamra, El El Zugm, Al Beida, Khadra, and Al Gaar) of Wadi El-Natrun (depression) region, Egypt.

1. Soil samples were collected randomly from reclaimed soil around the lakes

   Mud samples were collected at random from different sites inside and along the shore of lakes.

   Salt crust samples were collected at random from mineral formation present along the shores of the Lake.

   Water samples were collected in sterile bottles from different sites inside the lake by means sterile bottles.

2. Samples (soil, mud, and salt crusts) were put directly into a clean plastic bag as described [4].

3. At least five samples are taken at random from each place, then the five or more samples from each replication were brought into one composite sample, which was mixed thoroughly several times.

4. Samples (soil, mud, and salt crusts) were brought into the laboratory and kept in a cold place (5°C) till chemical and fungal analysis (**Figure 1**).

Figure 1.
Showing the reclaimed soil around Fasida Lake (1) and salt crusts of El El Zugm Lake (2).

2.2 Chemical analysis of soil samples

pH value: A pH meter (Orior Research Model GOHL Digital Ionalyzer) was used for the determination of soil pH. The electrode was immersed directly in the soil suspension with a ratio of 1:5 (w /v) [5].

Organic matter content (OM %): A semi-quantitative method was used for the determination of organic matter, which involves the heated destruction of all organic matter in the soil (Astem 2000). OM% is calculated as the difference between the initial and final sample weights divided by the initial sample weight times 100%.

Total soluble salts (TSS): The specific electrical conductance was measured in the soil extract using the conductance meter (YSI, model 35). The percentage TSS in the samples was estimated using this equation: % TSS in the dry sample - 0.064 × EC × extract ratio. The conversion factor to percentage salts (0.064) was fairly applied for solutions extracted from the soil [5].

Sodium and potassium (Na^+ & K^+): Flame photometer method [5], using Carl Zeiss flame photometer, was used for the determination of Na^+ and K^+ cations.

Carbonate and bicarbonate: Total carbonate and bicarbonate were determined directly in the soil through hydrochloric acid digestion [4].

Calcium and magnesium (Ca^{+2} & Mg^{+2}): The versene (disodium dihydrogen ethylene diamine tetraacetic acid) titration method as recommended [6] was employed for Ca^{+2} and Ca^{+2} + Mg^{+2} determinations.

Chloride (Cl^-): Soluble chloride was estimated by applying the silver nitrate titration method using potassium chromate as an indicator [4].

2.3 Isolation of fungi

From soil, mud, and salt crusts: The dilution plate method was used to enumerate different fungal species [7] and employed in this laboratory. At least five samples are taken at random from each place, and then the five or more samples from each replication were brought into one composite sample, which was mixed thoroughly several times.

From the air: Replicate plates of 9 cm diameter containing sterile agar media (five for each medium type) were exposed to the air for 15 minutes from January 2006–May 2007. The plates were sealed, brought back to the laboratory, then incubated at 28°C for 7–21 days, during which the developing fungi were identified and counted.

1. **Medium used for isolation of *osmophilic* and *osmotolerant fungi***

   Czapek Dox agar supplemented with 40% sucrose was used for isolation of osmophilic and osmotolerant fungi, from all sources investigated.

2. **Medium used for isolation of *halophilic* and *halotolerant fungi***

   Modified *Czapek Dox agar* medium (in which glucose, 10 g/l, replaced sucrose), supplemented with 10% sodium chloride, was used for isolation of halophilic and halotolerant fungi.

3. **Media used for isolation of *acidiphilic* and *aciditolerant fungi***

   Modified *Czapek Dox agar* media in which pH was adjusted at four or five using diluted HCl were used for isolation of *acidiphilic* and *aciditolerant fungi*.

4. **Media used for isolation of *alkaliphilic* and *alkalitolerant fungi***

   Modified *Czapek Dox agar* in which pH was adjusted at 10, 13 using NaOH were used for isolation of *alkaliphilic* and *alkalitolerant fungi*.

2.4 Identification of fungi

The identification of fungal genera and species (purely morphologically based on macroscopic and microscopic features).

Enzymatic activities of fungal isolates

Forty fungal isolates represented by ten species, commonly encountered from different sources at Wadi El-Natrun region, were screened for their abilities to produce six extracellular enzymes on solid media.

A: Cellulase production

Cellulase production was tested on medium as described by [8].

B: Protease production

The fungal proteolytic ability was tested using casein hydrolysis medium [9].

C: Lipase production

Lipolytic ability of fungal isolates was tested on the medium [10] with slight modification, in which tween 80 (Sorbitan polyoxyethylene monooleate) replaced tween 20.

D: Phosphatase production

The ability of fungal isolates to produce phosphatase enzymes was detected using phosphatase medium [11].

E: Pectinase production

The method was carried out as described by Hankin et al. [12].

F: Xylanase production

Modified *xylan agar* medium [13].

3. Results and discussion

3.1 Chemistry of Wadi El Natrun

3.1.1 Chemical analysis of soil samples collected from around Wadi El Natrun lakes

From the collective data of soil chemical analysis from the eight lakes investigated, it is obvious that soil samples collected from around Al Gaar lake possessed the highest values of pH (9.05 ± 0.6), moisture content (23.1 ± 11.0), total soluble salts (30.5 ± 16.8), potassium (0.3 ± 0.1), carbonate (0.3 ± 0.03), bicarbonate (0.5 ± 0.03), and chloride (3.6 ± 2.5) compared to those recorded from soil collected from the other 7 lakes of Wadi El-Natrun. On the other hand, other parameters showed their peaks in different lakes, for example, organic matter (2.0 ± 2.4) in El Zugm Lake, calcium ($0.2 \pm 9.7 \times 10^{-3}$) in Hamra Lake, magnesium (0.04 ± 0.2) in Al Beida, and sodium (11.1 ± 8.4 mg/g) in Umm Risha (**Table 1**) [14].

3.2 Chemical analysis of mud samples collected from Wadi El-Natrun lakes

From the collective data of mud chemical analysis from the eight lakes investigated, it is evident that mud samples collected from Hamra Lake showed the highest levels of pH (9.4 ± 0.3), organic matter (0.7 ± 0.7), sodium (38.3 ± 17.9), carbonate (0.5 ± 0.15), bicarbonate (0.76 ± 0.7), magnesium (0.2 ± 0.3), and chloride (16.4 ± 11.2). On the other hand, other parameters showed their peaks in different lakes, for example, Moisture content (44.7 ± 28.5), potassium (2.3 ± 4.5) in El Zugm Lake, total soluble salts (47.7 ± 26.2) in Fasida, and calcium (0.1) in Al Beida and Fasida (**Table 2**) [14].

3.3 Chemical analysis of salt crust samples collected from Wadi El-Natrun lakes

From the collective data of chemical analysis of the salt crusts collected from the eight lakes investigated, it is obvious that salt samples showed the highest values of moisture content (25.5 ± 19.9) and calcium (0.4 ± 0.4) in El Zugm Lake, of total soluble salts (88.7 ± 10.8), sodium (51.9 ± 21.2), potassium (0.6 ± 0.4) and chloride (20.6 ± 10.4) in Al Gaar, and carbonate (0.5 ± 0.3) and magnesium (0.4 ± 0.02) in Al Beida Lake. On the other hand, other parameters showed their peaks in other lakes, for example, pH (9.8 ± 0.43) in Umm Risha, organic matter (0.6 ± 0.7) in Hamra, and bicarbonate (2.1 ± 2.5) in Khadra (**Table 3**) [14].

3.4 Chemical analysis of water samples collected during spring 2007 from Wadi El-Natrun Lakes water

Chemical analysis revealed that water samples collected from Wadi El-Natrun Lakes were highly alkaline, with pH ranging from 8.4–9.5 and of high levels of total soluble salts, chlorides, sodium, and potassium. Water collected from El-Zugm Lake showed the highest levels of organic matter, sodium, calcium, magnesium, and chlorides among the eight lakes investigated. On the other hand, some parameters showed their peak in other lakes, for example, pH (9.4) and total soluble salts (87%) in Fasida (**Table 4**) [15].

4. Fungi in Wadi El-Natrun lakes

4.1 In soil around Wadi El-Natrun

Soil samples collected from different lakes during different seasons harbored the highest number of genera and species 48 genera, 137 species, and four varieties. The widest spectrum of species was recorded on the control medium (69 species +2 varieties) and the lowest on 10% NaCl medium (36).

	Al Gaar	Khadra	Al Beida	El Zugm	Hamra	Rosetta	Umm-Risha	Fasida
pH	9.05	8.2	8.7	8.5	8.7	8.4	8.5	8.3
OM (%)	0.3	1.2	1.4	2.0	1.4	1.3	1.7	0.7
MC (%)	23.1	3.1	6.4	3.9	3.0	4.5	8.9	5.4
TSS (mmol/L)	30.5	1.9	2.7	6.3	1.7	4.9	5.7	4.9
Na^+	4.9 ± 3.1	3.9 ± 3.4	9.9 ± 8.7	4.7 ± 2.7	7.4 ± 8.9	3.5 ± 2.3	11.1 ± 8.36	10.1 ± 9.7
K^+	0.3 ± 0.1	0.14 ± 0.05	0.15 ± 0.1	0.16 ± 0.08	0.25 ± 0.1	0.1 ± 0.01	0.13 ± 0.03	0.2 ± 0.3
CO_3^{-2}	0.3 ± 0.03	0.04 ± 0.02	0.01 ± 0.2	0.07 ± 0.07	0.26 ± 0.17	0.2 ± 0.3	0.1 ± 0.1	0.2 ± 0.3
HCO_3^-	0.5 ± 0.5	0.1 ± 0.07	0.2 ± 0.2	0.1 ± 0.1	0.1 ± 0.08	0.26 ± 0.3	0.07 ± 0.04	0.2 ± 0.2
Ca^{+2}	0.01 ± 0.009	0.03 ± 0.03	0.03 ± 0.02	0.01 ± 0.1	0.2 ± $9.7x10^{-3}$	0.02 ± 0.01	0.01 ± 0.01	0.008 ± 0.0
Mg + 2	0.023 ± 0.01	0.01 ± 0.01	0.04 ± 0.2	0.02 ± 0.01	0.03 ± 0.02	0.02 ± 0.01	0.03 ± 0.03	0.01 ± 1.4
Cl^-	3.6 ± 2.5	0.7 ± 0.1	3.4 ± 2.5	2 ± 2.2	1 ± 0.3	2.0 ± 2.5	0.9 ± 0.5	1.6 ± 0.8

OM and TSS are calculated as percentage of the samples analyzed; Na^+, K^+, $CO3^{-2}$, $HCO3^-$, Ca^{+2}, Mg^{+2}, and Cl are calculated as mg/ml.

Table 1.
Chemical analysis from soil samples collected from around Wadi El-Natrun lakes.

	Al Gaar	Khadra	Al Beida	El Zugm	Hamra	Rosetta	Umm-Risha	Fasida
pH	9.2	9.0	9.3	9.3	9.4	9.0	9.3	9.3
OM	0.5	0.3	0.6	0.3	0.7	0.3	0.4	0.23
MC	18.0	23.0	15	44.7	12.5	32.5	17.4	36.1
TSS	21	30.5	35.7	35.0	29.3	32.8	43.6	47.7
Na^{+}	31.4 ± 27.0	23.8 ± 24.7	23.8 ± 25.0	20.2 ± 18.1	38.3 ± 17.9	15.5 ± 4.7	27.1 ± 13.6	28 ± 9
K^{+}	0.6 ± 0.6	0.3 ± 0.12	0.55 ± 0.56	2.3 ± 4.5	0.5 ± 0.1	0.2 ± 0.07	0.2 ± 0.2	0.22 ± 0.7
CO_3^{-2}	0.02 ± 0.01	0.35 ± 0.03	0.4 ± 0.06	0.3 ± 0.04	0.45 ± 0.15	0.3 ± 0.02	0.4 ± 0.09	0.4 ± 0.07
HCO_3^{-}	0.3 ± 0.2	0.5 ± 0.5	0.5 ± 0.4	0.18 ± 0.1	0.76 ± 0.7	0.2 ± 0.15	0.4 ± 0.36	0.4 ± 0.3
Ca^{+2}	0.01 ± 0.01	0.01 ± 0.009	0.1 ± 0.2	0.03 ± 0.01	0.02 ± 0.02	0.07 ± 0.1	0.01 ± 6.5^{-3}	0.1 ± 0.2
Mg^{+2}	0.03 ± 0.02	0.02 ± 0.01	0.05 ± 0.06	0.02 ± 0.03	0.2 ± 0.3	0.04 ± 0.04	0.02 ± 0.02	0.1 ± 0.2
Cl^{-}	13.3 ± 11.0	8.5 ± 13.2	6.3 ± 3.9	7.7 ± 4.7	16.4 ± 11.2	5.7±	13.6 ± 7.9	13.8 ± 3.2

OM and TSS are calculated as percentage of the samples analyzed; Na^{+}, K^{+}, $CO3^{-2}$, $HCO3^{-}$, Ca^{+2}, Mg^{+2}, and Cl are calculated as mg/ml.

Table 2.
Chemical analysis of mud samples collected from Wadi El-Natrun lakes.

	Al Gaar	Khadra	Al Beida	El Zugm	Hamra	Rosetta	Umm-Risha	Fasida
pH	9.5	9.6	9.3	9.0	9.6	8.6	9.8	9.5
OM	0.17	0.1	0.2	0.1	0.6	0.085	0.09	0.3
MC	7.8	16.3	12.5	25.5	9.5	18.52	5.6	16.9
TSS	88.7 ± 10.8	88.6	80.9	86.45	86.5	88.2	87.2	88.7
Na^+	51.9 ± 21.2	41.1 ± 22.3	43.6 ± 20.8	50.1 ± 16.9	48.4 ± 16.0	43.8 ± 23.7	33.8 ± 20.1	39.9 ± 17.4
K^+	0.6 ± 0.4	0.5 ± 0.2	0.3 ± 0.2	0.5 ± 0.3	0.55 ± 0.1	0.38 ± 0.05	0.2 ± 0.12	0.4 ± 0.07
CO_3^{-2}	0.08 ± 0.05	0.4 ± 0.4	0.5 ± 0.3	0.4 ± 0.03	0.37 ± 0.07	0.38 ± 0.05	0.45 ± 0.16	0.38 ± 0.07
HCO_3^-	0.4 ± 0.4	2.1 ± 2.5	1.8 ± 2.1	0.35 ± 0.2	1.1 ± 0.6	0.35 ± 0.26	1.55 ± 1.6	0.77 ± 0.47
Ca^{+2}	0.03 ± 0.02	0.02 ± 0.01	0.02 ± 6^{-3}	0.4 ± 0.4	0.07 ± 0.08	0.15 ± 0.18	0.02 ± 0.009	0.05 ± 0.04
Mg^{+2}	0.05 ± 0.04	0.04 ± 0.02	0.4 ± 0.02	0.2 ± 0.2	0.2 ± 0.2	0.26 ± 0.28	0.03 ± 0.03	0.05 ± 0.04
Cl^-	20.6 ± 10.4	18.8 ± 13.9	19.5 ± 10.1	17.1 ± 8	22.4 ± 9.05	11.3 ± 5.0	19.4 ± 18.8	17.9 ± 7.1

OM and TSS are calculated as percentage of the samples analyzed; Na^+, K^+, $CO3^{-2}$, $HCO3^-$, Ca^{+2}, Mg^{+2}, and Cl are calculated as mg/ml.

Table 3.
Chemical analysis from salt samples collected from Wadi El-Natrun lakes.

	Al Gaar	Khadra	Al Beida	El Zugm	Hamra	Rosetta	Umm-Risha	Fasida
pH	8.8	9	9.0	8.4	9	9.1	9	9.5
OM	0.05	0.1	1.0	0.1	0.07	0.08	0.05	0.1
TSS	50	80	65	80.2	80	67	70.2	87
Na^+	17	22	22	44	40	20	40	13
K^+	0.2	0.3	0.2	0.2	0.3	0.3	0.1	0.2
CO_3^{-2}	0.2	1.1	1.3	0.2	0.5	0.3	1.0	0.2
HCO_3^-	0.22	2	0.23	0.2	1.0	0.11	1.2	0.19
Ca^{+2}	0.02	0.04	0.1	0.5	0.01	0.1	0.01	0.1
Mg^{+2}	0.03	0.05	0.02	0.4	0.03	0.3	0.03	0.09
Cl^-	12	14	10.5	22	20.1	12.5	23	15.2

OM and TSS are calculated as percentage of the samples analyzed; Na^+, K^+, $CO3^{-2}$, $HCO3^-$, Ca^{+2}, Mg^{+2}, and Cl are calculated as mg/ml.

Table 4.
Chemical analysis of water samples collected from Wadi El-Natrun lakes.

The genera *Aspergillus*, *Fusarium*, *Penicillium*, and *Emericella* were the most dominant with high proportions of propagules being recorded on all isolation media; however, *Stachybotrys* was also common but was not encountered on 10% NaCl, *Eurotium* was common on 40% sucrose and 10% NaCl media and *Acremonium* was common on alkaline media only.

Of *Aspergillus*, *Aspergillus terreus* followed by *A. niger*, *Aspergillus flavus*, and *A. fumigatus* gave the highest counts and frequencies on all isolation media; however, some other aspergilli were dominant on both acidic and alkaline media (*A. sydowii* and *A. ustus*), on 10% NaCl (*A. carneus* and *A. sydowii*), on 40% sucrose (*A. sydowii*), while *A. ochraceus* was dominant on the control medium as well as the salt and alkaline media.

Other most commonly encountered species comprised *Emericella nidulans*, *E. quadrilineata*, *Fusarium solani*, *Penicillium puberulum*, and *Acremonium furcatum*. On the other hand, *F. subglutinans*, *Penicillium chrysogenum*, and *Stachybotrys chartarum* were absent in NaCl medium, and *Acremonium strictum* was absent in 40% sucrose agar medium only.

It is worth mentioning that six out of the seven *Acremonium* species recorded from soil were isolated on the alkaline media with more propagules than on the other five isolation media used, however, five species were recorded on acidic media, two on 10% NaCl, and only one on 40% sucrose. This implies that this fungus prefers alkaline media rather than other media.

1. The current results show that *Acremonium fusidioides*, *Acrophialophora fusispora*, *Aspergillus deflectus*, *Cladosporium oxysporum*, *Cochliobolus tuberculatus*, *Curvularia penniseti*, *Eurotium amstelodami*, *E. repens*, *Fusarium nygamai*, *Gliocladium solani*, *Humicola insolens*, *Monodictys castaneae*, and *Scopulariopsis brevicaulis* were isolated from soil on one or both acidic media but not on alkaline media.

Moreover, some species could be isolated on medium of pH 10 (*Acremonium blochii*, *Microdochium nivale*, *Fusarium lateritium*, *Penicillium variabile*, *Mucor circinelloides*, *Penicillium verrucosum*, *Pseudoallescheria boydii*, *Scytalidium lignicola*, and *Trimmatostroma betulinum*) or pH 13 (*Aspergillus cremeus* and *Microascus trigonosporus*) or on both pHs (*A. hyalinulum*, *A. roseulum* and *Paecilomyces variotii*), however, these species were not isolated on both acidic media (adjusted at pH 5 and pH 4).

Some species were recorded only on NaCl medium namely *Cochliobolus monoceras*, *Scopulariopsis halophilica*, *S. carbonaria*, and *Ulocladium consortiale* or on 40% sucrose medium (*Aspergillus candidus* and *Gliocladium catenulatum*) (**Tables 5** and **6**) [14, 16].

4.2 In mud

Mud samples collected from different lakes during different seasons contributed much narrow spectrum of genera and species (13 and 48) compared to that recorded from soil (48 and 137 + 4 varieties). The widest spectrum of species was recorded on 40% sucrose and medium adjusted at pH 10 (25 species), and the narrowest on 10% NaCl (3).

Aspergillus was the most dominant genus possessing the highest propagules (over 75% of the total CFUs) on all isolation media; however, *Penicillium* was also dominant on 40% sucrose, acidic, and alkaline media while *Fusarium* was dominant on 40% sucrose and alkaline media, *Emericella* and *Eurotium* on 40% sucrose and *Acremonium* on alkaline media only.

Aspergillus showed its peak in spring 2007 in Al Gaar on all isolation media except on 10% NaCl medium in Fasida Lake.

Of *Aspergillus, A. terreus* followed by *A. fumigatus, A. flavus,* and *A. niger* were the most common on all isolation media; however, some other *Aspergilli* were dominant on the control, acidic and alkaline media (*A. ochraceus*), on 40% sucrose (*A. candidus, A. sydowii*) and on alkaline media (*A. carbonarius*).

1. Other most commonly encountered species comprised *F. solani* and *P. chrysogenum* on all media but not encountered on 10% NaCl medium. On the other hand, *E. nidulans* was absent on medium adjusted at PH10 and 10% NaCl medium while *P. puberulum* was absent on control and 10% NaCl media.

2. Interestingly, the three *Acremonium* species and some unidentified species recorded from mud were isolated on alkaline media with more propagules than on the other five isolation media used; however, two species were recorded on acidic media and only one on control, 40% sucrose and 10% NaCl media.

3. It is worthy to mention that some fungal species were recorded on only 40% sucrose (*A. candidus*, *E. nidulans* var. lata, *E. amstelodami*, *E. repens*, *Fennellia nivea*, *Fusarium sterilihyphosum*, and *Humicola fuscoatra*), acidic media (*H. insolens*, *Penicillium echinulatum*, *P. expansum*, and *P. janczewskii*), or alkaline media (*A. hyalinulum*, *A. fusispora*, *Myrothecium verrucaria*, *Fusarium semitectum*, *F. subglutinans*, *Penicillium brevicompactum*, and *P. griseofulvum*) but not on the other isolation media used (**Tables 5** and **6**) [17–19].

4.3 In salt crusts

Aspergillus (54.6–86.1% of the total propagules) followed by Penicillium (3.6–6.8%) were the most dominant genera with the highest propagules on all isolation media,

Fungal taxa	Soil	Mud	Salts	Water	Air
Acremonium Link	+	+	+	+	+
A. blochii (Matruchot) W. Gams	+				
A. furcatum F. & V. Moreau ex W. Gams	+	+	+	+	+
A. curvulum W. gams					
A. fusidioides (Nicot) Gams	+				
A. hyalinulum (Sacc.) W. Gams		+	+	+	
A. kiliense Grütz	+			+	
A. strictum W. Gams	+	+	+	+	
A. roseulum (G.S.M.) W. Gams	+		+		
Acremonium spp.	+	+	+	+	
A. fusispora (Saksena) Samson	+	+			
Alternaria Nees	+				
A. alternata (Fries) Keissler	+		+		+
A. chlamydospora Mouchacca					+
A. tenuissima (Kunze: Pers.) Wiltshire	+		+	+	+
Alternaria spp.	+	+	+		+
Aspergillus Mich. ex fr.	+	+	+	+	+
A. aculeatus Lizuka	+	+	+		
A. aegyptiacus Moubasher and Moustafa			+		
A. aureolatus Munt., Cvet. & Bata	+	+			
A. carneus (V. Tiegh.) Blochwitz	+				
A. candidus Link	+	+	+		
A. carbonarius (Bainier) Thom	+	+	+		
A. cremeus	+				
A. deflectus Fennell and Raper	+	+			
A. flavus Link	+	+	+	+	+
A. flavus var. *columnaris* Raper and Fennell	+				+
A. fumigatus Fresenius	+	+	+	+	+
A. japonicus Saito			+		
A. niger Van Tieghem	+	+	+	+	+
A. ochraceus Wilhelm	+	+	+	+	+
A. parasiticus Speare	+			+	
A. parvulus Smith	+	+			
A. phoenicis (Cda.) Thom		+	+		
A. pulverulentus (McAlpine) Thom	+				
A. puniceus	+		+		
A. sydowii (Baineir & Sartory) Thom and Church	+	+	+	+	+

Fungal taxa	Soil	Mud	Salts	Water	Air
A. terreus Thom	+	+	+	+	+
A. terreus var. *africanus* Raper and Fennell	+				
A. versicolor (Vuillemin) Tiraboschi	+				
A. ustus (Bainier) Thom and Church	+			+	
Botryodiplodia theobromae Patouillard	+				
Botryotrichum Saccardo and Marchal	+				
B. piluliferum Saccardo and Marchal	+				
B. atrogriseum	+				
Botrytis Micheli ex Pers.	+				
B. cinerea Persoon	+				
Botrytis sp.			+		
Chatomium Kunze	+			+	
C. globosum Kunze	+			+	
C. olivaceum Cooke and Ellis	+				
Chatomium spp.	+				
Cladosporium link	+	+	+	+	+
C. cladosporioides (Fres.) de Vries	+	+	+	+	+
C. herbarum (Pers.) Link ex S. F. Gray	+		+		+
C. oxysporum Berkeley and Curtis	+		+		+
C. sphaerospermum Penzig	+	+	+		+
Cladosporium spp.	+		+		+
Cochliobolus Drechsler	+	+	+	+	+
C. australiensis (Tsudal and Yeyama) Alcorn	+		+		
C. tuberculatus Sivanesan	+	+	+	+	+
C. spicifer Nelson	+				
Cochliobolus sp.					
Curvularia					
C. lunata var. aeria (Batista, Lima, and Vasconcelos) M. B. Ellis	+				
C. penniseti (Mitra) Boedijn	+				
Cunninghamella echinulata (Thaxt.) Thaxt. ex	+				
Blakeslee					
Cylindrocarpon sp.	+				
E.nigrum Link			+	+	+
Emericella Berkeley and Broome	+	+	+	+	+
E. acristata Fennell and Raper	+				
E. nidulans var. *lata* Thom and Raper	+	+	+		
E. nidulans(Eidam) Winter	+	+	+		

Fungal taxa	Soil	Mud	Salts	Water	Air
E. quadrilineata Thom and Raper	+	+	+	+	+
E. variecolor Berkeley and Broome	+		+		
Emericella spp.	+				+
Eurotium					
E. chevalieri Mangin	+		+	+	
E. amstelodami Mangin	+				
E. repens De Bary	+		+		+
Eurotium spp.	+				+
Fennellia Wiley and Simmons	+	+			
F. flavipes Wiley and Simmons		+			
F. nivea (Wiley and Simoons) Samson	+	+			
Fusarium Link	+	+	+	+	+
F. camptoceras Wellenweber and Reinking emend.	+				
Marasas & Logrieco					
F. chlamydosporum Wellenweber and Reinking	+				
F. equiseti (Corda) Saccardo	+				
F. lateritium Nees	+	+			
F. nygamai Burgess and Trimboli	+				
F. oxysporum Schlechtendahl emend. Snyder and	+		+		
Hansen					
F. poae (Peck) Wollenweber	+				
F. proliferatum (Matsushima) Nirenberg	+				
F. sambucinum Füchel	+		+		
F. semitectum Berkeley and Ravenel	+	+	+		
F. solani (Martius) Appel and Wollenweber emend.	+	+	+	+	+
Snyder and Hansen					
F. sporotrichioides Sherbakoff	+				
F. subglutinans (Wollenweber and Reinking)	+		+	+	
Nelson, Toussoun and Marasas					
F. sterilihyphosum Britz, Marasas, and Wingfield		+			
F. tricinctum (Corda) Saccardo	+				
F. tumidum Sherb.	+				
F. udum Butler			+		
F. verticillioides (Saccardo) Nirenberg	+	+	+		
Fusarium spp.	+				

Fungal taxa	**Soil**	**Mud**	**Salts**	**Water**	**Air**
Gliocladium Corda	+		+		
G. catenulatum Gilman and Abboll	+				
G. roseum Bainier	+		+		
G. solani (Harting) Petch	+				
Gliocladium spp.					
Graphium Corda	+				+
G. penicillioides Corda	+				+
Graphium spp.	+				
Humicola Traaen	+	+	+	+	
H. fuscoatra Traaen	+	+			
H. grisea Traaen	+	+	+	+	
H. insolens Cooney and Emerson	+	+	+		
Humicola spp.	+				
Hypomyces chrysospermus Tulasne	+				
Macrophomina phaseolina (Tassi) Goid	+				+
Memnoniella echinata (Riv.) Galloway					+
M. nivale (Fr.) Ces.	+				
M. trigonosporus Emmons et Dodge	+				+
Myrothecium Tode	+	+			
M. roridum Tode ex Steudel	+				
M. striatisporum Preston	+				
M. verrucoria (Albertini and Schweinitz) Ditmer ex	+	+			
Steudel					
M. castaneae (Wallr.) Hughes	+				
M. circinelloides Van Tiegh	+				
Neurospora Crassa Shear and Dodge	+				
Nigrospora					
N. sphaerica (Sacc.) Mason	+		+		+
N. oryzae					
Paecilomyces Bainier	+		+		
Paecilomyces lilacinus (Thom) Samson	+		+		
P. variotii Bainier	+		+		
Penicillium Link	+	+	+	+	+
P. aurantiogriseum Dierckx	+	+		+	
P. brevicompactum Dierchx		+			
P. chrysogenum Thom	+	+	+	+	+
P. citrinum Thom	+				

Fungal taxa	Soil	Mud	Salts	Water	Air
P. crustosum Thom		+	+		
P. duclauxii Delacroix	+	+	+		+
P. echinulatum Raper and Thom ex Fassatiová		+			
P. expansum Link	+	+		+	
P. funiculosum Thom	+	+	+		
P. griseofulvum Dierckx	+		+		
P. janczewskii Zaleski		+			
P. islandicum Sopp	+				
P. oxalicum Currie and Thom		+	+	+	
P. pinophilum Hedgcock	+	+			
P. puberulum Bainier	+	+	+	+	
P. purpurgenum Stoll	+	+	+		
P. variabile Sopp	+				
P. verrucosum Peyronel				+	
P. viridicatium Westling			+		
Pencillium spp.	+	+	+		+
Phialophora sp.	+				
Phoma Saccardo	+				
Phoma herbarum Westendorp	+	+	+		+
Phoma sp.	+				
Pochonia sp	+				
P. boydii (Shear) McGinnis *et al.*	+				
Rhizopus Ehrenberg	+				
R. oryzae Went and Gerlings	+				
R stolonifer (Ehrenb.) Lind	+				
Rhizopus sp.	+				
Scopulariopsis Bainier	+		+	+	
S. brevicaulis (Saccardo) Bainier	+				
S. brumptii Salvanet - Duval	+		+	+	
S. carbonaria Morton and Smith	+				
S. halophilica Tubaki	+				
S. sphaerospora	+				
Scopulariopsis spp	+				
S. lignicola Pesante	+				
Setosphaeria Leonard and Suggs	+				
S. holmii	+				
S. monoceras Alcorn	+				
S. rostrata Leonard	+				

Fungal taxa	Soil	Mud	Salts	Water	Air
Setosphaeria sp.	+				
Stachybotrys Corda	+				
S. chartarum (Ehrenb. Ex Lindt) Hughes	+		+		+
S. kampalensis Hansf.	+				
S. coccosporum Meyer and Nicot				+	
Stemphylium Wallroth					
S. botryosum Wallroth	+				+
Stemphylium spp.					
Talaromyces C. R. Benjamin	+				
Talaromyces helicaus (Raper and Fennell)	+				
Talaromyces sp.	+				
Thermoascus auranticus Miehe	+				+
Torula					
T. herbarum (Persoon) Link	+				+
Torula sp.					+
Trichoderma spp. (Persoon) Harz	+	+	+	+	
Trichothecium roseum (Persoon) Link ex Gray	+				
Trichurus Spiralis Hasselbring	+				
T. betulinum (Corda) Hughes	+				
Ulocladium Preuss	+			+	+
U. atrum Preuss	+				
U. botrytis Preuss	+			+	+
U. consortiale (Thümen) Simmons	+				
U. oudemansii	+				
Ulocladium spp.	+				
YEASTS	+			+	+
No. of genera (50)	48	13	18	16	21
No. of species (157 species + 4 varieties)	137 and 4 varieties	47 + 1 variety	59 + 1 variety	33	34 + 1 variety

(+) indicate the presence of the fungal species in this substrate.

Table 5.
Summarized data of fungal taxa recorded from various substrates in different lakes of Wadi El-Natrun.

however, both were not encountered on 10% NaCl. *Fusarium* and *Emericella* were common on two isolation media (pH 5 and 40% sucrose for *Emericella*, control and 40% sucrose for *Fusarium*) while *Acremonium* was common on alkaline media only.

Of *Aspergillus*, *A. flavus*, *A. niger*, *A. terreus*, and *A. fumigatus* were the most common on all isolation media, however, some other aspergilli were dominant on both acidic and alkaline media (*A. phoenicis*) or on 40% sucrose (*A. sydowii*).

Fungal taxa	Cz	40% sucrose	10% NaCl	pH 4	pH 5	pH 10	pH 13
Acremonium Link	+	+	+	+	+	+	+
A. blochii (Matruchot) W. Gams						+	
A. curvulum W. Gams	+						
A. furcatum F. and V. Moreau ex W. Gams	+	+	+	+	+	+	+
A. fusioides (Nicot) Gams				+	+		
A. hyalinulum (Sacc.) W. Gams						+	+
A. kiliense Grütz	+			+	+	+	
A. strictum W. Gams	+	+	+	+	+	+	+
A. roseulum (G.S.M.) W. Gams	+					+	+
Acremonium spp.	+	+		+	+	+	+
A. fusispora (Saksena) Samson				+	+		+
Alternaria Nees	+	+	+	+	+	+	+
A. alternata (Fries) Keissler	+	+	+	+	+	+	+
A. chlamydospora Mouchacca			+				
A. tenuissima (Kunze: Pers.) Wiltshire	+	+	+	+	+	+	+
Alternaria spp.	+	+		+	+		
Aspergillus Mich. ex Fr.	+	+	+	+	+	+	+
A. aculeatus Lizuka				+	+	+	
A. aegyptiacus Moubasher and Moustafa	+						
A. aureolatus Munt. Cvet. and Bata	+						
A. carneus (V. Tiegh.) Blochwitz		+	+				
A. candidus Link		+					
A. carbonarius (Bainier) Thom				+	+	+	
A. aculeatus					+	+	+
A. cremeus							+
A. deflectus Fennell and Raper	+			+			
A. flavus Link	+	+	+	+	+	+	+
A. flavus var. *columnaris* Raper and Fennell	+						
A. fumigatus Fresenius	+	+	+	+	+	+	+
A. japonicus Saito		+			+		
A. niger Van Tieghem	+	+	+	+	+	+	+
A. ochraceus Wilhelm	+	+	+	+	+	+	+
A. parasiticus Speare	+						
A. parvulus Smith	+						
A. phoenicis (Cda.) Thom	+						
A. pulverulentus (McAlpine) Thom	+						
A. puniceus	+						

Fungal taxa	**Cz**	**40% sucrose**	**10% NaCl**	**pH 4**	**pH 5**	**pH 10**	**pH 13**
A. sydowii (Bainier and Sartory) Thom and Church	+	+	+	+	+	+	+
A. terreus Thom	+	+	+	+	+	+	+
A. terreus var. *africanus* Raper and Fennell	+						
A. versicolor (Vuillemin) Tiraboschi		+					
A. ustus (Bainier) Thom and Church	+	+	+	+	+	+	+
B. theobromae Patouillard				+			
Botryotrichum Saccardo and Marchal				+		+	
B. piluliferum Saccardo and Marchal				+		+	
B. atrogriseum	+						
Botrytis Michel ex Pers.	+	+					
B. cinerea Persoon	+						
Botrytis sp.	+	+					
Chatomium Kunze	+	+	+	+	+	+	
C. globosum Kunze	+	+	+	+	+	+	
C. olivaceum Cooke and Ellis			+	+	+		
Chatomium spp.	+						
Cladosporium link	+	+	+	+	+	+	+
C. cladosporioides (Fres.) de Vries	+	+	+	+		+	+
C. herbarum (Pers.) Link ex S. F. Gray	+	+	+				
C. oxysporum Berkeley and Curtis	+	+	+			+	+
C. sphaerospermum Penzig	+	+	+	+	+	+	+
Cladosporium spp.	+	+					
C. Drechsler	+	+	+	+	+	+	+
C. australiensis (Tsudal and Yeyama) Alcorn	+	+		+	+	+	+
C. tuberculatus Sivanesan	+	+	+	+	+	+	+
C. spicifer Nelson	+						
Cochliobolus sp.						+	
Curvularia		+		+			
C. lunata var. *aeria* (Balista, Lima and Vasconcelos) M. B. Ellis		+					
C. penniseti (Mitra) Boedijn				+			
C. echinulata (Thaxt.) Thaxt. ex						+	
Blakeslee							
Cylindrocarpon sp.	+	+					
E. nigrum Link	+	+	+				
Emericella Berkeley and Broome	+	+	+	+	+	+	+
E. acristata Fennell and Raper	+	+					

Fungal taxa	**Cz**	**40% sucrose**	**10% NaCl**	**pH 4**	**pH 5**	**pH 10**	**pH 13**
E. nidulans var. *lata* Subramanian	+	+		+			
E. nidulans (Eidam) Vuillemin	+	+	+	+	+	+	+
E. quadrilineata (Thom and Raper) Benjamin	+	+	+	+	+	+	+
E. variecolor Berkeley and Broome	+			+		+	
Emericella spp.	+						
Eurotium Link	+	+	+	+	+	+	+
E. chevalieri Mangin	+	+	+	+	+	+	+
E. amstelodami Mangin		+	+	+	+		
E. repens De Bary		+					
Eurotium spp.	+	+					
Fennellia Wiley and Simmons			+	+		+	
F. flavipes Wiley and Simmons				+		+	
F. nivea (Wiley and Simoons) Samson		+	+				
Fusarium Link	+	+	+	+	+	+	+
F. camptoceras Wellenweber and Reinking				+		+	+
emend. Marasas and Logrieco							
F. chlamydosporum Wellenweber and Reinking	+			+	+	+	+
F. equiseti (Corda) Saccardo	+						
F. lateritium Nees		+	+			+	
F. nygamai Burgess and Trimboli					+		
F. oxysporum Schlechtendahl emend. Snyder	+	+	+	+	+	+	+
and Hansen							
F. poae (Peck) Wollenweber	+						
F. proliferatum (Matsushima) Nirenberg	+			+	+		+
F. sambucinum Füchel	+	+	+	+	+	+	+
F. semitectum Berkeley and Ravenel	+	+	+	+	+	+	+
F. solani (Martius) Appel and Wollenweber	+	+	+	+	+	+	+
emend. Snyder and Hansen							
F. sporotrichioides Sherbakoff		+					
F. subglutinans (Wollenweber and Reinking)	+	+		+	+	+	+
Nelson, Toussoun and Marasas							
F. sterilihyphosum Britz, Marasas, and		+					
Wingfield							
F. tricinctum (Corda) Saccardo	+	+					
F. udum Butler	+						
F. verticillioides (Saccardo) Nirenberg	+	+					
Fusarium spp.				+		+	+

Fungal taxa	**Cz**	**40% sucrose**	**10% NaCl**	**pH 4**	**pH 5**	**pH 10**	**pH 13**
Gliocladium Corda	+	+		+	+	+	+
G. catenulatum Gilman and Abboll		+					
G. roseum Bainier	+			+	+	+	+
G. solani (Harting) Petch				+	+		
Gliocladium spp.		+			+	+	
Graphium Corda		+			+	+	
G. penicillioides Corda		+			+	+	
Graphium spp.							
Humicola Traaen	+	+		+	+	+	+
H. fuscoatra Traaen	+			+	+	+	
H. grisea Traaen	+	+		+	+	+	+
H. insolens Cooney and Emerson		+		+	+		
Humicola spp.	+				+		
H. chrysospermus Tulasne				+			
M. phaseolina (Tassi) Goid	+			+	+	+	
M. echinata (Rivolta) Galloway	+						
M. nivale (Fr.) Ces.						+	
M. trigonosporus Emmons et Dodge	+						+
M. Tode	+	+		+	+	+	+
M. roridum Tode ex Steudel	+	+		+	+	+	+
M. striatisporum Preston	+						
M. verrucoria (Albertini and Schweinitz)	+	+		+	+	+	+
Ditmer ex Steudel							
M. castaneae (Wallr.) Hughes				+			
M. circinelloides Van Tiegh						+	
Neurospora Crassa Shear and Dodge		+	+	+	+	+	+
Nigrospora							
N. sphaerica (Saccardo) Mason		+	+	+	+	+	+
N. oryzae	+						
Paecilomyces Bainier							
P. lilacinus (Thom) Samson				+			
P. variotii Bainier		+				+	+
Paecilomyces spp.				+	+	+	
Penicillium Link	+	+	+	+	+	+	+
P. aurantiogriseum Dierckx		+		+	+	+	
P. brevicompactum Dierchx						+	
P. chrysogenum Thom	+	+		+	+	+	+

Fungal taxa	**Cz**	**40% sucrose**	**10% NaCl**	**pH 4**	**pH 5**	**pH 10**	**pH 13**
P. citrinum Thom		+	+			+	+
P. crustosum Thom		+			+		
P. duclauxii Delacroix	+	+		+	+	+	+
P. echinulatum Raper and Thom ex Fassatiová							
P. expansum Link		+		+	+	+	+
P. funiculosum Thom	+	+		+	+	+	+
P. griseofulvum Dierckx					+		
P. janczewskii Zaleski		+					
P. oxalicum Currie and Thom	+	+	+	+	+	+	+
P. pinophilum Hedgcock	+						
P. puberulum Bainier	+	+	+	+	+	+	+
P. purpurgenum Stoll		+		+	+	+	+
P. variabile Sopp						+	
P. verrucosum Peyronel						+	
P. viridicatium Westling				+	+	+	+
Pencillium spp.	+	+	+	+	+	+	+
Phialophora sp.	+						
Phoma Saccardo	+			+	+	+	+
Phoma herbarum Westendorp	+			+	+	+	+
Phoma sp.	+						
Pochonia sp.	+						
P. boydii (Shear) McGinnis *et al.*						+	
Rhizopus Ehrenberg		+		+		+	+
R. oryzae Went and Gerlings		+		+		+	+
R stolonifer (Ehrenb.) Lind							+
Rhizopus sp.	+						
Scopulariopsis Bainier							
S. brevicaulis (Sacc.) Bainier	+	+	+	+	+	+	
S. brumptii Salvanet - Duval		+	+	+	+		+
S. carbonaria Morton and Smith			+				
S. halophilica Tubaki			+				
S. sphaerospora							+
Scopulariopsis sp.							
S. lignicola Pesante	+	+	+				+
Setosphaeria Leonard and Suggs	+	+		+	+	+	+
S. holmii	+						
S. monoceras Alcorn			+				

Fungal taxa	Cz	40% sucrose	10% NaCl	pH 4	pH 5	pH 10	pH 13
S. rostrata Leonard	+	+		+	+	+	+
Setosphaeria sp.	+						
Stachybotrys Corda	+	+	+	+	+	+	+
S. chartarum (Ehrenb. Ex Lindt) Hughes	+	+	+	+	+	+	+
S. kampalensis Hansf.	+			+	+		
S. coccosporum Meyer and Nicot				+			
Stemphylium Wallroth	+	+					
S. botryosum Wallroth	+	+					
Stemphylium spp.							
Talaromyces C. R. Benjamin	+				+	+	
T. helicaus (Raper and Fennell)	+				+	+	
Talaromyces spp.	+						
Thermoascus aurantiacus Miehe		+		+			
Torula herbarum (Pers.) Link	+						
Torula sp.	+						
Trichoderma spp. (Persoon) Harz	+	+		+	+	+	+
T. roseum (Pers.) Link ex Gray		+					
T. Spiralis Hasselbring							+
T. betulinum (Corda) Hughes						+	
Ulocladium Preuss	+	+	+		+	+	+
U. atrum Preuss	+						
U. botrytis Preuss	+	+	+		+	+	+
U. consortiale (Thümen) Simmons			+				
Ulocladium spp.	+						
Yeasts					+	+	
No. of genera (50)	32	31	17	30	24	33	24
No. of species (157 species +4 varieties)	95	76	46	77	69	77	59

(+) indicate the (presence) of the fungal species in this medium.

Table 6.
Summarized data of fungal taxa recorded on different media Wadi El-Natrun from various substrates.

Other most commonly encountered species comprised *F. solani* and *P. chrysogenum* on all media but not encountered on 10% NaCl. *P. puberulum* was dominant on all media except the control and 10% NaCl media.

It is worth mentioning that out of five *Acremonium* species (four identified and one unidentified) recorded from salt, four were isolated on alkaline media while only three on the control medium and one was on 40% sucrose medium. *Acremonium* contributed higher propagules on both alkaline media (7.1% and 38.4 of the total CFUs) than on other media.

Some fungal species were recorded on 40% sucrose agar only (*A. candidus*, *Emericella variecolor*, and *F. nivea*), on acidic media (*Aspergillus puniceus*, *Cochliobolus australiensis*, *Fusarium camptoceras*, *Paecilomyces lilacenus*, and *Penicillium crustosum*), or on alkaline media (yeasts, *Cladosporium sphaerospermum*, *P. variotii*, *Scopulariopsis sphaerospora*, and *A. hyalinulum*) but not on other isolation media (**Tables** 5 and **6**) [17–19].

4.4 In water

Aspergillus and *Acremonium* followed by *Penicillium* were the most dominant genera possessing the highest proportions of propagules on all isolation media except on 10% NaCl. On the other hand, only species of the genera *Scopulariopsis* and *Acremonium* were isolated on 10% NaCl medium.

Aspergillus showed its peak in Al Beida during winter 2007 on both acidic and alkaline media while in spring 2007 on control medium (from Khadra Lake) and on 40% sucrose (from El Zugm Lake).

Of *Aspergillus, A. terreus* followed by *A. flavus* and *A. niger* were the most common on all isolation media. On the other hand, *A. ochraceus* was dominant in acidic media only.

Other most commonly encountered species, *P. chrysogenum* and *P. puberulum* were encountered on all media but not on 10% NaCl medium.

Some species were isolated on one medium but not on the others: *S. halophilica* (on 10% NaCl), *E. quadrilineata* (on 40% sucrose), *Staphylotrichum coccosporum* (on medium adjusted at pH 4), and *A. hyalinulum* (on alkaline media) (**Tables** 5 and **6**) [15].

4.5 In air

Aeromycobiota were represented by 21genera and 35 species with the widest spectrum of species being recorded on 40% sucrose medium (28) and the lowest on 10% NaCl medium (18).

The dematiaceous hyphomyceteous genera *Cladosporium* and *Alternaria* were the most dominant followed by *Aspergillus* and possessed the highest number of propagules on all isolation media.

Cladosporium cladosporioides, *Alternaria tenuissima*, *A. alternata* followed by *A. flavus*, *A. niger*, and *A. terreus* were the most common in all isolation media.

Other most commonly encountered species were related to dematiaceous hyphomycetes and these are *C. oxysporum, Epicoccum nigrum, S. chartarum,* and *Ulocladium botrytis* on all isolation media. *Stemphylium botryosum* was also dominant in control and 40% sucrose media, but was not encountered in 10% NaCl medium (**Tables** 5 and **6**) [17].

5. Enzymes produced by the most common fungi

Total of 40 isolates of the most commonly encountered species from different sources, lakes, and isolation media were tested for their capabilities of producing cellulase, protease, lipase, phosphatase, xylanase, and pectinase enzymes. The following observations could be outlined:

Most isolates had the capabilities of producing cellulase (96%), protease (86.8%), lipase (92.3%), and phosphatase (100%) but with different degrees; however, only three out of 20 isolates tested were xylanolytic (15%) and only one out of 38 was pectinolytic.

Total of 36 isolates showed high-producing abilities of either phosphatase (27 isolates), lipase (21), cellulase (11), protease (7) and xylanase (2) on different screening media (**Table** 7).

Some of these isolates were high producers for more than one enzyme, on one or more of the screening media.

5.1 Of the high cellulase producers

- Some isolates produced cellulase on one medium only, for example, the control medium (*Alternaria alternata* AUMC No5666), medium adjusted at PH4 (*A. terreus* AUMC No5675, *C. cladosporioides* AUMC Nos 5681 and 5683), or medium supplemented with 10% NaCl (*E. nidulans* AUMC No5685, *F. solani* AUMC No5691, and *C. australiensis* AUMC No5694).

- Some isolates produced cellulase on the control medium, medium supplemented with 10% NaCl in addition to the acidic (*E. nidulans* AUMC No 5687) or on control, 10% NaCl and alkaline media (*C. cladosporioides* AUMC No. 5684).

- Some isolates produced cellulase on the control, acidic, alkaline, and NaCl media (*E. nidulans* AUMC Nos 5686 and 5689) (**Table** 7) [20, 21].

5.2 Of the high protease producers some isolates showed this property on either

- Control medium (three isolates belong to *C. cladosporioides* AUMC Nos 5681 and 5683 and *M. verrucaria* AUMC No5697)

- Acidic media (1, *M. verrucaria* AUMC No 5699)

- Medium supplemented with 10% NaCl (1, *C. australiensis* AUMC No 5694)

- Both the control and the acidic media (two isolates assigned to *E. nidulans* AUMC Nos 5686 and 5689) (**Table** 7) [20, 21].

5.3 Of the 21 highly lipase-producing strains, some provided this character on

- The control medium (nine isolates, *A. flavus* AUMC No. 5669 and 5670, *A. terreus* AUMC No. 5672 and 5674, *C. cladosporioides* AUMC No 5683, *M. verrucaria* AUMC No. 5697 and *P. chrysogenum* AUMC Nos 5700, 5702, and 5704)

- The acidic media (four isolates, *C. cladosporioides* AUMC5 No. 5680, *C. australiensis* AUMC No. 5696, *M. verrucaria* AUMC No. 5699, and *P. chrysogenum* AUMC No. 5703).

- Alkaline media (*F. solani* AUMC No. 5692)

- Both the control and the acidic media (*E. nidulans* AUMC Nos 5686 and 5688, *F. solani* AUMC No. 5690, and *C. australiensis* AUMC No. 5695)

- Acidic and alkaline media (*E. nidulans* AUMC No. 5687)
- Control, acidic, alkaline, and salt media (*A. terreus* AUMC Nos 5676 and 5677) (**Table** 7) [20, 21].

5.4 The high phosphatase production has been proved on one (or more) medium types

- *F. solani* AUMC No. 5693 (on the control medium)
- *C. cladosporioides* AUMC No. 5680 (on 10% NaCl medium)
- *A. flavus* AUMC No. 5671 and *P. chrysogenum* AUMC No. 5702 (on acidic medium)
- *F. solani* AUMC No. 5692 and *M. verrucaria* AUMC No. 5698 (on alkaline media)
- *C. cladosporioides* AUMC No. 5681 and *E. nidulans* AUMC No. 5688 (on the control and the acidic media)
- *C. australiensis* AUMC No. 5694 and *P. chrysogenum* AUMC No. 5703 (on the control and the alkaline media), *E. nidulans* AUMC No. 5685 (on the control and NaCl media)
- *Alternaria alternata* AUMC No. 5668 (on acidic and alkaline media)
- *A. terreus* AUMC No. 5674 and 5675 and *C. cladosporioides* AUMC No. 5684 (on acidic and salt media)
- *A. flavus* AUMC No. 5669 (on alkaline and NaCl media)
- *A. terreus* AUMC Nos 5672 and 5676, *Chaetomium globosum* AUMC No. 5679, and *P. chrysogenum* AUMC No. 5704 (on the control, acidic, and NaCl media)
- *Alternaria alternata* AUMC No. 5667, *C. australiensis* AUMC No. 5696, and *P. chrysogenum* AUMC No. 5700 (on the control, alkaline, and NaCl media)
- *Alternaria alternata* AUMC No. 5665 and *A. flavus* AUMC No. 5670 (on acid, alkaline, and NaCl media)
- *E. nidulans* AUMC No. 5687 (on the control, acidic, alkaline, and NaCl media) (**Table** 7) [20, 21].

5.5 The highly xylanolytic strains were demonstrated only by *E. nidulans* AUMC Nos. 5685 and 5688

(**Table** 7) [20, 21].

Species	Source	Lake name	Isolation medium	No.	Cellulase	Protease	Lipase	Phosphatase	Xylanase
Alternaria alternata	Soil	Al Gaar	Cz (pH 13)	3469				pH 4, pH 10,10%NaCl	
	Soil	El Zugm	Cz (10% Nacl)	4001	C				
	Mud	Al Gaar	Cz (40% S)	952				C, pH 10,10%NaCl	
	Water	Khadra	Cz (PH5)	3869				pH 4, pH 10	
A. flavus	Soil	Umm Risha	Cz (pH 13)	3776			C	pH 10,10%NaCl	
	Soil	khadra	Cz (pH 4)	3059			C	pH 4, pH 10,10%NaCl	
	Soil	Hamra	Cz (40% S)	5				pH 4	
A. terreus	Air		Cz	205			C	C,pH 4,10%NaC l	
	Mud	khadra	Cz (pH 3)	3035					
	Soil	El Zugm	Cz (pH 3))	3437			C	pH 4,10%NaCl	
	Soil	Hamra	Cz (pH 3)	3913	pH 4			pH 4,10%NaCl	
	Water		Cz (pH 3)	3300			C, pH 4, pH 10,10%NaCl	C, pH 4, 10%NaCl	
	Salts	Umm Risha	Cz (pH 13)	3310			C, pH 4, pH 10,10%NaCl		
Chatomium	Soil	El Zugm	Cz (PH5)	4205					
	Soil	Al Beida	Cz (PH10)	3939				C, pH 4, 10%NaCl	
C. cladosporioides	Soil	Umm Risha	Cz (pH 13)	3783			pH 4	10%NaCl	
	Soil	UmmRisha	Cz (pH 4)	3779	pH 4	C		C, pH 4	
	Soil		Cz (pH 4)	3789	pH 4	C	C		
	Soil	Al Gaar	Cz (40% S)	1240	C,pH 10,10			pH 4, 10%NaCl	
E. nidulans	Soil	El Zugm	Cz (10% Nacl)	4060	10% NaCl			C, 10%NaCl	c
	Soil	Umm Risha	Cz (pH 4)	3339	C, pH 4, pH 10,10%	C, pH 4	C, pH 4		
	Soil	Al Gaar	Cz (pH 13)	3563	C, pH 4,		pH 4, pH 10	C, pH 4,	
					10%NaCl			pH 10,10%NaCl	

Species	Source	Lake name	Isolation medium	No.	Cellulase	Protease	Lipase	Phosphatase	Xylanase
	Soil		Cz (pH 4)	3003			C, pH 4	C, pH 4	c
	Water	Al Gaar	Cz (Ph13)	3594	C, pH 4,	C,			
					pH 10,10%	pH 4			
					NaCl				
	ce	name	medium						
F. Solani	Soil	Al Beida	Cz (pH 4)	3877			C, pH 4		
	Soil	Hamra	Cz (pH 13)	3946	10%NaCl				
	Soil	Umm Risha	Cz (10% Nacl)	4046			pH 13	PH 10	
	Salts	Khadra	Cz (40% S)	76				C	
Cochliobolus austaliensis	Soil	Hamra	Cz (pH 4)	3560	10%NaCl	10% NaCl		C, pH 10	
	Soil	Hamra	Cz (pH 13)	3598			C, pH 4	pH 10,10%NaCl	
	Salts	Hamra	Cz (pH 13)	3810			pH 4	C, pH 10,10%NaCl	
Myrothium verrucoria	Soil	Hamra	Cz (pH 4)	3557		C	C		
	Soil	Khadra	CZ (pH 13)	3950				pH 10	
	Salts	Hamra	Cz (PH10)	3590		pH 4	pH 4		
P. chrysogenum	Soil	Al Gaar	Cz (pH 4)	3079			C	C, pH 10,10%NaCl	
	Soil	Al Beida	Cz (pH 13)	3681					
	Soil	Umm Risha	Cz (40% S)	753			C	pH 4	
	Water		Cz (pH 4)	3858			pH 4	C, pH 10	
	Mud	Khadra	Cz (PH5)	3155			C	C, pH 4,10%NaCl	

Screening media: C = control, pH 4 = medium adjusted at pH 4, pH 10 = medium adjusted at pH 10, pH 13 = medium adjusted at pH 13, and 10% NaCl = medium supplemented with 10% NaCl.

Table 7.
The highly producing isolates for cellulose, protease, lipase, phosphatase, and/or xylanase enzymes on different screening media.

6. Conclusions

Survey of mycobiota of Wadi El-Natrun depression, western desert of Egypt gave in general, 159 species; in addition, four species varieties assigned to 50 genera were recovered during the current investigation. The widest spectra of species were recorded in the genera *Aspergillus* (22 species +2 varieties), *Penicillium* (19), *Fusarium* (17), and *Acremonium* (8). The widest spectrum of species was recorded in El Zugm Lake (82 species) while the lowest was in Fasida (51). Also, the control medium contributed the widest spectrum of species (95 species) while 10% NaCl medium had the lowest (46 species), with the wider spectrum also being recorded in winter and spring seasons and the narrowest during summer.

Author details

Hassan Abdel Motagly Abdel Mougod Gouda[1*], Abdel-Aal Hassan Moubasher[2,3], Mady Ahmed Ismail[2] and Nammat Abd el Gowad Hussein[2]

1 Plant Pathology Research Institute, Agricultural Research Center, Giza, Egypt

2 Department of Botany and Microbiology, Faculty of Science, Assiut University, Assiut, Egypt

3 Assiut University Mycological Centre (AUMC), Assiut University, Assiut, Egypt

*Address all correspondence to: mycologist2010@yahoo.com

References

[1] Zahran MA, Willis AJ. The Vegetation of Egypt. London: Chapman and Hall; 1992. p. 424

[2] Satyanarayana T, Raghukumar C, Shivaji S. Extremophilic microbes: Diversity and perspectives. Current Science. 2005;**89**:1-10

[3] Hendry P. Extremophiles: There's more to life environ. Chem. 2006;**3**:75-76

[4] Johnson LF, Curl EA. Methods for Research on the Ecology of Soil-Borne Plant Pathogens. Minneapolis, MN: Burgress Publishing Company; 1972

[5] Jachson ML. Soil Chemical Analysis. London: Constable and Co.; 1958

[6] Williams V, Twin S. Flame photometric method for sodium, potassium and calcium. In: Paech K, Tracev MV, editors. Modern Methods of Plant Analysis. Vol. V. Berlin: Springer-Verlag; 1960. pp. 3-5

[7] Schwarzenbach G, Biedermann, W.: Complexons. XAlkaline earchcomplexes of O, O-dihydroxy azo days. Helvetica Chimica Acta. 1948;**31**:678-687 (in German)

[8] Eggins H, Pugh PJF. Isolation of cellulose decomposing fungi from soil. Nature. 1962;**193**:94-95

[9] Paterson RRM, Bridge PD. Biochemical methods for filamentous fungi. In: IMI Technical Handbooks No. 1, Wallingford. UK: CAB International; 1994

[10] Ullman U, Blasins C. A simple medium for the detection of different lipolytic activity of microorganisms. Zentrabl Bakteriol Hyg II Abt A. 1974;**2**: 264-267

[11] Gochenaur SE. Fungi of a Long Island oak-birch forest II. Population dynamics and hydrolase patterns for the soil Penicillia. Mycologia. 1984;**76**:218-231

[12] Hankin L, Zucker M, Sands M. Improved solid medium for the detection and enumeration of pectolytic bacteria. Applied Microbiology. 1971;**22**:205-209

[13] Nakamura S, Wakabayashi K, Horikoshi K. Purification and some properties of an alkaline xylanase from akaliphilic bacillus sp. strain 41 M-1. Applied and Environmental Microbiology. 1993;**59**:2311-2316

[14] Gouda HAA. Studies on xerophilic, acidiphilic and alkaliphilic fungi in Wadi El-Natrun. [M.SC. thesis] Department of Botany, Faculty of Science, Assiut University, Assiut, Egypt. 2009. pp. 613

[15] Moubasher AH, Ismail MA, Hussein NA, Gouda AAH. Terrestrial fungi tolerating the hypersaline water of Wadi El-Natrun Lakes, Egypt. Journal of Basic & Applied Mycology (Egypt). 2013; **4**:47-58

[16] Moubasher AH, Ismail MA, Hussein NA, Gouda AH. Osmophilic/osmotolerant and halophilic/halotolerant mycobiota of soil of Wadi El-Natrun region, Egypt. Journal of Basic & Applied Mycology (Egypt). 2015;**6**:27-42

[17] Gouda HA, Moubasher AH, Ismail MA, Hussein NA. Osmophilic/osmotolerant and halophilic/halotolerant fungi from mud, salt crusts, and air collected from Wadi El-Natrun lakes. Journal of Multidisciplinary Sciences. 2020;**2**(2):30-40

[18] Gouda HA, Moubasher AH, Ismail MA, Hussein NA. Acidophilic and acidotolerant fungi in mud and salt

crusts collected from Wadi El- Natrun lakes. Part 1. Journal of Agriculture, Food and Environment. 2020;**1**(2):41-51

[19] Gouda HA, Moubasher AH, Ismail MA, Hussein NA. Alkaliphilic and alkalitolerant fungi in mud and salt crusts collected from Wadi El- Natrun lakes, part 2. Journal of Agriculture, Food and Environment. 2020;**1**(2):52-62

[20] Moubasher AH, Ismail MA, Hussein NA, Gouda AH. Enzyme producing capabilities of some extremophilic fungal strains isolated from different habitats of Wadi El-Natrun, Egypt part 1: Protease, lipase and phosphatase. European Journal of Biological Research. 2021;**6**(2):92-102

[21] Moubasher AH, Ismail MA, Hussein NA, Gouda AH. Enzyme producing capabilities of some extremophilic fungal strains isolated from different habitats of Wadi El-Natrun, Egypt part 2: Protease, lipase and phosphatase. European Journal of Biological Research. 2016;**6**(2):103-111